The Architecture of Cognition

The Architecture of Cognition

Rethinking Fodor and Pylyshyn's Systematicity Challenge

edited by Paco Calvo and John Symons

The MIT Press
Cambridge, Massachusetts
London, England

MIT Press books may be purchased at special quantity discounts for business or sales promotional use. For information, please email special_sales@mitpress.mit.edu or write to Special Sales Department, The MIT Press, 55 Hayward Street, Cambridge, MA 02142.

This book was set in Stone Sans and Stone Serif by Toppan Best-set Premedia Limited, Hong Kong. Printed and bound in the United States of America.

Library of Congress Cataloging-in-Publication Data is available.

ISBN: 978-0-262-02723-6

10 9 8 7 6 5 4 3 2 1

For Anabel, Brendan, Hortensia, Marie, Paquilo, and Zoe

Contents

Preface

Is human thought systematic? How can we best explain it? The present volume aims to explore a variety of conceptual and empirical strategies for responding to these two questions. Twenty-five years after Jerry Fodor and Zenon Pylyshyn originally challenged connectionist theorists to explain the systematicity of cognition, our task in this volume is to reassess and rethink systematicity in the post-connectionist era.

In their seminal "Connectionism and Cognitive Architecture: A Critical Analysis" (*Cognition* 28: 3–71, 1988), Fodor and Pylyshyn argued that the only way for connectionist theory to explain the systematicity of thought is by implementing a classical combinatorial architecture. Connectionist explanations, they claimed, are destined to fail, managing at best to inform us with respect to details of the neural substrate. Explanations at the cognitive level, they argue, simply must be classical insofar as adult human cognition is essentially systematic. It is difficult to overstate the importance of Fodor and Pylyshyn's argument in cognitive science. In fact, it is not easy task to find an introductory text that does not give a central role to the "systematicity challenge."

However, a quarter of a century later, we inhabit a post-connectionist world, where the disagreement is not between classical and connectionist models, but rather between cognitivism writ large and a range of methodologies such as behavior-based AI, ecological psychology, embodied and distributed cognition, dynamical systems theory, and nonclassical forms of connectionism, among others. Thus, it is worth revisiting the initial challenge to connectionist theory, as originally formulated, in order to understand how the debate looks in this new context.

The twenty-fifth anniversary of the publication of Fodor and Pylyshyn's critical analysis provides a suitable occasion to revisit the challenge. To this end, we organized a workshop entitled "Systematicity and the Post-connectionist Era: Taking Stock of the Architecture of Cognition" in the tiny village of San José in a beautiful corner of Almería, Spain. We are very grateful to the co-organizers, Ángel García, Toni Gomila, and Aarre Laakso, and to the participants for several days of highly productive discussion.

Participants in the workshop were asked to focus on systematicity from the perspective of nonclassical approaches in cognitive science. We addressed the following questions: Can we identify novel lines of response from, say, ecological psychology, embodied and distributed cognition, or neurobiologically plausible neural network theory? Would such strategies face the same conceptual challenges as previous connectionist responses? In what way might an implementation-level orientation in embodied cognitive science serve to inform psychological explanation? Is there reason to rethink the claim that thought is systematic? What is the empirical evidence for or against the systematicity of thought? How does the systematicity of human thought relate to human and nonhuman systematic behaviors? What areas of research, other than language, can throw light on the systematicity argument? A number of contributors to this volume (Ken Aizawa, Anthony Chemero, Alicia Coram, Fernando Martínez-Manrique, Brian McLaughlin, Steven Phillips, Michael Silberstein, and David Travieso) presented preliminary versions of their chapters at the workshop. The topics addressed cut across the cluster of disciplines that constitute contemporary cognitive science.

Ensuing discussion and informal interaction continued well beyond the workshop itself, forming the basis for the present volume. To improve the diversity of reactions to the systematicity challenge, we subsequently invited an additional set of essays from a number of researchers who had not been at the workshop (Gideon Borensztajn, Willem Zuidema, and William Bechtel; Jeff Elman; Stefan Frank; Edouard Machery, Gary Marcus; Randall O'Reilly, Alex Petrov, Jonathan Cohen, Christian Lebiere, Seth Herd, and Seth Kriete; Bill Ramsey). The present volume is the finished product of this joint effort. We are confident that it provides a representative sample and an overview of some of the most important developments in the scientific literature on the systematicity of cognition.

We are thankful to many people for their work on this three-year project. First of all, we are very grateful to the contributors. They have worked in a highly collaborative spirit, reviewing each other's work and responding to one another in a way that has helped to unify and consolidate the volume. The sharp focus of this volume has meant that in many of the papers there is some degree of overlap with respect to introductory discussions of systematicity. As editors, we encouraged contributors to set the stage as they wished. In our view, their divergent emphases with respect to the history of the debate and the conceptual landscape of the problem inform their respective responses to the systematicity challenge. While the essays can profitably be read by themselves, or in whatever order the reader chooses, the careful reader will appreciate the organizational principle that underlies this collection of essays into unlabeled blocks, as they appear in parts I through IV in the table of contents. The thematic arrangement is not accidental and the reader may wish to read through from beginning to end in order to find the system for herself.

This project was made possible thanks to the financial support of a number of agencies. We gratefully acknowledge the support of Fundación Séneca (Agencia de Ciencia y Tecnología de la Región de Murcia), who funded us through project 11944/PHCS/09, and to the Ministry of Economic Affairs and Competitiveness of Spain, for its support through Project FFI2009-13416-C02-01. We also thank the European Network for the Advancement of Artificial Cognitive Systems, Interaction and Robotics for additional support.

We are grateful to Philip Laughlin at MIT Press for his patient and encouraging support throughout this process. On a personal note, we wish to thank the generous people of San José for warmly welcoming our motley crew of scholars and scientists. The staff at Hotel Doña Pakyta and Hotel El Sotillo, and especially Oti, the "commander-in-chief," provided the coziest of atmospheres for our workshop. We cannot forget Paco and Pipe, at Bar El Duende, and Amada, at Bar Mónsul, where many participants at the workshop came to understand what philosophers mean by perennial questions.

San José, Almería
Lawrence, Kansas
Easter, 2013

I

1 Systematicity: An Overview

John Symons and Paco Calvo

Introduction

In 1988, Jerry Fodor and Zenon Pylyshyn published "Connectionism and Cognitive Architecture: A Critical Analysis." Their article presented a forceful and highly influential criticism of the explanatory relevance of neural network models of cognition. At the time, connectionism was reemerging as a popular and exciting new field of research, but according to Fodor and Pylyshyn, the approach rested on a flawed model of the human mind. Connectionism is the view that the mind can be understood in terms of an interconnected network of simple mechanisms. Its proponents contend that cognitive and behavioral properties can be modeled and explained in terms of their emergence from the collective behavior of simple interacting and adaptive mechanisms. According to Fodor and Pylyshyn, connectionist approaches neglect an essential feature of thought—its systematic nature. On their view, the basic psychological fact that thoughts are intrinsically related to other thoughts in systematic ways becomes inexplicable if one denies that representations are structured in a syntactically and semantically classical combinatorial manner.[1] Connectionism, they argued, inevitably fails to provide a meaningful explanation of cognition insofar as it confuses the intrinsically systematic nature of thought with a system of associations.[2] Connectionism might shed some light on the way that cognitive architectures happen to be implemented in brains, but the explanation of cognition does not take place at the level of biology or hardware. A cognitive architecture must be systematic to the core in order to shed light on the intrinsically systematic character of cognition.

One prominent message of their article, that a cognitive architecture must explain systematicity in order to explain human cognition, came to be called *the systematicity challenge*. The meaning and implications of the challenge can be interpreted in more than one way, and it quickly generated a vigorous response in philosophy and cognitive science. The ensuing debate resulted in an enormous literature from a variety of perspectives.[3] Philosophers and scientists on all sides of the issue generally agree that the paper helped to sharpen central questions concerning the nature of explanation in cognitive science and that Fodor and Pylyshyn have encouraged the scientific community to think carefully about what the goals of cognitive science should be. Nevertheless, in cognitive science and philosophy, opinion is sharply divided concerning the role of systematicity in a mature science of mind.

The criticism of connectionism in the 1988 paper harks back to Fodor's earlier arguments for the idea that cognition should be understood by analogy with classical computational architectures as a system of rules and representations. In the 1970s, Fodor had argued that the language of thought hypothesis explained the systematic features of thought. On his view, all thoughts that have propositional content are representational in nature, and these representations have syntactical and semantic features that are organized in a way that is similar to the transformational rules of natural languages. Insofar as tokens of mental representation figure in thoughts and insofar as we can judge those thoughts to be true or false, they must be organized in a language-like way (Fodor 1975). Fodor presents the language of thought hypothesis as the best way to account for a range of features of our psychology. The three basic explananda that Fodor highlights in his work are:

(a) The productivity of thought: we have an ability to think and understand new thoughts and previously unheard sentences.
(b) The systematicity of thought: to genuinely understand a thought is to understand other related thoughts.
(c) The principle of compositionality: the meaning of sentences results from the meanings of their lexical parts.

Fodor and Pylyshyn's criticism of connectionism is shaped by Fodor's early articulation of the language of thought hypothesis and by their view that competing explanatory strategies miss what is distinctively cognitive about cognition. Their 1988 article applied a challenging philosophical argument to a lively and ongoing scientific controversy. By any measure, their paper has served as a focal point for one of the most active debates in the philosophy of cognitive science over the past twenty-five years.

In our view, the scientific landscape has changed in ways that call for a fresh look at this influential set of arguments. Most obviously, the quarter-century since Fodor and Pylyshyn's paper has seen the development of new approaches to connectionism that depart in a number of important respects from the modeling strategies that were the direct target of their criticism in the 1980s. More generally, the broader scientific context has changed in ways that are relevant to the kinds of explanations that are available to cognitive science. Fodor and Pylyshyn presented an argument about what should count as the right kind of explanation of cognition. In the intervening years, a range of scientific and philosophical developments have supported alternative approaches to explanation in the study of cognition. Dynamical, embodied, situated, ecological, and other methodologies are no longer exotic or marginal options for cognitive scientists. At the other end of the spectrum, the majority of researchers in neuroscience adopt highly reductionist approaches to the brain, focusing increasingly, and very fruitfully, on the cellular and subcellular details.[4]

Systematicity tends to be conflated with classical computational kinds of cognitive architecture and traditional research in artificial intelligence. However, the contemporary landscape with respect to artificial intelligence has also shifted in important ways. Researchers can choose from a range of classical, hybrid, and nonclassical architectures, along with a growing set of noncognitive architectures inspired by developments in robotics. Much has changed since Fodor and Pylyshyn published their article, and this volume is intended as a way of taking stock of one of the most important debates in the history of cognitive science from a contemporary perspective. The question of what counts as a good explanation of cognition clearly has not been settled decisively.

Setting the Stage for the Systematicity Challenge

It is helpful to locate the systematicity challenge in the context of the development of cognitive science in the second half of the twentieth century. To begin with, it is important to note that Fodor and Pylyshyn were not the first to challenge the explanatory status of nonclassical approaches, including network models, for the study of cognition. They were well aware of the historical context and pointed back to what they saw as the decisive defeat suffered by advocates of perceptrons, behaviorism, and the like. They claimed that the discussion of what the correct architecture of cognition looks like "is a matter that was substantially put to rest about thirty years ago; and the arguments that then appeared to militate decisively in favor of the Classical view appear to us to do so still"

(Fodor and Pylyshyn 1988, 3). Indeed, many of the central conceptual questions were already coming into focus in the early 1960s.

Regarding the big-picture philosophical concerns, the relative status of associationism and rationalism, for example, were in play in throughout the early history of cognitive science. These concerns came into focus, for example, in debates over the explanatory status of perceptron models. The inability of single-layered perceptrons to learn how to approximate non-linearly separable functions was one of the reasons that popular scientific opinion had turned against neural networks two decades earlier. The combination of Chomsky's (1959) review of Skinner's *Verbal Behavior* and Minsky and Papert's (1969) critique of Rosenblatt's (1958) perceptrons encouraged Fodor and Pylyshyn to see the fight against network models and behaviorism as having been settled decisively in favor of classicism and rationalism.

The neural network approach was delayed but not derailed. In addition to the development of well-known parallel distributed processing strategies that culminated in the two volumes of Rumelhart and McClelland's (1986) *Parallel Distributed Processing: Explorations in the Microstructure of Cognition*, developments in the 1970s and early 1980s include Stephen Grossberg's (1982) adaptive resonance theory and self-organizing maps and Kunihiko Fukushima's (1980) neocognitron, among others. Although neural network approaches may not have been as prominent during this period, Grossberg and others were developing increasingly sophisticated formal techniques that allowed researchers to sidestep many of the objections from the 1960s.

Several well-known milestones marked the revival of popular interest in connectionism during the late 1970s and early '80s.[5] The most prominently cited is the development of the backpropagation learning algorithm for multilayered perceptrons (Rumelhart, Hinton, and Williams 1986) that permitted researchers to address problems that had previously been regarded as intractably difficult challenges for network modelers. The debate over the acquisition of the past tense in English played a historically important role in this regard. Heated discussion as to how much of the developmental psycholinguistic data could be accounted for by statistical means alone continued throughout the 1980s and '90s. The phenomenon to be explained takes the following form. As they learn the past tense of English, children develop their abilities in a familiar "U-shaped developmental profile" (Berko 1958; Ervin, 1964). Initially, they correctly produce the past tense forms of regular and irregular verbs. However, they soon go through a period where they make predictable errors in the inflection of verbs, for example, "goed" and "falled" instead of "went" and "fell." Errors of this kind, at this stage in the developmental process, are widely understood as being the result of the overgeneralization of rules. After this decline in their performance, children begin to correct these overgeneralization errors and their performance improves.

According to connectionists, explanations of the characteristic U-shaped pattern of acquisition can be provided by network models that mimic the pattern of acquisition with a reasonable level of accuracy. (See, e.g., Plunkett and Juola 1999.) Against the connectionists, Pinker and Prince (1988) maintained that researchers would need to resort to more abstract symbol-and-rule mechanisms if they were ever to model the phenomenon adequately. At this point, the trade-offs were becoming clear. The kind of approach championed by Pinker, Prince, and others offered a simple and precise solution to the narrow challenge of explaining the U-shaped developmental profile. Connectionist models performed in less precise ways, but had the virtue of being applicable to a broad range of cases rather than being a tailor-made system of symbols and rules that applied narrowly to the specific problem of learning the English past tense. In addition to being more general, connectionists claimed that their models provided the required level of generalization and productivity, without the unparsimonius proliferation of modules that came with symbol-based approaches. It was in this atmosphere of excitement about the prospects of connectionist theory that Fodor and Pylyshyn published their article. As for the debate over the English past tense, we continue to see a lack of consensus as to whether statistical mechanisms embedded in artificial neural networks are genuinely explanatory (Ramscar 2002; Pinker and Ullman 2002).

The terms of this disagreement echo Fodor and Pylyshyn's discussion of what it means to give an explanation of a cognitive phenomenon. Research on language acquisition in infants (Marcus, Vijayan, Bandi Rao, and Vishton 1999; Seidenberg and Elman 1999) and speech processing in adults (Peña et al. 2002; Endress and Bonatti 2007; Laakso and Calvo 2011) are areas where many of the central themes of the original debate continue to be explored.[6] The set of considerations that were raised in debates over the past tense have spread throughout the field of psycholinguistics and the rest of the cognitive sciences, with contemporary versions of the dispute being held between "probabilistic" (Griffiths, Chater, Kemp, Perfors, and Tenenbaum 2010) and "emergentist" (McClelland, Botvinick, Noelle, Plaut, Rogers, Seidenberg, and Smith 2010) models of thought.

While the so-called great past tense debate was affected directly by Fodor and Pylyshyn's criticisms, their argument was aimed at a very general set of questions rather than at the specifics of a particular debate in science. They were unmoved by the fact that this or that area of cognition could actually be modeled statistically. On their view, a simulation that mimics some feature of human cognition is not an explanation of that feature. Their criticism applied equally, for example, to single-layered and multi-layered perceptrons; they were relatively unconcerned with the intricacies of particular battle lines in particular cases. Their criticisms were fundamentally philosophical, targeting the underlying associative character of neural network processing per se. In their view, the "parallel distributed processing" of the 1980s was basically equivalent to the kind of associationism that philosophers would associate with Locke or Hume. Insofar as it claims to be a theory of cognition, connectionism is simply associationism dressed up in the jargon of vectorial patterns of activation, matrices of weighted connections, and gradient descent learning. Any claim to being neurobiologically plausible was also irrelevant to their criticisms of connectionism. Fodor and Pylyshyn's chief concern was the very nature of human cognition, rather than the details of how cognition happened to be implemented in the nervous system. Insofar as connectionist theory echoed British empiricist philosophy of mind, the inferential treatment of classical cognitive science took the side of the rationalists. Thus, while they sided with Pinker and Prince (1988) against Rumelhart and McClelland (1986) in the dispute over the English past tense, Fodor and Pylyshyn had their eye on traditional philosophical questions. They famously conclude their essay by acknowledging that these debates have a venerable heritage in the history of philosophy as well as the more recent history of psychology: "We seem to remember having been through this argument before. We find ourselves with a gnawing sense of déjà vu" (Fodor and Pylyshyn 1988, 70).

By acknowledging the historical precedents for the debate, Fodor and Pylyshyn certainly did not mean to imply that the situation is a stalemate or that this debate exemplifies perennial problems in philosophy that will never be resolved. In their view, there is a clear winner: when it comes to cognition, rationalism is, as Fodor (1975) had earlier claimed of the language of thought hypothesis, *the only game in town*.

The Systematicity of Thought

At the heart of the systematicity debate is a basic disagreement concerning what, precisely, needs to be explained in the science of mind. For example, in the case of language, is it the actual linguistic parsing and production performed by cognitive agents or their competence-level characterization in generative linguistic terms (Chomsky 1965)? What is the correct level of description that captures the distinctively cognitive core of the phenomenon? Another basic difference between the classicists and the connectionists was their attitude toward the autonomy of psychology relative to the details of implementation (Fodor 1974). Fodor and Pylyshyn's strategy was to focus on what they took to be a clear-cut and pervasive set of phenomena that they regarded as uncontrovertibly cognitive and to downplay questions of implementation and performance.

Fodor and Pylyshyn focused on the productive and systematic features of thought along with its inferential coherence. Notwithstanding constraints on human hardware capacities, the productivity of thought refers to our capacity to entertain or grasp an indefinitely large number of thoughts. Thought processes, on the other hand, are systematic to the extent that our capacity to entertain or grasp a thought appears to be intrinsically connected with our capacity to entertain or grasp a number of other semantically related thoughts. Likewise, thought exhibits inferential coherence. Our capacity to follow a pattern of inference appears intrinsically connected to our capacity to draw certain other inferences. The productivity, systematicity, and inferential coherence of thought strongly suggest that mental representations possess a constituent structure without which it is difficult to come to terms with the interconnections among thought-related capacities.[7]

According to Fodor and Pylyshyn, an inference to the best explanation should lead us to regard these explananda as involving operations performed on a stock of representations that can be combined and recombined in accordance with a set of rules. On their view, unless we postulate syntactic and semantic combinatorial relations and unless thought is compositional, the way we are endowed with these abilities remains a mystery. Unstructured connectionist networks would, at first sight, lack the resources to explain the productivity, systematicity, or inferential coherence of thought.

Fodor and Pylyshyn developed a number of parallel lines of argument in support of the superiority of a symbol-and-rule-based approach. In the case of productivity, as Fodor and Pylyshyn observe, the existence of structured representational schemata is inferred from the fact that our competence does not seem to be finite; we appear to be able to entertain an indefinitely large number of thoughts. Arguments from productivity invite a number of responses. Calling the very existence of a competence–performance divide into question was one common line of response. A less direct approach that was favored by some scientists was to simply consider performance and competence separately, approaching one side of the divide and not the other on methodological grounds (see Elman, this volume; Frank, this volume). Although productivity has been a central topic that figured in responses to the 1988 paper, it strikes us that the core of Fodor and Pylyshyn's argument is the notion that our capacity to think is not *punctate*; once we have entertained a thought, we have the resources to deal with others that are semantically related. This is the critical datum that compels us, on their view, to see thought as requiring structured representational schemata.

The simplicity of the argument's starting point is powerful and compelling. It is intuitively obvious that a speaker's capacity to understand native sentences of her language is related to her understanding of a number of other semantically related sentences. Pathologies aside, it is difficult to imagine that someone could understand the sentence "John loves Mary" without having *ipso facto* the resources to produce or understand "Mary loves John."[8]

Imagine attempting to learn a language using only a phrase book. Punctate understanding is what one gets when a phrase book is one's only access to a foreign language. What you can say and understand depends on the particular phrases you happen to look up. This contrasts sharply with the way competent speakers of a language understand sentences. The question is whether this intuitive starting point is sufficient to license Fodor and Pylyshyn's claim that plausible models of human thought must have a classical combinatorial structure.

According to Fodor and Pylyshyn, the systematicity of thought argument against connectionism as an architectural hypothesis of cognition runs as follows:

(i) It is a fact of psychology that thought is systematic insofar as our thoughts are intrinsically related to one another in such a way that having one thought means having the capacity to access an indefinitely large set of other thoughts. So, to take the canonical illustration, someone who can think JOHN LOVES MARY must have the capacity to think MARY LOVES JOHN.

(ii) An explanation of systematicity requires syntactic and semantic constituency relations among mental representations and a set of processes that are sensitive to such internal structure such as those provided by the language of thought (LOT) hypothesis.

(iii) Connectionist theory posits neither syntactic and semantic constituency relations among mental representations, nor any set of processes that are sensitive to the internal structure of mental representations.

(iv) Therefore, connectionism is unable to account for the systematicity of thought.

It should be noted that the fact that LOT provides the explanatory framework required is meant to imply that it *guarantees* the phenomenon and not merely that it is compatible with its occurrence. The explanatory framework is meant to satisfy the sense that the cognitive architecture and cognition itself share a common core. On Fodor and Pylyshyn's view, a genuine explanation involves a robust constraint on acceptable architectures: the demand is that the model accords with systematicity in a way that is not merely the product of an exercise in data-fitting (see Aizawa 2003). Instead, Fodor and Pylyshyn are eager to emphasize that systematicity follows from classical architecture as a matter of nomological necessity and is not simply a contingent fit between the architecture and the explanandum. Of course, this does not mean that it is conceptually impossible that other architectures might play the same explanatory role or even that some other architecture might also have a necessary connection with the explanandum. Indeed, there may be psychological theories that play an equivalent role, but in order to satisfy the Fodorian demand it is not enough to show that the framework is compatible with the explanandum. Insofar as the systematicity of thought is understood to be a basic psychological fact, or even a psychological law (McLaughlin 2009), it would not suffice for a neural modeler to hit upon a configuration of weights that *happens to* allow a connectionist network to mimic the cognitive explanandum.[9] It would need to be shown how it follows of necessity from that architecture, as is supposedly the case for LOT (but see Phillips, this volume). Fodor and Pylyshyn's empirical bet is that we will not be able to find successful explanations of the systematicity of thought that do not involve full-fledged compositional semantics.

Cognitive Architecture

For Fodor and Pylyshyn, the issue is whether connectionism can be understood to serve as an explanatory *cognitive* theory, rather than as a high-level description of the underlying neural substrate of cognition. But this raises some core issues with regard to the very notion of "cognitive architecture."

A cognitive architecture provides the theoretical framework that constrains and aims at explaining the putative mental capacities of a physical system. However, fixing precisely what is meant by a "cognitive architecture" is a basic conceptual problem in itself. Fodor and Pylyshyn adopted Newell's (1980, 1982) distinction between a knowledge level and a physical-symbol-system level in their treatment of neural networks. Newell noted that "given a symbol level, the architecture is the description of the system in whatever system-description scheme exists immediately next below the symbol level" (1990, 81). Appealing in a physical symbol system to the level *immediately below* the symbol level implies a clear division of labor in a cognitive architecture. Clearly, this division of labor is congenial with the autonomy of psychology as championed by Fodor (1974). According to Fodor and Pylyshyn:

The architecture of the cognitive system consists of the set of basic operations, resources, functions, principles, etc. (generally the sorts of properties that would be described in a "user's manual" for that architecture if it were available on a computer), whose domain and range are the *representational states* of the organism. (1988, 5)

Although they make no reference to David Marr in their essay, the latter promoted a particular account of explanation and a distinction of levels that was common currency in the 1980s, and which has helped to frame the discussion of the architecture of cognition ever since. What Fodor and Pylyshyn (and Newell) are after is exemplified by Marr's (1982) well-known tripartite approach to the description of cognitive systems in terms of the computational, the algorithmic, and the implementational. We may thus read them as endorsing a Marrian view of the appropriate analysis of levels. Their aim is to explain the systematicity of thought *algorithmically*, once the phenomenon has been defined at the computational level, and with details of substrate implementation being entirely left aside.

Fodor and Pylyshyn's critical analysis targets neural network modeling only insofar as it presumes to address the cognitive level of explanation. It is in this *top-down* spirit that their challenge is to be read. Fodor and Pylyshyn argue that accounting for the systematicity of thought is only achieved to the extent that connectionist models import the structural features of classical combinatorial processes. However, insofar as this is the case, no alternative algorithm or explanatory framework is actually being provided. On this view, *structured* neural networks are relegated to the status of "implementational connectionism" (Pinker and Prince 1988). Thus, classicists will happily concede that connectionism may well be able to unearth details of the neural substrate that allow for the physical implementation of structured sets of mental representations. Clearly, having the right story about the neural hardware would be a welcome supplement to psychological theorizing. The target for the classicist is possession of an algebraic, symbolic, and rule-based explanation of systematicity, on the one hand, and a neurological understanding of how the capacity to remain structurally sensitive to compositional processes is implemented in the brain and nervous system, on the other.

Let's consider the trade-off between neural structure and cognitive function. Traditionally, the discussion has taken the form of determining what the correct level of description or explanation is (Broadbent 1985). The tendency of both the "eliminative connectionists" and the classicists (Pinker and Prince 1988) has been to focus exclusively on either neural structure or cognitive function. On the cognitive side, the mission has been to explain systematicity in terms of the set of "operations, resources, functions and principles" that Fodor and Pylyshyn regard as governing the representational states of a physical system. By contrast, the eliminative connectionist tendency has been to focus on the structure and processes of the brain and nervous system rather than on cognition.

The idea that there are cognitive mechanisms that explain the systematicity of the human mind, and that any empirically adequate theory should incorporate them, is not at issue. Instead, disagreements center on the question of the relationship between the cognitive and the biological levels of analysis. Although Marr's understanding of the methodological relationship between computational, algorithmic, and implementational levels was highly influential for cognitive science, many philosophers and scientists have argued for a reevaluation of his tripartite methodological framework (Symons 2007). Nowadays, most cognitive scientists find it difficult to accept, without significant qualification, the top-down recommendation that understanding the goal of computations should take priority over the investigation of the implementational level. Marr understood investigations of the computational level to involve the determination of the problem that a system was forced to solve. However, the classical hierarchy of (autonomous) computational, algorithmic, and implementational levels of analysis is not uncontroversially assumed by contemporary cognitive scientists. Neurobiological constraints have become centrally important to the characterization of the computational level. In general, there is a growing tendency to merge top-down and bottom-up considerations in the determination of the architecture of cognition.[10] In this way, contemporary treatments of the correct architecture of cognition concern the relationship between implementation structure and cognitive function.

Another important feature of these debates is the central place given to representation in classicist arguments against connectionism. This topic is far too subtle and complex for an introductory overview, but it is important to note that Fodor and Pylyshyn presented their challenge to connectionism in representationalist terms:

> If you want to have an argument about *cognitive* architecture, you have to specify the level of analysis that's supposed to be at issue. If you're *not* a Representationalist, this is quite tricky since it is then not obvious what makes a phenomenon cognitive. (1988, 5)

It is not at all clear to us how one ought to read this appeal to representationalism in the contemporary context. The connectionist of the 1980s would not see this as particularly problematic insofar as neural network modeling was generally speaking committed to some sort of representational realism, typically in the form of context-dependent subsymbols. But it is no longer the case that contemporary connectionists would accept the kind of representationalist view of the mind that Fodor and Pylyshyn assumed. At the very least, the classicist would need to provide a more developed argument that the denial of standard representationalism is equivalent to some form of noncognitivist behaviorism.[11] While the challenge is not directly aimed at nonrepresentational connectionists, this does not prevent nonrepresentational connectionism from having something to say about systematicity.

Of course, Fodor and Pylyshyn assume that the etiology of intentional behavior must be mediated by representational states. As far as explanation is concerned, this assumption should not be understood as placing an extra, asymmetrical burden of proof on someone wishing to provide nonrepresentationalist accounts of the phenomenon in question. Calvo, Martín, and Symons (this volume) and Travieso, Gomila, and Lobo (this volume) propose neo-Gibsonian approaches to systematicity. While Fodor and Pylyshyn were committed representationalists, neo-Gibsonians can take the systematicity of thought or the systematicity of behavior as an explanandum for cognitive and noncognitive architectural hypotheses alike, irrespective of whether those hypotheses include representations.

Twenty-Five Years Later: Taking Stock of the Architecture of Cognition

Systematicity arguments have figured prominently in discussions of cognitive architecture from the heyday of connectionism in the 1980s and '90s to the advent of a "post-connectionist" era in the last decade. Many of the aforementioned concerns are either explored or cast in a new light in the chapters of this volume.

The present volume represents a collective effort to rethink the question of systematicity twenty-five years after Fodor and Pylyshyn's seminal article in light of the wide variety of approaches in addition to connectionist theory that are currently available. As more implementational details are being honored, artificial neural networks have started to pay closer attention to the neurobiology (Borensztajn et al., this volume; O'Reilly et al., this volume). Overall, what we find is a greater sensitivity in the scientific literature to the aforementioned trade-off between structure and function. Memory, for instance, is not understood as the general configuration of a weight matrix, but instead is modeled by means of specific, neurobiologically plausible attractor networks in which components and their activities are organized with great precision and with considerable effort to maintain biological plausibility.

We may then consider whether networks of these kinds either meet the systematicity challenge or change the terms of the debate in significant ways. In what sense does Fodor and Pylyshyn's critical analysis extend to nonclassical forms of connectionism? When they objected to the possibility of having "punctate minds," they were thinking of models like those one finds in the work of Hebb, Osgood, or Hull; connectionism, according to Fodor and Pylyshyn, was *nothing but old wine in new bottles*. Whether or not they are correct in their judgment that nothing of substance had changed in the period between the precursors of connectionism and the 1980s,[12] superficially at least, there seem to be many reasons to reassess the debate today in light of the widespread use of "nonclassical connectionist" approaches. Fodor and Pylyshyn would acknowledge that connectionist networks have grown in sophistication, but that the basic principles remain the same. Had they written their paper in 2013, the argument would have probably been that at the cognitive level, the architecture of the mind is not a "nonclassical connectionist parser" (Calvo Garzón 2004), and that nonclassical connectionism may at best provide an implementational account of thought. The basic lines of the argument still apply, in spite of impressive scientific developments. This, in itself, is an interesting feature of the debate. Fodor and Pylyshyn are presenting an argument for a particular conception of explanation. Connectionists, no matter how sophisticated, have simply missed what is most important and interesting about cognition. The fact that connectionists are likely to have grown very weary of responses of this kind does not mean that Fodor and Pylyshyn are simply wrong.

In addition to connectionism, a constellation of methodologies and architectures have entered the field, many of which explicitly tackle basic questions concerning the nature of explanation in cognitive science. Modern post-connectionist viewpoints include dynamical, embodied, and situated cognitive science, the enactive approach, and neo-Gibsonian approaches. By focusing their criticism on connectionism, Fodor and Pylyshyn limited the range of hypotheses under consideration in their original article. Throughout his career, Fodor has engaged directly with alternative conceptions of psychological explanation.[13] However, it is fair to say that the Fodorian side of the systematicity debate has maintained (rightly or wrongly) a very fixed picture of what counts as a genuinely cognitive explanation.

The present volume represents the state of play in 2013. Aizawa (chapter 3, this volume) argues that, if one is to pay close attention to what ecological psychology, enactivism, adaptive behavior, or extended cognition actually say, it is unclear what the dividing line between cognition and behavior is—either because these methodologies are behaviorally inclined in themselves, downplaying their relevance to questions concerning cognition, or (even worse) because they identify or conflate cognition itself with behavior. These constitute new challenges to the systematicity arguments in the post-connectionist era. Of course, if that is the case, the rules of the game may no longer be clear. One way or another, Aizawa concludes, the post-connectionist era brings about "tough times to be talking systematicity."

In the immediate aftermath of Fodor and Pylyshyn's article, classicists and connectionists focused primarily on the dichotomy between context-free versus context-dependent constituency relations. Context-dependent constituents that appear in different thoughts as syntactically idiosyncratic tokens were first discussed by Smolensky (1987) and Chalmers (1990). This idea is revisited by Brian McLaughlin (chapter 2, this volume), who takes issue with Smolensky and Legendre's most recent views as presented in *The Harmonic Mind*. According to McLaughlin, Smolensky and Legendre's integrated connectionist symbolic architecture is only able to explain systematicity and productivity, despite its hybridity, by collapsing into a full-fledged LOT model.

Gary Marcus (chapter 4, this volume) defends the idea that the mind has a neurally realized way of representing symbols, variables, and operations over variables. He defends the view that the mind is a symbol system against eliminativist varieties of connectionism. On Marcus's view, minds have the resources to distinguish types from tokens and to represent ordered pairs and structured units, and have a variety of other capacities that make the conclusion that minds are symbol systems unavoidable. In his view, connectionist architectures have proved unable to exhibit this set of capacities. Connectionism therefore continues to be subject to the same sorts of considerations that Fodor and Pylyshyn raised twenty-five years ago. Marcus, however, identifies one particular capacity where classicism has not been able to succeed, namely, in the representation of arbitrary trees, such as those found in the treatment of syntax in linguistics. Unlike computers, humans do not seem to be able to use tree structures very well in mental representation. While we can manipulate tree structures in the abstract, our actual performance on tasks requiring manipulation of tree-like structures is consistently weak.

Marcus appeals for an integrative approach to problems of this kind, arguing that the symbolic and statistical features of mind should be modeled together. Nevertheless, on Marcus's view, the human mind is an information-processing system that is essentially symbolic, and none of the developments in the years since Fodor and Pylyshyn's paper should shake that conviction.

Fodor and Pylyshyn focused on rules as a source of systematicity in language. In chapter 5, Jeff Elman points out that the intervening years have seen an increased interest in the contribution of lexical representations to the productivity of language. The lexicon was initially thought to be a relatively stable set of entities with relatively little consequence for cognition. Elman notes that the lexicon is now seen as a source of linguistic productivity. He considers ways in which systematicity might be a feature of the lexicon itself, and not of a system of rules. He proceeds to provide a model of lexical knowledge in terms of performance and distributed processing without positing a mental lexicon that is independently stored in memory.

Frank (chapter 6, this volume) considers that neural network success in accounting for systematicity cannot rely on the design of toy linguistic environments, and he scales up simple recurrent networks (SRNs; Elman 1990) by using more realistic data from computational linguistics. In his chapter, he sets out to compare empirically a connectionist recurrent neural network with a probabilistic phrase-structure grammar model of sentence processing in their systematic performance under more or less realistic conditions. As Frank reports, the performance of both models is strikingly similar, although the symbolic model displays slightly stronger systematicity. In his view, nevertheless, real-world constraints are such that in practice performance does not differ susbtantially across models. As a result, the very issue of systematicity becomes much less relevant, with the litmus test residing in the learning and processing efficiency of the models in dispute, in their flexibility to perform under adverse conditions, or in the explanation of neuropsychological disorders. This range of phenomena, among others, is what in Frank's view "getting real about systematicity" boils down to. Interestingly enough, systematicity, as presented by Fodor and Pylyshyn, may not be that relevant after all.

Overall, what this type of neural modeling hints at, regardless of the level of realism involved, is that it is the nonlinear dynamics that result from a word's effect on processing that counts. Constituency may be understood in terms of dynamical basins of attraction, where convergence toward stability is compatible with dynamic states involved in combinatorial behavior being transient. It is then a step forward that falls between SRNs and other connectionist networks, and dynamical systems theory.[14] In this way, if SRNs exploit grammatical variations as dynamical deviations through state space, the explanatory framework and formal tools of dynamic systems theory (Port and van Gelder 1995) provides yet a more solid avenue of research. The working hypothesis is that there is no need to invoke information-processing concepts and operations, with combinatorial behavior grounded in sensorimotor activity and the parameter of time. However, although there is a trend to replace the symbols and rules of classical models with quantities, different types of attractors and their basins may furnish different dynamical means of implementing combinatorial structure. Thus, monostable attactors (globally stable controllers inspired in cortical circuitry and which hold single basins of attraction; Buckley et al. 2008), for instance, may hint toward different sets of solutions than those in terms of the trajectories that get "induced by sequences of bifurcations ('attractor chaining')" (van Gelder 1998). Either way, the dynamical setting of monostable attractors, attractor basins, or attractor chaining points toward alternative ways to understand cognition and its temporal basis. Other connectionist proposals that have exploited some of the toolkit of dynamic systems theory use articulated attractors (Noelle and Cottrell 1996), including the deployment of wide enough basins of attraction to capture noisy patterns, and stable enough attractors to remember input indefinitely. But humans do not implement arbitrarily long sequential behavior. If compositionality is to be modeled, it seems it have to depend on other sort of resources, than memory resources per se.[15]

Of course, connectionist "structure-in-time" models are incomplete in a number of respects, most notably in being disembodied and in the fact that the vectorial representations they make use of cannot be taken as theoretical primitives. If we pay attention to ontogenesis in the developmental psychology experimental literature, we find that models point toward the decentralization of cognition (Thelen et al. 2001). It is then a step forward to move from SRNs and dynamical systems theory to an embodied and situated cognitive science (Calvo and Gomila 2008; Robbins and Aydede 2008). Hotton and Yoshimi (2011), for instance, exploit "open" (agent-*cum*-environment) dynamical systems to model embodied cognition with dynamic-based explanations of perceptual ambiguity and other phenomena. According to embodied cognitive science, we should be phrasing *cognitively* the following question: what is it that adults *represent* from the world that allows them to behave systematically and productively? This rendering of the situation presupposes an answer where a connectionist or dynamical phase space obtains stable representational states (regardless of whether they are context-dependent or collapse into context-free states). But embodied constituents are not hidden manipulable states, but rather states that change continuously in their coupling with the environment. A system converges to stable states from nearby points in phase space as a result of external conditions of embodiment of the system and endogenous neurally generated and feedback-driven activity. The appropriate question may then be what sensory-to-neural continuous transformations permit adults to exhibit a combinatorial behavior.

In contrast with connectionist, more or less traditional lines of response, Coram (chapter 11, this volume) focuses on the extended theory of cognition, a framework that has proved valuable in informing dynamic systems models of the mind. Putative explanations of systematicity reside in the wider cognitive system, something that includes language and other structures of public representational schemes. Coram compares this extended shift to a strategy that Pylyshyn has tried out in his research on imagistic phenomenon, and argues that embodied and embedded cognitive science need not redefine the phenomenon of systematicity itself, but rather can account for it in its classical clothes with some revisions to the concept of representation at play. Her proposed explanation combines extended explanatory structures with internal mechanisms.

The systematic features of visual perception appear to be an example of a non-linguistic forms of systematicity (Cummins 1996). Following this thread, in chapter 16, Calvo, Martín, and Symons show how systematicity may also emerge in the context of simple agents, taking a neo-Gibsonian perspective to the explanation of this form of systematicity. The objective of this chapter is to provide an explanation of the emergence of systematic intelligence per se rather than providing a defense of a particular cognitive architecture. To this end, Calvo, Martín, and Symons examine marginal cases of behavioral systematicity in the behavior of minimally cognitive agents like plants and insects, rather than beginning with the linguistically mediated cognition of adult human beings, with the intention to provide a basis for understanding systematicity in more sophisticated kinds of cognition.

Other authors such as Travieso, Gomila, and Lobo (chapter 15, this volume) favor a dynamical, interactive perspective and discuss the alleged systematicity of perception as illustrated by the phenomenon of amodal completion (see Aizawa, this volume). According to Travieso et al., amodal completion emerges globally out of context-dependent interactions and cannot be explained compositionally. They further discuss systematicity in the domain of spatial perception, and argue that although systematic dependencies are not found in perception in general, a Gibsonian ecological approach to perception that recurs to higher-order informational invariants in sensorimotor loops has the potential to explain a series of regularities that are central to perception, despite remaining unsystematic. Research on sensory substitution and direct learning serves to make their case. One way or another, it seems that the fact that extracranial features (bodily or environmental) play a constitutive role for the sake of cognitive processing is compatible both with cognition being extended and with cognition reducing to behavior while the latter is accounted for directly in neo-Gibsonian terms.

As we've already mentioned, probably every single aspect of Fodor and Pylyshyn's argument has been questioned. If from the connectionist corner, rejoinders included buying into constituent structure with an eye to unearthing implementational details of an otherwise LOT cognitive architecture or developing an alternative form of context-dependent constituent structure, more recent post-connectionist responses include variations such as the development of spiking neural network and plasticity implementational models (Fernando 2011) or realistic linguistic settings to feed SRNs without resorting to constituency internalization. It might be possible to bypass classical compositionality by individuating neural network internal clusters in a hierarchical manner (Shea 2007). Work with a SINBAD (set of interacting backpropagating dendrites; Ryder 2004) neural model that exploits cortical hierarchies for the purpose of allowing for increasing generalization capabilities by scaffolding variables as we move cortically away from the periphery of the system may be read in this light. The Hierarchical Prediction Network (HPN) of Borensztajn, Zuidema, and Bechtel (chapter 7, this volume) points in this direction, contributing a notable combination of functional abstraction with substrate-level precision.

In particular, Borensztajn et al., inspired by Hawkins's Memory Prediction Framework, and departing from the emergentist view of unstructured connectionist modeling, explore dynamic binding (Hummel and Biederman 1992) among processing units, and develop a neurobiologically plausible version that is structured and can thus account for systematicity. Encapsulated representations that result from the hierarchical organization of the cortex enable categories to play causal roles. In their chapter, they elaborate on how encapsulated representations can be manipulated and bound into complex representations, producing rulelike, systematic behavior. This, Borensztajn et al. argue, does not make their proposal implementational, because despite the fact that encapsulated units can act as placeholders, the value of encapsulated representations only gets set in the system's interaction with an external environment. As to how to combine encapsulated representations into complex representations, Borensztajn et al. show how temporary linkages between representations of the sort allowed by dynamic binding (Hummel and Biederman 1992) may deliver the goods.

If a hierarchical category structure can play causal roles and account for systematicity by treating constituents as "substitution classes," with an eye to exploiting encapsulated representations for the purpose of respecting compositionality but without retaining classical constituents, Phillips and Wilson (chapter 9, this volume) rely on "universal constructions" for the same purpose. According to them, neither classicism, nor connectionism, nor dynamicism, among other methodologies, for that matter, has managed thus far to fully explain the systematicity of human thought in a way that is not ad hoc. They propose instead a *categorial* cognitive architecture: a category theoretic (Mac Lane 2000) explanation based on the concept of a "universal construction" that may constitute the right level of description to inform empirical sciences. Whereas substitution classes do the trick in hierarchical prediction networks, their model, relying on a formal theory of structure, relates systematically maps of cognitive processes that are structurally preserved, allegedly meeting Fodor and Pylyshyn's challenge.

On the other hand, the SAL framework of O'Reilly, Petrov, Cohen, Lebiere, Herd, and Kriete (chapter 8, this volume) provides a synthesis of ACT-R (Adaptive Control of Thought—Rational; Anderson and Lebiere 1998) and the Leabra model of cortical learning (O'Reilly and Munakata, 2000) and is a plea for pluralism in the cognitive sciences in the form of a biologically based hybrid architecture, where context-sensitive processing takes place first, on the ground of evolutionary and online-processing considerations. Their systems neuroscience approach is then aimed at illuminating how partially symbolic processing, to the extent that human performance happens to approximate a degree if systematicity, is the result of complex interactions, mainly in the prefrontal cortex/basal ganglia (PFC/BG) system.

Although Fodor and Pylyshyn presuppose architectural monism to be the default stance, a commitment to some form of pluralism is shared by a number of authors in this volume. In fact, strategies inspired by "dual-process" theories have gained increasing support in recent years (Evans and Frankish 2009). The dual-process working hypothesis is that the architecture of cognition is split into two processing subsystems, one older than the other, evolutionary speaking. Whereas the former puts us in close relation to our fellow nonhuman animals (e.g., pattern-recognition), the latter system is in charge of abstract reasoning, decision making, and other competencies of their ilk. Gomila et al. (2012), for instance, adopt a dual-process framework to argue that systematicity only emerges in the restricted arena of the newer subsystem. In their view, thought's systematicity is due to the fact that human animals are *verbal* (Gomila 2012). Language, in the external medium, underlies our ability to think systematically (see also Coram, this volume; Travieso et al., this volume). Martínez-Manrique (chapter 12, this volume), in turn, argues that connectionist rejoinders to Fodor and Pylyshyn's challenge have never been satisfactory. Systematicity, nevertheless, need not be a general property of cognition, and in that sense there is some room to maneuver. Martínez-Manrique motivates a variety of conceptual pluralism according to which there are two kinds of concepts that differ in their compositional properties. Relying in part on the dual-process approach, he suggests a scenario of two processing systems that work on different kinds of concepts. His proposal boils down to an architecture that supports at least two distinct subkinds of concepts with different kinds of systematicity, neither of which is assimilable to each other.[16]

Architectural pluralism retains, in this way, at least partially, a commitment to representations, but there are other options. In his contribution, Ramsey (chapter 10, this volume) contends that the fact that systematicity is a real aspect of cognition should not be seen as bad news for connectionism. Failure to explain it does not undermine its credibility since the mind need not have only one cognitive architecture. Ramsey is again calling for some form of architectural pluralism via dual-process theories that would allow connectionist theory to shed its distinctive light on those aspects of cognition that remain unsystematic. However, his proposal is radical enough to allow for the vindication of connectionism despite not just their inability to account for systematicity, but also their not constituting a representational proposal.[17]

In chapter 13, Edouard Machery applies Fodor and Pylyshyn's early criticism of connectionist models to neo-empiricist theories in philosophy and psychology. After reviewing the central tenets of neo-empiricism, especially as presented in the work of Jesse Prinz and Lawrence Barsalou, Machery focuses on its characterization of occurrent and non-occurrent thoughts, he argues that amodal symbols are necessary conditions for non-occurrent thoughts. On Machery's view, if occurrent thoughts are individuated by their origins then some feature of the architecture of cognition other than the contingent history of the learning process is needed to account for thought's inferential coherence. Machery's chapter reflects the continuing influence that systematicity arguments have in contemporary philosophy of psychology.

Still more radical departures come from systems neuroscience, a field that, by integrating the scales of specific neural subsystems with constraints of embodiment for cognition and action (Sporns 2011), provides a further twist in the tale. In fact, from the higher point of view of vastly interconnected subnetworks, the brain as a complex system appears to some authors to defy a part-whole componential reading. With the focus placed in the brain-body-environment as a complex system, Chemero (chapter 14, this volume) calls our attention to a *radical embodied cognitive science* (2009) that suggests that cognitive systems are interaction dominant to some extent, and that this requires that we fully revisit Fodor and Pylyshyn's notion of systematicity. He describes a number of examples and argues that interaction dominance is inconsistent with the compositionality of the vehicles of cognition. Since compositionality underlies the phenomenon of systematicity, cognition happens not to be systematic at least to the extent that cognitive systems are interaction dominant. In addition, Silberstein (chapter 17, this volume) combines systems neuroscience and psychopathology to shed light on theories of standard cognitive functioning. In particular, he proposes to make some empirical progress by studying the effects on cognition and behavior when inferential coherence fails to obtain in patients with schizophrenia. According to Silberstein, the fact that the absence of dynamical subsymbolic properties of biological neural networks correlates with the breakdown in systematic inferential performance tells against a symbol-and-rule approach.

One can imagine a variety of approaches to the evaluation of theories of mind in addition to the systematicity criterion. So, for example, behavioral flexibility, faithfulness to developmental considerations, or performance in real time could all constitute plausible functional constraints on the architecture of cognition.[18] It goes without saying that there are many more approaches that could potentially shed light on these questions than those presented in the pages of this book. Our purpose here has been to consider a sample of nonclassical connectionist empirical and theoretical contestants in light of the conceptual challenge that Fodor and Pylyshyn articulated. The central question for readers is whether the symbol-and-rule stance is required for genuine explanations in cognitive science. In the late 1980s, it was clear to Fodor and Pylyshyn what the answer should be. It may be the case twenty-five years later that their verdict would remain the same, but assessing whether their original arguments continue to have the same force for the range of approaches included in this volume is something we leave to the reader to judge.

Acknowledgments

Many thanks to Alicia Coram for very helpful comments. This research was supported by DGICYT Project FFI2009-13416-C02-01 (Spanish Ministry of Economy and Competitiveness) to PC, and by Fundación Séneca-Agencia de Ciencia y Tecnología de la Región de Murcia, through project 11944/PHCS/09 to PC and JS.

Notes

1. The most thorough examination of the arguments associated with the systematicity challenge is Kenneth Aizawa's 2003 book, *The Systematicity Arguments*.

2. As David Chalmers wrote, their trenchant critique "threw a scare into the field of connectionism, at least for a moment. Two distinguished figures, from the right side of the tracks, were bringing the full force of their experience with the computational approach to cognition to bear on this young, innocent field" (1990, 340).

3. At the time of this writing, "Connectionism and Cognitive Architecture: A Critical Analysis" has been cited over 2,600 times, according to Google Scholar.

4. See, e.g., John Bickle's (2003) defense of the philosophical significance of developments in cellular and subcellular neuroscience.

5. For an account of the history of connectionism, see Boden 2006, ch. 12.

6. See Marcus 2001 for a critical appraisal from the classicist perspective.

7. According to Fodor (1987), constituents appear in different thoughts as syntactically identical tokens. "The constituent 'P' in the formula 'P' is a token of the same representational type as the 'P' in the formula 'P&Q', if 'P' is to be a consequence of 'P&Q'" (Calvo Garzón 2000, 472).

8. On pathological cases, see Silberstein, this volume.

9. After all, neural networks are universal function approximators (Hornik, Stinchcombe, and White 1989). Thus, since they are Turing equivalent (Schwarz 1992), the worry is not whether they can compute, but rather whether they can compute "systematicity functions" without implementing a classical model in doing so.

10. Clark (2013) and Eliasmith (2013) are recent illustrations.

11. Ramsey (2007), for example, has recently argued that only classical cognitive science is able to show that a certain structure or process serves a representational role at the algorithmic level. Connectionist models, Ramsey argues, are not genuinely representational insofar as they exploit notions of representation that fail to meet these standards (but see Calvo Garzón and García Rodríguez 2009).

12. One could argue that already in the 1980s they were working with an incomplete picture of the state of network theory. They make no reference to Stephen Grossberg's adaptive resonance approach to networks, for example.

13. See, e.g., Fodor and Pylyshyn's (1981) response to Gibson.

14. As a matter of fact, the line between connectionist and dynamicist models of cognition is anything but easy to draw (see Spencer and Thelen 2003).

15. In the case of some "structure-in-time" models, such as "long short-term memory" models (LSTM; Schmidhuber Gers and Eck 2002), the implementational outcome is more clearly visible. Long short-term memories are clusters of nonlinear units arranged so that an additional linear recurrent unit is places in the middle of the cluster, summing up incoming signals from the rest. The linear unit allows the system to maintain a memory of any arbitrary number of time steps, which apparently would make the model collapse into our original context-free versus context-dependent dichotomy. In addition, the linear units that LSTM models employ are unbiological.

16. Interestingly, as Martínez-Manrique discusses, if cognitive processes happen not to be systematic in Fodor and Pylyshyn's sense, a nonclassical systematicity argument may be run by analogy to their systematicity argument.

17. This is something Ramsey has argued for elsewhere (Ramsey 2007): that cognitive science has taken a U-turn in recent years away from representationalism and back to a form of neobehaviorism (see Calvo Garzón and García Rodríguez 2009 for a critical analysis).

18. For other constraints and further discussion, see Newell 1980 and Anderson and Lebiere 2003.

References

Aizawa, K. 2003. *The Systematicity Arguments*. Dordrecht: Kluwer Academic.

Anderson, J. R., and C. Lebiere. 1998. *The Atomic Components of Thought*. Mahwah, NJ: Erlbaum.

Anderson, J. R., and C. Lebiere. 2003. The Newell test for a theory of cognition. *Behavioral and Brain Science* 26:587–637.

Berko, J. 1958. The child's learning of English morphology. *Word* 14:150–177.

Bickle, J. 2003. *Philosophy and Neuroscience: A Ruthlessly Reductive Account*. Dordrecht: Springer.

Boden, M. A. 2006. *Mind as Machine: A History of Cognitive Science*. Oxford: Oxford University Press.

Broadbent, D. 1985. A question of levels: Comment on McClelland and Rumelhart. *Journal of Experimental Psychology: General* 114:189–192.

Buckley, C., P. Fine, S. Bullock, and E. A. Di Paolo. 2008. Monostable controllers for adaptive behaviour. In *From Animals to Animats 10: The Tenth International Conference on the Simulation of Adaptive Behavior*. Berlin: Springer.

Calvo Garzón, F. 2000. A connectionist defence of the inscrutability thesis. *Mind and Language* 15:465–480.

Calvo Garzón, F. 2004. Context-free versus context-dependent constituency relations: A false dichotomy. In *Compositional Connectionism in Cognitive Science: Proceedings of the American Association for Artificial Intelligence Tech. Report FS-04-03*, ed. S. Levy and R. Gayler, 12–16. Menlo Park, CA: AAAI Press.

Calvo Garzón, F., and A. García Rodríguez. 2009. Where is cognitive science heading? *Minds and Machines* 19:301–318.

Calvo, P., and A. Gomila, eds. 2008. *Handbook of Cognitive Science: An Embodied Approach*. Amsterdam: Elsevier.

Chalmers, D. 1990. Why Fodor and Pylyshyn were wrong: The simplest refutation. In *Proceedings of the 12th Annual Conference of the Cognitive Science Society*, 340–347. Hillsdale, NJ: Erlbaum.

Chemero, A. 2009. *Radical Embodied Cognitive Science*. Cambridge, MA: MIT Press.

Chomsky, N. 1959. Review of B. F. Skinner's *Verbal Behavior*. *Language* 35:26–58.

Chomsky, N. 1965. *Aspects of the Theory of Syntax*. Cambridge, MA: MIT Press.

Clark, A. 2013. Whatever next? Predictive brains, situated agents, and the future of cognitive science. *Behavioral and Brain Sciences* 36 (3):1–73.

Cummins, R. 1996. Systematicity. *Journal of Philosophy* 93:591–614.

Eliasmith, C. 2013. *How to Build a Brain: A Neural Architecture for Biological Cognition*. New York: Oxford University Press.

Elman, J. L. 1990. Finding structure in time. *Cognitive Science* 14:179–211.

Endress, A. D., and L. L. Bonatti. 2007. Rapid learning of syllable classes from a perceptually continuous speech stream. *Cognition* 105 (2):247–299.

Ervin, S. M. 1964. Imitation and structural change in children's language. In *New Directions in the Study of Language*, ed. E. H. Lenneberg, 163–189. Cambridge, MA: MIT Press.

Evans, J. S. B. T., and K. Frankish, eds. 2009. *In Two Minds: Dual Processes and Beyond*. Oxford: Oxford University Press.

Fernando, C. 2011. Symbol manipulation and rule learning in spiking neuronal networks. *Journal of Theoretical Biology* 275:29–41.

Fodor, J. 1974. Special science, or the disunity of science as a working hypothesis. *Synthese* 28:97–115.

Fodor, J. 1975. *The Language of Thought*. Cambridge, MA: Harvard University Press.

Fodor, J. 1987. *Psychosemantics*. Cambridge, MA: MIT Press.

Fodor, J., and Z. W. Pylyshyn. 1981. How direct is visual perception? Some reflections on Gibson's "Ecological Approach." *Cognition* 9:139–196.

Fodor, J., and Z. W. Pylyshyn. 1988. Connectionism and cognitive architecture: A critical analysis. *Cognition* 28:3–71.

Fukushima, K. 1980. Neocognitron: A self-organizing neural network model for a mechanism of pattern recognition unaffected by shift in position. *Biological Cybernetics* 36 (4):193–202.

Gomila, A. 2012. *Verbal Minds: Language and the Architecture of Cognition*. Amsterdam: Elsevier.

Gomila, A., D. Travieso, and L. Lobo. 2012. Wherein is human cognition systematic? *Minds and Machines* 22 (2):101–115.

Griffiths, T. L., N. Chater, C. Kemp, A. Perfors, and J. B. Tenenbaum. 2010. Probabilistic models of cognition: Exploring the laws of thought. *Trends in Cognitive Sciences* 14:357–364.

Grossberg, S. 1982. *Studies of Mind and Brain: Neural Principles of Learning, Perception, Development, Cognition, and Motor Control.* Dordrecht: Kluwer Academic.

Hornik, K., M. Stinchcombe, and H. White. 1989. Multilayer feedforward networks are universal approximators. *Neural Networks* 2:359–366.

Hotton, S., and J. Yoshimi. 2011. Extending dynamical systems theory to model embodied cognition. *Cognitive Science* 35:444–479.

Hummel, J. E., and I. Biederman. 1992. Dynamic binding in a neural network for shape recognition. *Psychological Review* 99:480–517.

Laakso, A., and P. Calvo. 2011. How many mechanisms are needed to analyze speech? A connectionist simulation of structural rule learning in artificial language acquisition. *Cognitive Science* 35:1243–1281.

Mac Lane, S. 2000. *Categories for the Working Mathematician,* 2nd ed. New York: Springer.

Marcus, G. F. 2001. *The Algebraic Mind: Integrating Connectionism and Cognitive Science.* Cambridge, MA: MIT Press.

Marcus, G. F., S. Vijayan, S. Bandi Rao, and P. M. Vishton. 1999. Rule learning by seven-month-old infants. *Science* 283:77–80.

Marr, D. 1982. *Vision: A Computational Investigation into the Human Representation and Processing of Visual Information.* New York: Freeman.

McClelland, J. L., M. M. Botvinick, D. C. Noelle, D. C. Plaut, T. T. Rogers, M. S. Seidenberg, and L. B. Smith. 2010. Letting structure emerge: Connectionist and dynamical systems approaches to cognition. *Trends in Cognitive Sciences* 14: 348–356.

McLaughlin, Brian P. 2009. Systematicity redux. *Synthese* 170:251–274.

Minsky, M. L., and S. A. Papert. 1969. *Perceptrons.* Cambridge, MA: MIT Press.

Newell, A. 1980. Physical symbol systems. *Cognitive Science* 4:135–183.

Newell, A. 1982. The knowledge level. *Artificial Intelligence* 18:87–127.

Newell, A. 1990. *Unified Theories of Cognition.* Cambridge, MA: Harvard University Press.

Noelle, D., and G. W. Cottrell. 1996. In search of articulated attractors. In *Proceedings of the 18th Annual Conference of the Cognitive Science Society.* Mahwah, NJ: Erlbaum.

O'Reilly, R. C., and Y. Munakata. 2000. *Computational Explorations in Cognitive Neuroscience: Understanding the Mind by Simulating the Brain.* Cambridge, MA: MIT Press.

Peña, M., L. Bonatti, M. Nespor, and J. Mehler. 2002. Signal-driven computations in speech processing. *Science* 298:604–607.

Pinker, S., and M. T. Ullman. 2002. The past and future of the past tense. *Trends in Cognitive Sciences* 6 (11):456–463.

Pinker, S., and A. Prince. 1988. On language and connectionism: Analysis of a parallel distributed processing model of language acquisition. *Cognition* 28:73–193.

Plunkett, K., and P. Juola. 1999. A connectionist model of English past tense and plural morphology. *Cognitive Science* 23 (4):463–490.

Port, R., and T. van Gelder. 1995. *Mind as Motion*. Cambridge, Mass.: MIT Press.

Ramscar, M. 2002. The role of meaning in inflection: Why the past tense does not require a rule. *Cognitive Psychology* 45 (1):45–94.

Ramsey, W. 2007. *Representation Reconsidered*. New York: Cambridge University Press.

Robbins, P., and M. Aydede. 2008. *The Cambridge Handbook of Situated Cognition*. Cambridge: Cambridge University Press.

Rosenblatt, F. 1958. The perceptron: A probabilistic model for information storage and organization in the brain. *Psychological Review* 65 (6):386–408.

Rumelhart, D. E., G. E. Hinton, and R. J. Williams. 1986. Learning internal representations by error propagation. In *Parallel Distributed Processing: Explorations in the Microstructure of Cognition*, vol. 1: *Foundations*, D. E. Rumelhart, J. L. McClelland, and the PDP Research Group, 318–362. Cambridge, MA: MIT Press.

Rumelhart, D. E., and J. L. McClelland. 1986. On learning the past tenses of English verbs. In *Parallel Distributed Processing: Explorations in the Microstructure of Cognition*, vol. 1: *Foundations*, D. E. Rumelhart, J. L. McClelland, and the PDP Research Group. Cambridge, MA: MIT Press.

Rumelhart, D. E., and J. L. McClelland, and the PDP Research Group. 1986. *Parallel Distributed Processing: Explorations in the Microstructure of Cognition*, vol. 1: *Foundations*. Cambridge, MA: MIT Press.

Ryder, D. 2004. SINBAD neurosemantics: A theory of mental representation. *Mind & Language* 19 (2):211–240.

Schmidhuber, J., F. Gers, and D. Eck. 2002. Learning nonregular languages: A comparison of simple recurrent networks and LSTM. *Neural Computation* 14 (9): 2039–2041.

Schwarz, G. 1992. Connectionism, processing, memory. *Connection Science* 4: 207–225.

Seidenberg, M. S., and J. L. Elman. 1999. Do infants learn grammar with algebra or statistics? *Science* 284:435–436.

Shea, N. 2007. Content and its vehicles in connectionist systems. *Mind and Language* 22:246–269.

Smolensky, P. 1987. The constituent structure of connectionist mental states: A reply to Fodor and Pylyshyn. *Southern Journal of Philosophy* 26:137–163.

Spencer, J. P., and E. Thelen eds. 2003. Connectionist and dynamic systems approaches to development. *Developmental Science* (Special Issue) 6:375–447.

Sporns, O. 2011. *Networks of the Brain*. Cambridge, MA: MIT Press.

Symons, J. 2007. Understanding the complexity of information processing tasks in vision. In *Philosophy and Complexity: Essays on Epistemology, Evolution, and Emergence*, ed. C. Gershenson, D. Aerts, and B. Edmonds, 300–314. Singapore: World Scientific.

Thelen, E., G. Schöner, C. Scheier, and L. Smith. 2001. The dynamics of embodiment: A field theory of infant perseverative reaching. *Behavioral and Brain Sciences* 24:1–86.

van Gelder, T. 1998. The dynamical hypothesis in cognitive science. *Behavioral and Brain Sciences* 21:615–665.

2 Can an ICS Architecture Meet the Systematicity and Productivity Challenges?

Brian P. McLaughlin

It has been a quarter of a century since the publication of Jerry Fodor and Zenon Pylyshyn's "Connectionism and Cognitive Architecture: A Critical Analysis." Their seminal paper presents several related challenges to the hypothesis that the cognitive architecture of beings with the ability to think is a connectionist architecture.[1] None concern computational power. There are kinds of multilayered connectionist networks that are Turing equivalent—that can compute all and only the same functions as a universal Turing machine. The challenges are explanatory challenges: challenges to explain certain facts about the abilities of thinkers[2] by appeal to a connectionist architecture that is not implementation architecture for a language of thought (LOT) architecture.[3] The challenges pose the following dilemma for the view that the computational architecture underlying the ability to think is connectionist. If connectionism cannot adequately explain the facts in question, then it fails to offer an adequate theory of the computational architecture of thinkers; and if it explains them by appeal to a connectionist architecture that implements a LOT architecture,[4] then it fails to offer an alternative to the hypothesis that the computational architecture of thinkers is a LOT architecture.

The literature on the challenges is vast, and the reactions to them have been many and varied. I will make no attempt to canvass them here. Moreover, I will be concerned only with the most fundamental challenges posed by Fodor and Pylyshyn: the challenges to explain the systematicity and productivity of thought by appeal to a connectionist architecture that is not an implementation architecture for a LOT architecture. I believe that connectionism can offer an adequate alternative to the LOT hypothesis only if it can meet these challenges. And I believe they have not been met.

After some preliminary discussion, I will examine an attempt by Paul Smolensky and Géraldine Legendre (2006) to meet them. Their attempt is, without question, the best to date. In fact, it is the only attempt to meet the systematicity challenge that Fodor and Pylyshyn actually intended to pose, a challenge that has been badly misunderstood in the literature.[5] I will argue that they do not succeed.

1 The Systematicity and Productivity Challenges

Fodor and Pylyshyn (1988) claim that our thought abilities are productive. By that they mean we can, in principle, think an unbounded number of thoughts. That claim is of course an idealization since our limited life spans, to note just one factor, prevents that. Because some connectionists (e.g., Rumelhart and McClelland 1986, 191) reject idealizations to unbounded cognitive abilities, Fodor and Pylyshyn appeal to the systematicity of thought, which involves no such idealization, to pose a related challenge to connectionism. The idea that thought is systematic is that any being able to have a certain thought would be able to have a family of related thoughts. Abilities to have thoughts are never punctate; they come in clusters. That idea can be captured by saying that they come in pairs, where it is understood that a given thought ability can be a member of more than one pair. The idea that they come in pairs is that it will be true (at least *ceteris paribus*) and counterfactual supporting that a thinker has one member of a pair if and only if the thinker has the other. A paradigm example from the literature of two systematically related thought abilities is the ability to think the thought that *Sandy loves Kim* and the ability to think the thought that *Kim loves Sandy*. One can of course think that *Sandy loves Kim* without thinking that *Kim loves Sandy*. But the claim is that any being able to think the one thought would be able to think the other. By "think the thought that," Fodor and Pylyshyn do not mean "believe that" or "think that." They sometimes mean "entertain the thought that," but, more generally, they mean "mentally represent in thought that." The systematicity claim is thus that a thinker has the ability to mentally represent in thought that *Sandy loves Kim* if and only if the thinker has the ability to mentally represent in thought that *Kim loves Sandy*.

Fodor and Pylyshyn (1988) claim that thought abilities come in clusters *because* members of the clusters are intrinsically connected. What it is for thought abilities to be intrinsically connected is for them to be complex abilities that are constituted by at least some of the same abilities. The ability to mentally represent that *Sandy loves Kim* and the ability to mentally represent that *Kim loves Sandy* both involve the ability to mentally represent Kim, the ability to mentally represent Sandy, and the ability to mentally represent one individual as loving another. These are conceptual abilities. The intrinsic connectedness idea thus seems to be that systematically related thought abilities are *constituted* by the same conceptual abilities.[6] Let us call the thesis that abilities to mentally represent in thought that something is the case are constituted by conceptual abilities "the conceptual constitution thesis."[7] The systematicity challenge might be formulated as follows: to explain, by appeal to a connectionist architecture that does not implement a LOT, counterfactual-supporting generalizations asserting the copossession of mental representational abilities that are constituted by the same conceptual abilities.

Appeal to the conceptual constitution thesis is not question begging in the current dialectical context, for the conceptual constitution thesis does not imply that there is a LOT.[8] Of course, if the challenge is formulated in this way, then connectionists that reject the conceptual constitution thesis will reject the challenge. They would, then, face a different challenge. They would have to explain abilities to mentally represent in thought without appeal to the hypothesis that such abilities are constituted by conceptual abilities.

The systematicity challenge can be formulated without appeal to the conceptual constitution thesis and, indeed, even without appeal to conceptual abilities. Counterfactual-supporting generalizations correlating the relevant pairs of representational abilities—low-level psychological laws[9]—can be identified by appeal to sentence schemata. That is, we can specify schemata such that the English sentences that are instances of them are such laws (or express such laws)—systematicity laws. In the effort to characterize the systematicity of thought in the least contentious way, this has been the most common practice in the literature that followed Fodor and Pylyshyn 1988.[10]

The grammar of predicate logic is completely understood. Although the grammar of English is not, enough is known about it that it is uncontroversial that predicate logic schemata have instances in English. Consider, then, this sentence schema:

(S1) A thinker is able to mentally represent in thought that aRb if and only if the thinker is able to mentally represent in thought that bRa.[11]

Our earlier paradigm example is an instance of this sentence schema, as are no end of other counterfactual supporting generalizations in English. If we avoid using the horseshoe (using instead "if ... then," which is not captured by the horseshoe), we can appeal to the following sentence schema to identify an unbounded number of other systematicity laws expressible in English, without making any assumptions about the grammar of English beyond that there are English sentences with no truth-functional structure (ones that can be substitution instances of "P" and "Q"):

(S2)   A thinker is able mentally represent in thought that *if P then Q* if and only if the thinker is able to mentally represent in thought that *if Q then P*.

Many other sentence schemata can be formulated to identify further systematicity laws. But the two schemata in question serve to illustrate the strategy for identifying systematicity laws. In what follows, we will only have occasion to appeal to our paradigm example of systematicity.

As proponents of the *representational* theory of mind, LOT theorists maintain that mental abilities to represent (nomologically) require possession of mental representations. That is a nontrivial thesis to maintain, because from the fact that we mentally represent things, it does not logically follow that we have mental representations. As proponents of the *computational* theory of mind, LOT theorists maintain that mental abilities consist in subpersonal-level abilities to compute cognitive functions. As proponents of *LOT*, they maintain that the ability to compute cognitive functions consists in possession of a certain kind of mental symbol system and algorithms for constructing and manipulating the symbols. This last thesis is, of course, one that non-implementational computational connectionists dispute.

We will focus here exclusively on the cognitive architecture of human beings (as do Smolensky and Legendre [2006]). A human LOT architecture will include a finite base of atomic mental symbols, and algorithms that can construct molecular symbols from them. The grammar of LOT will be such that it will generate a mental symbol that means that ψ, for all and only the ψs such that a normal human being has the capacity[12] to mentally represent in thought that ψ.[13] On this view, conceptual representations with complete contents (propositional contents) are sentences in LOT.

The LOT explanation of productivity is that the algorithms for generating molecular mental symbols from atomic ones involve recursive procedures by means of which an unbounded number of mental symbols with propositional content could be constructed. As we noted, the claim that thought is systematic, unlike the productivity claim, does not imply that we have an unbounded competence. In a nutshell, the LOT explanation of the systematic relations among pairs of mental representational abilities is that the abilities in such pairs will share a computational architectural basis; that is, there will be an architectural basis that is a basis for both abilities. That basis will consist in the architectures including certain atomic mental symbols and an algorithm for constructing mental sentences with either of the contents in question from those atomic symbols. Since both abilities share such an architectural basis, anyone who has the one ability would *ipso facto* have the other.

2   Can Distributed Connectionist Representations Have Symbolic Structure?

The hypothesis that there are mental representations with syntactic structures thus figures essentially in how LOT theorists explain the systematicity and productivity of thought. Connectionists that aim to meet the systematicity and productivity challenges must either show us how to explain systematicity and productivity without appeal to representations that have syntactic structures or else show us how a connectionist architecture can include such representations without implementing a LOT. Connectionists have not told us how to explain either systematicity or productivity without appeal to representations with syntactic structure. So, the question arises whether connectionist representations can have syntactic structure.

Tim van Gelder tells us

the best answer to the received question about Connectionist representations [Are they syntactically structured?] is that they are *neither* syntactically structured *nor* not syntactically structured. Neither answer characterizes them acceptably. We need to set aside that particular vocabulary, which is embedded in and effectively limited to the Classical framework [the LOT framework], and develop a new set of conceptual tools which will be sufficient to comprehend the emerging connectionist alternative. (1991, 372; emphasis in the original)

That cannot be the best answer. The reason is that that answer is equivalent to a contradiction: connectionist representations are not syntactically structured and are not not syntactically structured. I take it that van Gelder did not intend to make an assertion that flouted the principle of noncontradiction, but meant instead to deny an instance of the law of the excluded middle, in particular, "Either connectionist representations have syntactic structure or they do not." Classical logic embraces the law of excluded middle, and so requires us to accept that claim. So do some nonclassical logics, for example, quantum logic. But nonclassical logics typically do not embrace the law of excluded middle. So perhaps in addition to rejecting classical computationalism van Gelder means also to reject classical logic, and indeed any logic that embraces the law of excluded middle. Classical logic and Zermelo–Fraenkel set theory provide the foundations of classical mathematics, the mathematics deployed throughout the sciences. But perhaps the revolution that van Gelder envisions is supposed to extend to classical mathematics as well, and so to all of applied mathematics in science.

In any case, a problem remains for van Gelder even if he eschews logics that include the law of excluded middle. The problem is that in virtually every nonclassical logic, one cannot deny an instance of the law of excluded middle on pain of contradiction. Not only in classical logic and quantum logic, but also in intuitionistic logic, weak three-valued logic, strong three-valued logic, and relevance logics, one cannot deny, on pain of contradiction, that connectionist representations either have or do not have syntactic structure. The pain of contradiction is that a contradiction entails every claim, and, so, entails, for instance, that the Moon is made of green cheese.

Paraconsistent logics are logics in which a contradiction does not entail every claim. Perhaps van Gelder meant after all to make a claim that flouted the principle of noncontradiction, but would try to escape the pain of contradiction by embracing a paraconsistent logic. Paraconsistent logics mainly differ just in their treatment of conditionals, and van Gelder is not making a conditional claim. So, let's consider the logic of paradox, the leading paraconsistent logic (Priest 2006). If the logic of paradox is the correct logic, then it would not be the case that in denying that connectionist representations either have or do not have syntactic structure, van Gelder is committed to the claim that the Moon is made of green cheese. But although he would not be committed to that claim, he will still have to say that connectionist representations have syntactic structure. And he will also have to say that connectionist representations do not have syntactic structure. Let it suffice to note that anyone who says both of those things has said too much.

So, we can return in good conscience to our question: Can connectionist representations have syntactic structure?

Connectionist architectures can have either of two kinds of representations: local or distributed.[14] A local connectionist representation is an individual unit that is activated or that has a certain level of activation (if units have levels of activation).[15] As Fodor and Pylyshyn (1988) point out, although in diagrams of connectionist networks individual units are sometimes labeled using symbols with syntactic structures, a local connectionist representation does not have a syntactic structure. It can, at best, only represent a syntactic structure. (In the toy network models in which they are labeled by symbols with syntactic structures, they represent syntactic structures by the modeler's stipulations.) Local connectionist representations are atomic representations: they do not contain other representations as constituents, as proper parts, and so they lack constituent structures.[16] A distributed connectionist representation is an *n*-tuple of units exemplifying a certain *n*-adic pattern of individual unit activation (at a time or throughout an interval of time).[17] Fodor and Pylyshyn (1988) claim that distributed representations too lack syntactic structure, and so are atomic. They maintain that although distributed representations can represent syntactic structures, they cannot themselves have syntactic structure.

It is generally acknowledged that local connectionist representations cannot have syntactic structure. But the claim that distributed connectionist representations cannot have syntactic structure remains vigorously disputed.

In a series of articles and technical reports, some single authored, some coauthored, starting in the mid-1980 and continuing through the 1990s,[18] Smolensky and various coauthors have championed the view that distributed representations can have syntactic structure. This work culminated in 2006 in Smolensky's massive two-volume book, coauthored with Legendre, entitled *The Harmonic Mind: From Neural Computation to Optimality-Theoretic Grammar*; the two volumes total 1,174 pages. The centerpiece of the book is Smolensky's Integrated Connectionist/Symbolic architecture (ICS), for which he was awarded in 2005 the David E. Rumelhart Prize for Contributions to the Theoretical Foundations of Human Cognition. In what follows, I will focus on the discussions in these two volumes, which present the most mature formulation of Smolensky's views about cognitive architecture.[19]

Smolensky and Legendre (2006, vols. 1 and 2) embrace a computational theory of mind and acknowledge that certain mental representations are systematically related, that our system of mental representation is productive, and that the explanation of systematicity and productivity must appeal to syntactically structured representations. But they maintain that systematicity and productivity can be explained by appeal to a connectionist architecture that has distributed representations with syntactic structure yet does not implement a LOT architecture. Indeed, they maintain that all of the challenges that Fodor and Pylyshyn pose for connectionism can be met by appeal to such an architecture (2006, vol. 1, 101). Our focus will remain on systematicity and productivity.

3   The Promise of an ICS Architecture

I will now present, mainly in their own words, Smolensky and Legendre's overall position on human cognitive architecture. They state:

> The search for a coherent computational architecture for the mind/brain has led cognitive science into a major crisis. ... The computational architecture of the *brain* seems to be connectionist, while the most successful explanation of the *mind* has depended on a very different architecture, symbolic computation. (2006, vol. 1, 101)[20]

And they say that "the fundamental tension that frames the research presented in this book" (2006, vol. 1, 31) is raised by the following questions:

> What type of computation is cognition? Connectionist "brain" computation—massively parallel numerical processing? Or symbolic "mind" computation—rule governed manipulation of combinatorially complex, discrete, abstract symbol structures? (ibid.)

On their view,

> The answer is this: it is both. What makes a mind/brain, what gives rise to human cognition, is a complex dynamical system that is a massively parallel numerical computer at a lower level of formal description—a level closer to the biophysical—and, at the same time, a rule-governed processor of discrete symbolic structures at a higher level of description—a level considerably more abstract than the biophysical, but nonetheless governed by formal laws, and related to the lower level description by precisely defined mathematical mappings. (ibid.)

They say, moreover:

> The connectionist computer is a description of the mind/brain at a lower level, while the symbolic computer is a description *of one and the same system* at a higher, more abstract level. (ibid.; Italics theirs)[21]

They thus appear here to posit a cognitive symbolic computer in "the mind/brain."

Smolensky posits not only mental symbols with syntactic structure, but a mental symbol system with a compositional semantics (2006, vol. 2, 544–545).[22] Further, he tells us:

> To address the problem of cognitive productivity [and systematicity], ICS adopts the combinatory strategy. ... Elements of a cognitive domain—for example, scenes or

sentences—are mentally represented using complex symbol structures. New inputs are represented as novel combinations of familiar input constituents; their outputs are generated by creating a corresponding novel combination of familiar output constituents. (2006, vol. 2, 556)

He thus states: "the combinatorial strategy explains the productivity of cognition, both in PSA [Purely Symbolic Architectures] and ICS" (ibid. 543). And he maintains the same for systematicity. He justifies this appeal to the combinatorial strategy in the same way that LOT theorists do. Speaking for all of the authors, Smolensky says:

The authors of this book have adopted this solution because we believe it has led to tremendous progress, and that no other comparably promising alternative currently exists. (ibid., 556)

Given the presentation thus far, it may very well appear that an ICS architecture is an implementation architecture for a symbolic (LOT) architecture.[23] If it is, then appeal to it will not answer any of Fodor and Pylyshyn's challenges. The reason, of course, is that their challenges are to explain certain features of cognitive abilities by appeal to a connectionist architecture that does not implement a symbolic architecture.

Smolensky and Legendre, however, are very explicit that an ICS architecture is not an implementation architecture for a symbolic architecture. They say:

ICS does not represent either an eliminativist or an implementationalist position on the relation between connectionist and symbolic computation ..., but a novel intermediate position. (2006, vol. 1, 33)

They thus tell us:

The ICS theory developed in this book aims to break the deadlock between the eliminativists—who claim that symbols have no place in a science of cognition—and the implementationalists—who maintain that symbolic computation provides all we need for a cognitive theory, with neural networks "merely implementing" symbolic theories. ICS theory, we argue, takes us closer to a satisfactory computational characterization of a mind/brain, assigning both symbolic and connectionist computation essential roles that preserve the strength of the former for mental explanation, and the advantages of the latter for reducing cognition to neurally plausible elementary computations. (ibid., 101)

4 Symbols and Cognitive Processes

Given that an ICS architecture includes a symbol system with a grammar and a compositional semantics and, moreover, the combinatorial strategy is deployed in explaining cognitive phenomena such as systematicity and productivity, why is not an ICS architecture (at best) a connectionist architecture that implements a symbolic (LOT) architecture?

A hint at the answer is given in the following passage:

> In the work presented in this book, ideas and techniques developed in mathematical physics and in computer science are brought together to construct a formal characterization of a mind/brain as a computational system that is parallel and numerical when described at the lower level, but, when described at the higher level, is symbolic in crucial respects. ... The "psychological reality" of the rules and symbols in the new computational architecture is somewhat reduced relative to traditional symbolic theories of mind. Yet the role of symbols and rules in explaining the crucial properties of higher cognition is an essential one. For this reason, the new architecture we develop here is called the Integrated Connectionist/Symbolic Cognitive Architecture (ICS): in this architecture, connectionist and symbolic computational descriptions each play an essential role in overall cognitive explanation. (2006, vol. 1, 33)

Despite playing an essential role in explaining "the crucial properties of higher cognition"—which, by their own lights, includes the systematicity and productivity of thought—an ICS architecture is supposed not to be an implementation architecture for a symbolic architecture because "the psychological reality" of the rules and symbols is "somewhat reduced relative to traditional symbolic theories of mind." It is the reduced psychological reality of symbols and rules for manipulating them that prevents an ICS architecture from being an implementation architecture.

But what is meant by saying that their psychological reality is "somewhat reduced"? Are there degrees of psychological reality? Attempting to answer the last question in the affirmative is not a promising way to go. But a few pages after the passage last quoted, Smolensky and Legendre offer an explanation of why an ICS architecture is not an implementation architecture that makes no appeal to the idea of reduced psychological reality. They say that an ICS architecture is not such an implementation architecture

> because only symbolic *representations*, and not symbolic *algorithms* for manipulating them, are claimed to be cognitively relevant. (2006, vol. 1, 36)

The key idea is that although an ICS architecture includes mental symbols (indeed syntactically structured mental symbols), it does not include symbolic algorithms. All of its algorithms are connectionist. Smolensky sometimes puts this by saying: "symbols are computationally relevant, but not symbolic algorithms" (2006, vol. 2, 519). Symbols are computationally relevant in that our cognitive systems represent the arguments and values of the functions it computes. But the algorithms by which it computes cognitive functions are not symbolic algorithms.

An implementation architecture for a symbolic architecture must include symbolic algorithms, not as fundamental algorithms, of course, but as algorithms derivative from the fundamental nonsymbolic algorithms of the architecture. (By fundamental algorithms, I mean basic algorithms, algorithms that are not executed by means of executing other algorithms.) If ICS architectures contain no symbolic algorithms at all, not even as derivative algorithms, then it is indeed the case that no ICS architecture is an implementation architecture for a symbolic one.

Of course, then, ICS will not yield at least one of its promises. Smolensky and Legendre say, you will recall: "The connectionist computer is a description of the mind/brain at a lower level, while the symbolic computer is a description *of one and the same system* at a higher, more abstract level" (2006, vol. 1, 31). If our cognitive system is an ICS system, then it will not be a connectionist computer at a lower level of description and a symbolic computer at a higher level of description. The reason is that it will not be a symbolic computer at any level of description, since it will not include symbolic algorithms at any level of description. There can be no symbolic computer without symbolic algorithms.

It is one question whether an ICS architecture includes symbolic algorithms among its basic or fundamental algorithms. It is another question whether it includes symbolic algorithms as derivative algorithms—algorithms whose steps are executed by executing nonsymbolic algorithms. Why does an ICS architecture not include symbolic algorithms as derivative algorithms? Smolensky and Legendre do not draw the distinction in question. What they do is insist that symbols are not "process relevant" (2006, vol. 2, 518). Indeed, it is the process irrelevance of symbols that Smolensky and Legendre have in mind when they say: "The 'psychological reality' of the rules and symbols in the new computational architecture [ICS architecture] is somewhat reduced relative to traditional symbolic theories of mind" (2006, vol. 1, 33). In a symbolic architecture, symbols are process relevant; in an ICS architecture, they are not.

By saying that symbols are "not process relevant," they mean that symbols are not cognitive process relevant. They do not mean to maintain that symbols do not participate in any causal processes whatsoever, and so are epiphenomena—devoid of any causal effects. Were that the case, one wonders what reason there would be for positing them. What they mean by saying that symbols are cognitive process irrelevant in an ICS architecture is that the algorithmic processes that can occur in an ICS architecture never consist in the manipulation of symbols. The algorithms such processes execute are thus not symbolic algorithms. In fact, they are connectionist algorithms. If it is indeed the case that an ICS architecture includes no symbolic algorithms even as derivative or nonbasic algorithms, then an ICS architecture is indeed not an implementation architecture for a symbolic architecture. It is not a symbolic computer at any level of description.

If, however, an ICS architecture includes no symbolic algorithms even as derivative or nonbasic algorithms, then, in what sense is the combinatorial strategy deployed in explanations of cognitive phenomena? In deploying the combinatorial strategy, LOT theorists appeal to algorithms that manipulate and construct symbols. How can the combinatorial strategy be deployed without such algorithms? To address that question properly, we will have to examine Smolensky and Legendre's discussions of ICS architectures and computational explanations of cognitive abilities in considerable detail.

5   Principle 1 of ICS Architectures

Smolensky and Legendre characterize ICS architectures by appeal to four principles. They tell us: "The first two principles concern how connectionist computation realizes symbolic computation" (2006, vol. 1, 74). Only these two principles need concern us here, since only they bear on why an ICS architecture is supposed to not be an implementation architecture for a symbolic architecture.[24]

Smolensky and Legendre tell us: "In integrating connectionist and symbolic computation, the most fundamental issue is the relation between the different types of representations they employ" (2006, vol. 1, 65). The different types of representations in question are atomic symbols and symbol structures, on the one hand, and distributed representations on the other. The issue is how they are related.

The first principle is supposed to address that issue. Smolensky and Legendre say:

The ICS hypothesis concerning this relation is our first principle, P_1, informally stated in (1).

(1)   **P1. Rep$_{ics}$:** Cognitive representation in ICS

Information is represented in the mind/brain by widely distributed activity patterns—activation vectors—that, for central aspects of higher cognition, possess global structure describable through the discrete structures of symbolic cognitive theory. (2006, vol. 1, 65)

There is much to discuss here.

By "the discrete structures of symbolic cognitive theory," they mean of course the atomic symbols and the symbolic structures that symbolic cognitive theory (LOT theory) posits. It is, however, important not to conflate being distinct with being discrete. Let us first address distinctness, then discreteness. LOT theory posits a system of distinct mental symbols, and so of distinct mental representations. But representational connectionism posits distinct representations too, either local or distributed. ICS (by stipulation) contains only distributed connectionist representations, and so contains no local representations. So, let us consider distributed connectionist representations. Given that types of distributed connectionist representations are identical to certain types of patterns of activation over groups of units, if two different types of patterns of activation over even exactly the same units are both types of distributed representations, then they are distinct (i.e., non-identical). Thus, ICS too posits a system of distinct representations.

What, though, about discreteness? There is a sense in which symbolic theory is committed to denying that all symbols are discrete. It posits molecular symbols. Molecular symbols are in a sense not discrete, since two molecular symbols can both share a symbol as a proper part. Symbolic structures are molecular symbols, and so distinct symbolic structures need not be discrete. Perhaps Smolensky and Legendre have in mind that distinct atomic symbols are discrete. Atomic symbols do not have other symbols as parts, and so do not share symbols as parts. That, Smolensky and Legendre might maintain, makes them discrete. It is, however, a substantive issue whether distributed representations are discrete in that sense. Fodor and Pylyshyn (1988) maintain that distributed representations are discrete in that sense, and so cannot have syntactic structure. Smolensky and Legendre may hold that distributed representations are not discrete since distinct ones can share subsymbolic (subrepresentational) parts, for example, a certain individual unit at a certain level of activation. An individual unit at a certain level of activation will not be a symbol or representation in an ICS architecture, because such an architecture contains no local representations. Distributed representations are indeed not discrete in that sense. But atomic symbols in a symbolic architecture need not be discrete in that sense either. Smolensky says at one point that symbolic architectures use "symbols all the way down" (2006, vol. 2, 506). But, given that symbols are representations, that is not in general the case. (Like

Smolensky and Legendre, I am using "symbols" in such a way that every symbol is a representation.) Two distinct atomic symbols can share sub-symbols (subrepresentations) as parts. I have in mind here the familiar distinction in symbolic theory between bits and bytes, and the fact that the individual bits need not be representations, and so need not be symbols. Atomic symbols can be thus distributed over constituents that are not symbols, but rather subsymbols. It should be noted, moreover, that it is compatible with the LOT hypothesis that atomic symbols are, or are con-stituted by, or are realized by[25] distributed connectionist representations. Indeed, one of those disjuncts will be the case in a LOT architecture that is implemented by a connectionist architecture.

Smolensky and Legendre say that distributed activity patterns "possess global structure describable through the discrete structures of symbolic cognitive theory" (2006, vol. 1, 65). The locution "describable through" is awkward. Presumably, it means "describable by means of" or "describable through the use of." But if our concepts are classical cognitive symbols, then Smolensky and Legendre's claim is trivial, since anything whatsoever that is in principle describable by us is in principle describable by (describ-able through the use of) a symbol in our cognitive system. As their subse-quent discussion makes clear, however, what they mean is that the global structure is describable by means of the terms of a symbolic cognitive theory in the sense that it is describable *as* a symbol system. Thus, they say, more perspicaciously:

A major contribution of the research program presented in this book is the demon-stration that a collection of numerical activity vectors can possess a kind of global structure that is describable as a symbol system. (2006, vol. 1, 66)

If a collection of numerical activity vectors possesses a kind of global struc-ture that is describable as a symbol system, then that kind of global struc-ture is a symbol system. (This is just an instance of the general principle that if anything is describable as F, then it is F.[26]) Thus, they seem to hold that a collection of numerical activity vectors can be a symbol system.

I noted earlier that Smolensky and Legendre take mental representations to be distributed representations. I also noted that a distributed connec-tionist representation is a group of (an n-tuple of) units exemplifying a certain ordered pattern of individual unit activation. That is a complex state of certain units within a network. But in the remark quoted above (from 2006, vol. 1, 66), they seem to take a collection of activity vectors to be a symbol system. Activity vectors are n-tuples of numbers, each number indexing the activation value of an individual unit in an n-tuple of units.[27] But, then, is a mental representation (symbol or symbol structure) in a cognitive ICS architecture supposed to be a pattern of activity over an n-tuple of units, or is it supposed to be an activity vector?

It is important to distinguish three kinds of things. First, there are "the structures of symbolic cognitive theory." They are mental symbols with syntactic structures and their symbolic constituents, including their ultimate symbolic constituents, atomic symbols. Second, there are distributed representations: groups of n units exhibiting an n-adic pattern of activation levels. A group of n units having a certain n-adic pattern of activation levels is, as I noted, a kind of state. Let us call such states "distributed activation states." (This is my term, not theirs.) Smolensky and Legendre maintain that the distributed activation states of the ICS architecture that is our cognitive architecture will be biophysical states of our brain. More specifically, they take it that the units will be neurons and the activation levels might be, for example, the firing rates of the neurons. (They do not, however, commit to the activation levels being firing rates; they are noncommittal about such details of the neural implementation.) They thus take it that a distributed activation state in the ICS architecture that is our cognitive architecture will be a complex state of a group or population of neurons. Third, there are activation vectors. Vectors are mathematical entities. An activity (or activation) vector is an n-tuple of numbers, where the individual numbers, the components or elements of the n-tuple, give a coordinate in a certain dimension of an n-dimensional vector space and the n-tuple as a whole gives the coordinates of a point in an n-dimensional vector space (2006, vol. 1, 161). The vector (1.0, –1.0., 0.8), for instance, gives the coordinates of a point in a three-dimensional vector space. What makes a numerical vector an activity vector is that the individual numbers of the n-tuple represent (index) the activation values of the corresponding individual units in the n-tuple of units that, at those respective levels of activation, comprise the distributed activation state. We represent points in activation space by the use of numerical vectors. Activity vectors represent distributed activation states by a point in a vector space.[28] So, then, three kinds of things are invoked in the characterization of an ICS architecture: (1) the structures of symbolic cognitive theory; (2) distributed activation states; and (3) activation vectors. The question I asked at the end of the preceding paragraph is essentially just whether, in a cognitive ICS architecture, mental symbols and symbol structures are supposed to be certain distributed activation states or instead certain activation vectors. They cannot be both, since activation vectors are not distributed activation states. The latter are represented by the former, not identical with them.

In answer to my question, Smolensky and Legendre maintain that mental representations are distributed activation states, certain kinds of biophysical states of the brain. Mental representations are not identical with points in a vector space. Rather, they are distributed activation states that can be represented by points in a vector space. They maintain that a system of distributed activation states can be a symbol system, because of its global structure. Activation vectors come into their picture in this way: they appeal to activation vectors to explain how such a system can be a symbol system.

As the foregoing discussion illustrates, Smolensky and Legendre have a tendency to run together talk of activation vectors with the activation states that they represent. This is encouraged by their use of "activity pattern," an expression they sometimes use to mean (what I have called) a distributed activation state and sometimes use to mean an activity vector that represents such a state. In their defense, it should be noted that it is common even in the literature in physics, the branch of science from which they draw their mathematics, to run together talk of vectors and the states they represent. To illustrate, it is common even in the physics literature to talk of superimposing states. But superposition is in the first instance a mathematical operator on vectors (Byrne and Hall 1999). Any state that can be represented by a vector can be represented by the super-position of two vectors. Confusing vectors with the states they represent is analogous to a use-mention confusion. It is like confusing "oil" with oil. Such confusions can of course lead to other mistakes; indeed, the history of philosophy is rife with use-mention mistakes. But the slide from talk of vectors and to talk of states that one finds in the physics literature is typi-cally harmless, leading physicists to no confusions. The slide is sometimes harmless in Smolensky and Legendre's discussions. But it is not always harmless.

To cite an example of harm, Smolensky raises the following problem about fully distributed representations, which he says Jerome Feldman posed to him in personal communication in 1983:

(3)   The two-horse problem

If the representation of *one* horse fills up an entire network, how is it possible to represent *two* horses? (2006, vol. 1, 160)

That is a superb question, and a telling one, since "fully distributed" rep-resentations are supposed to involve all of the units in a network (or in some portion of a network, a portion over which they are fully distributed).

Of course, one pattern of activation over the units of a network might represent a horse, while a different pattern of activation over those very same units represents a different horse (or even represents two horses or something else entirely). The units in the network may exhibit the first activation pattern at one time t, and thereby represent one horse; and those same units may exhibit the second activation pattern at another time t^*, and thereby represent another horse. But I take it that Feldman's question is this: if a representation of something fills up an entire network, then how is it possible for the network to also represent something else at the same time?

Smolensky immediately reframes the question this way:

More prosaically: if the representation of a symbol **A** is distributed throughout a certain portion of a network, how can the representation of a second symbol **B** also be distributed thought that same portion of the network—as is necessary for the distributed representation of even an extremely simple symbol structure like the set {**A**, **B**}? (2006, vol. 1, 160)

Here there is a shift from horses to symbols, but Feldman's question concerns the representation of anything by a fully distributed representation in a network. The telling shift is when Smolensky says, immediately following that remark, as he begins his answer: "Given that the representational medium we are working in is a space of activation vectors..." His full answer need not concern us at the moment since it would lead us into a discussion of a matter to be addressed later (the role of superposition in the mapping of symbol structures to vectors) after some background has been provided. It suffices to note for now that the full answer concerns vectors in a vector space. But Feldman's question is not about vectors in a vector space as a representational medium. His question is about network units with activation values as a representational medium.[29] A distributed representation is supposed to be a group of units exhibiting a pattern of activation values. If distributed representations are fully distributed, and so distributed over every unit in a network, then the network can be in only one such state at time. The reason is simple: a unit can have only one activation value at a time. For a group of units to have more than one activation pattern at a time, it would have to be the case that at least one unit in the group has more than one activation value at that time. But that is impossible. Smolensky misunderstands Feldman's question, and as a result misses its very telling point about fully distributed representations.

6   Smolensky and Legendre's Explication of Principle P₁

Smolensky and Legendre explicate principle P_1 in a series of steps, which
they label a–d. I shall now discuss them in turn.

a. In all cognitive domains, when analyzed at a lower level, mental representations
are defined by the activation values of connectionist units. When analyzed at a higher level, these representations are distributed patterns of activity—activation vectors.
For central aspects of higher cognition domains, these vectors realize symbolic structures. (2006, vol. 1, 66)

I will say no more here about shifts from talk of distributed activation
states to talk of activation vectors. I want to focus now on the last claim,
the claim that vectors realize symbolic structures.

It is important to note that in subprinciple (a), Smolensky and Legendre
are not using "realize" in its familiar sense from the functional theory of
mind. In the functional theory of mind, a realizer is a realizer of a functional role. A functional role includes causal role, a role as a cause and a
role as an effect. A realizer of a functional role is just an occupant of that
role, that is, a state or event that plays the role. If there is more than one
occupant of the functional role, then there is multiple realization. It is
quite natural to think that a distributed activation state would realize a
symbol by occupying a certain functional role within the cognizer (and
perhaps also in relation to the cognizer's environment—so that the causal
role is a "wide" role, rather than one narrowly limited to causal transactions within the cognizer). Indeed, if we want a naturalistic account of
how a distributed activation state could be a representation, the functionalist view is the only view now available; there are no other extant views to
which to appeal. On the functionalist view, a distributed activation state
can be a representation in a system by occupying a certain functional role
within that system (a role perhaps extending to include portions of the
environment). This is a filler-functionalist (or realization-functionalist)
account of what makes a distributed activation state a representation. I will
return to this later. For now, the point to note is just that Smolensky and
Legendre are not using the functionalist notion of realization in (a). And
in the last sentence of (a), they indeed mean that vectors realize symbol
structures.

When in (a), they say vectors realize symbol structures, they are using
"realize" in the mathematical sense of there being a mapping, a function
from symbolic structures to vectors in a vector space. Here is the definition
of a connectionist realization of symbol structures:

Definition. A **connectionist realization** (or **representation**) of the symbolic structures in a set S is a mapping ψ from S to a vector space V:

$\Psi: S \rightarrow V$. (2006 v.1, 277)

The term "representation" appears in parentheses in the above definition, because symbolic structures are represented by vectors in a vector space. The vectors that realize them (in the mapping sense) represent them.

To see how the mapping from symbolic structures to vectors is done, we need to look in turn at (b) and (c) of their explication of P_1. Claim (b) characterizes symbolic structures in terms of structural roles and fillers of those roles, and so in terms of variable binding.

b. Such a symbolic structure **s** is defined by a collection of **structural roles** $[r_i]$ each of which may be occupied by a **filler** f_i, **s** is a set of constituents, each a **filler/role binding** f_i/r_i. (2006, vol. 1, 66)

They thus say: "A structure is a set of **bindings** of various structural **roles** to their **fillers**" (2006, vol. 1, 168).

Smolensky tells us: "At highest level of abstraction, the sentence generator can be functionally described by" binary tree structures (2006, vol. 2, 507). A binary tree is a symbolic structure that includes "a left child" and "a right child." A binary tree can have another binary tree as a left child or as a right child. And a binary tree can have a leaf as a left child or as a right child. Leaves are childless. Figure 2.1 shows a diagram of binary tree structure.

Smolensky and Legendre characterize binary trees in terms of two roles and their fillers (the symbols that occupy the roles). The two roles are,

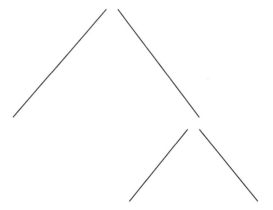

Figure 2.1

respectively, the roles of being the left child of and being the right child of. The leaves of a tree are atomic symbols. The binary tree structure above would have a leaf as its left child and a binary tree as its right child, one with a leaf as its left child and a leaf as its right child. Fillers of roles can thus be atomic symbols or be binary trees. One tree can be embedded in another. A tree can be a constituent of another tree. This feature of binary trees makes for recursion.

Smolensky and Legendre maintain that symbolic structures—binary trees—are realized by activity-vectors, by which they mean that binary trees can be mapped into activity-vectors. Clause (c) presents a mathematical formula for the mapping, where "Σ" represents the mathematical operation of superposition and "$\otimes$" represents the tensor product operation:

c. The connectionist realization of s [a symbolic structure] is an activity vector

$$\mathbf{s} = \Sigma_i \; \mathbf{f}_i \otimes \mathbf{r}_i$$

which is the sum of the vectors realizing the filler/role bindings. In these **tensor product representations**, the patterns realizing the structure as a whole is the superposition of patterns realizing all of its constituents. And the patterns realizing a constituent is the tensor product of a pattern realizing the filler and a pattern realizing the structural role it occupies. (2006, vol. 1, 66)

These filler/role bindings are called "tensor product bindings," because "the binding $f/\mathbf{r}$ of a filler $\mathbf{f}$ to a role $\mathbf{r}$ is realized as a vector $\mathbf{f}/\mathbf{r}$ that is the tensor product of a vector $\mathbf{f}$ realizing f with a vector $\mathbf{r}$ realizing $\mathbf{r}$: $\mathbf{f}/\mathbf{r} = \mathbf{f} \otimes \mathbf{r}$" (2006, vol. 1, 169).

Smolensky tells us:

In a connectionist realization of symbolic computation, the activation vectors realizing the atomic symbols, and those realizing the role vectors, are (linearly) independent. (2006, vol. 1, 173)

And he says:

Here, it is understood that the patterns realizing the atomic symbols lie in one vector space and must be independent within that space, while the patterns realizing the roles lie in a different vector space and must be independent within that space. (2006, vol. 1, 174)

As he notes: "for n linearly independent patterns to exist, there must be at least n nodes in the network" (2006, vol. 1, 280).

Here is (d), the final part of Smolensky and Legendre's explication of P_1:

d. In higher cognitive domains such as language and reasoning, mental representations are recursive: the fillers or roles of s have themselves the same type of internal structure as **s**. And these structured fillers **f** or roles r in turn have the same type of tensor product realization as **s**. (2006, vol. 1, 67)[30]

So, then, Smolensky and Legendre claim that binary trees can be mapped into vectors in a vector space in the way described in (a)–(d).[31] Binary trees are realized by activity-vectors in the mathematical sense that there is such a mapping from binary trees to vectors.

Let us look at how the mapping procedure is supposed to work in a simple case. Consider the two roles: r_0 (the role of being the left child of) and r_1 (the role of being the right child of). As Smolensky points out, the roles can be "realized by the vectors (1 0) and (0 1), respectively: in their own two-dimensional vector space (the **role space**), these vectors are indeed independent" (2006, vol. 1, 174). The vectors (1 2) and (3 5) are also in the two-dimensional vector space and are also linearly independent. We could instead map r_0 to (1 2) and r_1 to (3 5). That mapping would be mathematically equivalent to his mapping. But the vectors (1 0) and (0 1) are the standard or normal bases of a two-dimensional vector space, which can be preferable for explanatory purposes along the epistemic dimension of explanation. (The normal or standard bases of a three-dimensional vector space are the vectors: (1 0 0), (0 1 0), and (0 0 1). For a four-dimensional space, they are: (1 0 0 0), (0 1 0 0), (0 0 1 0), and (0 0 0 1). It should be clear how to generalize to find the standard bases of any n-dimensional vector space.) He also points out that if there are, say, five atomic symbols, **A, B, C, D, E**, they can be realized by (mapped to) linearly independent vectors in a five-dimensional vector space, the atomic symbol filler space. (If there are six atomic symbols, the atomic symbol filler space would have to be at least a six-dimensional vector space, and so on.) And, for illustrative purposes, he maps these five atomic symbols, respectively, to the standard bases of five-dimensional vector space. He maps atomic symbol **A** to (1 0 0 0 0), **B** to (0 1 0 0 0), **C** to (0 0 1 0 0), **D** to (0 0 0 1 0), and **E** to (0 0 0 0 1) (2006 v.1, 174).

Consider now a simple binary tree. Suppose that atomic symbol **A** fills the role of being the left child of the binary tree, and so fills role r_0; and that atomic symbol **B** fills the role of being the right child of the tree, and so fills r_1. Then, **A** in that role is mapped to the tensor product of **A** and r_0, namely, $A \otimes r_0$. Given the above role and atomic symbol vector assignments, $A \otimes r_0$ is the vector (or tensor): (1 0 0 0 0 0 0 0 0 0). **B** in the role of the right child of the binary tree is mapped to $B \otimes r_1$. Given the above assignments, $B \otimes r_1$ is the vector (or tensor): (0 0 0 0 0 0 1 0 0 0). $(A \otimes r_0) + (B \otimes r_1)$ is the superposition of those two tensors. It is the vector (1 0 0 0 0 0 1 0 0 0). A binary tree with **A** as its left child and **B** as its right child is thus mapped to that vector, which represents a point in a ten-dimensional vector space.

7 A Problem Flagged

Smolensky calls this mapping procedure "the crucial innovation" used in response to Fodor and Pylyshyn's dilemma (2006, vol. 2, 514). We have just seen how the mapping works for a simple binary tree that includes no other tree as a constituent. But a binary tree can contain another tree as a constituent, which makes for recursion; and that is essential for explaining productivity. It is also essential for explaining typical cases of systematicity. Thus, consider, again, that anyone who is able to represent that *Kim loves Sandy* is able to represent that *Sandy loves Kim*. Or, as Smolensky states his example of systematicity,

if a cognitive system can entertain the thought expressed by *Sandy loves Kim*, then it can entertain the thought expressed by *Kim loves Sandy*. (2006, vol. 2, 513)

The binary tree for *Kim loves Sandy* and the binary tree for *Sandy loves Kim* both have binary trees as their right child; they thus each contain an embedded tree. Using bracket notation, rather than an equivalent binary tree diagram, if "S" stands for Sandy, "K" stands for Kim, and "L" stands for loves, then the structure for *Sandy loves Kim* is [L, [S,K]]. The structure for *Kim loves Sandy* is [L, [K,S]].

Binary trees with binary trees as constituents pose a problem for the mapping procedure. Smolensky and Legendre's (c) tells us that the sum of the fillers in roles in any binary tree is supposed to be arrived at by superposition. Given the tensor product idea and the fact that a binary tree can contain another binary tree as a constituent, an alarm will go off for readers familiar with linear algebra. (No doubt that alarm has already gone off for such readers.) Superposition is defined only for tensors of the same rank, and vectors in the same vector space (vectors with the same number of components). Applying the formula for mapping trees to vectors, namely, $s = \Sigma_i\ f_i \otimes r_i$, where "$\Sigma$" is supposed to represent superposition, would require us, in the case of trees that contain trees as constituents, to superimpose tensors of different ranks. That is impossible. Thus, the would-be mapping procedure fails since it fails for trees with trees as constituents.

Unsurprisingly, Smolensky is aware of the problem, though it receives little attention in the volumes. He suggests two possible solutions that I will discuss in section 10 below. For now, let us simply spot Smolensky that there is a mapping procedure from binary trees to vectors using some sort of addition and the tensor product operation.

8 Principle P_2 of ICS Architectures

Smolensky and Legendre state P_2 as follows:

P_2. **Proc$_{ics}$**: Cognitive processing in ICS

Information is processed in the mind/brain by widely distributed connection patterns—weight matrices—which, for central aspects of higher cognition, possess global structure describable through symbolic expressions for recursive functions of the type employed in symbolic cognitive theory. (2006, vol. 1, 71)

They spell this out as follows:

Alg$_{ics}$ (HC,W): Recursive processing

Central aspects of many higher cognitive domains, including language, are realized by weight matrices with recursive structure. That is, feed-forward networks, and recurrent networks realizing cognitive functions have weight matrices with these respective forms:

Feed-forward: $W = I \otimes \underline{W}$ Recurrent: $W = \underline{W} \otimes R$

In either case, $\underline{W}$ is a finite matrix of weights that specifies the particular cognitive function. I and R are recursive matrices for feed-forward and recurrent networks, these are simply-defined unbounded matrices that are fixed—the same for all cognitive functions. (2006, vol. 1, 71–72)

As Smolensky notes, the primitive operations of Lisp, a programming language that he calls "the quintessential programming language for symbol manipulation" (2006, vol. 1, 182), "include the operations of extracting the left or right child of a pair (ex$_0$, ex$_1$; traditionally 'car' and 'cdr'), attaching a left or right child (**sef**), testing pairs of equality (**equal?**), assigning definitions to function names (**define**)" (2006, vol. 2, 509). Binary Lisp trees, he maintains, can be mapped into vectors. And the functions computed by these Lisp operations, he says, can be computed by matrix multiplication operations.

For present purposes, we need not examine P_2 in detail.[32] The key point of the principle that concerns us is just this: The algorithms in ICS are supposed to be, one and all, connectionist algorithms. Taken out of context, the claim that algorithms in an ICS architecture compute matrix multiplication operations might well suggest that the architecture includes matrix multiplication operations involving the manipulation of numerals (symbols for numbers). But an ICS architecture is supposed to include no symbolic algorithms, and so no numeral manipulation algorithms, not even ones that involve parallel computing in the sense of more than one numeral-multiplication operation being performed at the same time. The reason is that *that* sort of parallel processing is not ruled out by LOT theory. LOT theory does not imply that our cognitive system, or indeed even a given module within our cognitive system, can execute only one symbolic algorithm at a time. (It should go without saying that LOT is not Lisp, as I have now said twice. Indeed, we have yet to determine what the algorithms of our LOT architecture are. That is one of the major research projects in LOT theory.) In an ICS architecture, there are no local representations, only distributed representations. A unit with a certain activation value thus never represents a number since it never represents anything. Matrix multiplication functions are computed in ICS. But they are computed by connectionist algorithms involving the spread of activation through the network, not by numeral-manipulation algorithms, run serially or in parallel, since such algorithms are symbolic algorithms. To repeat: the algorithms of an ICS architecture are supposed to be, one and all, connectionist algorithms.

9   Why Symbols in ICS Are Supposed to Be Process Irrelevant

ICS is supposed to include no symbolic algorithms, because symbols are not cognitive process relevant. Why not? To answer that question, we must first consider Smolensky's distinction between three levels of description of a computational system, since his notion of process-relevance is defined by appeal to the three levels.

The levels are the functional level (the f-level), the computational level (the c-level), and the neural level (the n-level). Smolensky tells us:

a. The f-level: describes the *functions* computed
b. The c-level: describes the *algorithms* that compute this function[33]
c. The n-level: describes the *neural* (or other physical) processes that realize the algorithm. (2006, vol. 2, 516)

(In [c] above, "realize" is again used in the mathematical sense, not in the functionalist sense.) Smolensky notes that his three levels were inspired by David Marr's (1982) well-known three levels of analysis: what Marr called "the computational," "the algorithmic," and "the implementational" levels. But he also notes that his tripartite distinction is different from Marr's tripartite distinction.

Although there are indeed differences, it is nevertheless useful by way of explicating Smolensky's levels to compare them with Marr's. Smolensky notes: "It is the job of the f-description to specify a *function*" (2006, vol. 2, 537). His f-level, the functional level, is essentially Marr's (1982) computational level. At that level, we specify the cognitive functions that the architecture can compute. Smolensky says that

"computational level" refers to the highest level in Marr 1982 but to the middle level here. I will in fact generally avoid "computational level" in favor of **c-level**, which emphasizes the technical nature of the level terminology developed here; the label "computational" for the c-level should be construed as a rough gloss, intended more as a mnemonic than as a meaningful label. The glosses "functional level" and "neural level" for **f-level** and **n-level** should be taken in the same spirit. (2006, vol. 2, 519)

Still, he says, "it is the job of the c-level description to specify an *algorithm*" (2006, vol. 2, 537). So, the c-level is related to Marr's algorithmic level. But, as we will see shortly, the comparison is complicated by the fact that Smolensky splits the c-level into two levels: a "higher c-level" and a "lower c-level."[34] The lower c-level is the level at which the algorithms by which functions are computed are specified. The lower c-level, rather than the c-level itself, is essentially Marr's algorithmic level. The n-level is essentially Marr's implementational level. Smolensky says: "It will be called the n-level, in anticipation of the cognitive case in which it is the neural level" (2006, vol. 2, 520). Although he remarks at times that ICS is neurally plausible (see, e.g., 2006, vol. 1, 101, and 2006, vol. 2, 514), he is purposefully noncommittal about the details of the n-level, with the exception that he ventures the very substantive claim that units (or nodes) in the ICS architecture that is our cognitive architecture will be neurons. ICS is thus supposed to have a neural instantiation. As I emphasized earlier, the distributed representations of that ICS architecture are supposed to be biophysical states, specifically states of populations of neurons.

In ICS, the c-level is split

into two sublevels, providing two different ways of decomposing a state of the mind/brain. At the higher sublevel, a state is decomposed into superimposed patterns, corresponding to the decomposition of a symbol structure into its constituents. At the lower sublevel, a state is decomposed into individual unit activations, corresponding to the physical decomposition of a brain state at the neural level. (2006, vol. 2, 503)

The decomposition at the higher c-level is "decomposition of vectors using filler vectors and role vectors, their binding into constituents, via tensor product, and their combination via superposition (summation)" (2006, vol. 2, 566). The decomposition at the lower c-level is decomposition into individual units at levels of activation and weighted patterns of connectivity in which those units participate.

So, there are, in effect, four levels to consider: the f-level, the higher c-level, the lower c-level, and the n-level. Smolensky says that there is a realization (a mapping) from the f-level to the higher c-level, a realization from the higher c-level to the lower c-level, and a realization from the lower c-level to the n-level. There is thus a realization from the f-level to the n-level.

But a realization can fail to be an isomorphism. As Smolensky notes: "An isomorphism maps each part of one system to a part of another system in such a way as to preserve the relationships among the parts with each system" (2006, vol. 2, 514). Smolensky tells us the f-level is isomorphic to the higher c-level. The mapping procedure from symbols and symbol structures (binary trees and their constituents) to activity vectors described earlier is supposed to yield that isomorphic mapping. Thus, for instance, a binary tree with atomic symbol **A** as its left child and atomic symbol **B** as its right child is mapped to $A \otimes r_0 + B \otimes r_1$. That vectorial description is what Smolensky has in mind as the higher c-level decomposition. Thus, he tells us:

$s = A \otimes r_0 + B \otimes r_1$ expresses a decomposition of s into **A**- and **B**-constituents. This decomposition is isomorphic to the decomposition of a symbol structure $s = [A\ B]$ into its **A**- and **B**-constituents. (2006, vol. 2, 515)

Given our earlier assignments, $A \otimes r_0 + B \otimes r_1$ is the vector (1 0 0 0 0 0 1 0 0 0). At the lower c-level, that vector is decomposed into activation values of each of the components of (in this case) a 10-tuple of units, as well as the patterns of connectivity of each of those units with other units in the network. The lower c-level is supposed to be isomorphic to n-level. So, the higher c-level is isomorphic to the f-level, and the lower c-level is isomorphic to the n-level. There is a realization from the higher c-level to the lower c-level, but it is not an isomorphism; the higher c-level is not isomorphic to the lower c-level.

Let us look at a detailed statement of why he says the higher and lower c-levels are not isomorphic:

Because of distributed representations, however, the higher and lower orders are not isomorphic with each other: the two modes of decomposition cannot be put in one-to-one correspondence. ... A higher-level part like $A \otimes r_o$ corresponds to many lower-level parts; the activation values of the individual units in the pattern of activity $A \otimes r_o$. A lower-level part, the activation of unit k, corresponds to many higher-level parts, since it is part of the pattern defining many constituent vectors (e.g., both $A \otimes r_o$ and $B \otimes r_1$). The higher- and lower-level decompositions crosscut one another. (2006, vol. 2, 515)

Given that the higher c-level is isomorphic to the f-level but not isomorphic to the lower c-level, it follows that the f-level is not isomorphic to the lower c-level. Given that the lower c-level is isomorphic to the n-level, the f-level is not isomorphic to the n-level.

According to Smolensky, it is because the f-level is not isomorphic with the n-level that symbols in an ICS architecture—distributed representations—are not cognitive process relevant. He tell us that for cognitive process relevance, there must be an

isomorphism with the n-level, for that is the level at which the theory makes contact with such observable quantities as reaction times and capacity limitations. For example, in order for the number of steps in an algorithm to necessarily predict the actual time required to perform a computation, the step-by-step structure of the algorithm must be isomorphic to the moment-by-moment structure of the physical system realizing that algorithm. (2006, vol. 2, 516)

He says more generally:

A main point of the preceding sections is that algorithms can only meet the demands of models of mental processes if they are coupled with hypotheses about how the structure internal to the algorithm maps onto the physical device implementing the algorithm. (2006, vol. 2, 563)

He tells us,

The internal causal structure, described by the connectionist network description, has no corresponding description at the higher, "symbolic" level. ICS employs symbolic explanations for cognitive *functions*, but connectionist explanation for cognitive processes. (2006, vol. 2, 516)

According to Smolensky,

The general notion of **process-relevance** can be characterized as in (3).

(3)   Process-relevance of a computational description ...

In order for a computational description to account for the time, space, or other resource requirements of a process in a physical system, that description must be isomorphic to the n-level description of the system, with respect to the structural decomposition relevant to a given resource. (2006, vol. 2, 516)

Symbols and symbol structures are not process-relevant since there is no isomorphic structural decomposition to the relevant sources, namely, n-level processes.

The following passage nicely summarizes his position:

> The higher sublevel of the ICS c-level is isomorphic to the symbolic f-level account: symbols are computationally relevant. But only the lower c-sublevel is process-relevant, and at this level there are connectionist but not symbolic algorithms. The reduction from the higher to the lower c-sublevel is accomplished by a realization mapping that is *not* an isomorphism; this mapping *does* allow a reduction from symbolic structures to connectionist units—but, because representations are distributed [i.e., distributed activation states], it is a holistic, not a part-by-part, reduction. This mapping is formal, as required for a fully computational reduction, but the lack of part-by-part correspondence means symbolic decompositions do not map onto process-relevant decompositions. (2006, vol. 2, 519)

To minimize suspense, I hereby note that I will not challenge Smolensky's claim that the f-level and higher c-level fail to be isomorphic to the n-level in an ICS architecture. But the issue of whether an ICS architecture can include symbolic algorithms as derivative (nonbasic) algorithms remains. I will address it in section 11. First, however, in section 10, I will discuss the problem mentioned earlier with the procedure for mapping binary trees with embedded trees to vectors, since that will provide useful background for the discussion in 11. In section 12, I will conclude by putting in a plug for implementation connectionism.

10   The Mapping of Binary Trees into Vectors Reexamined

In the final chapter of volume 2, Smolensky tells us:

> In this chapter, I argue that ICS furnishes a unified cognitive architecture ... [that] resolves the alleged dilemma posed by Fodor and Pylyshyn (1988). The crucial innovation is tensor networks, or **tensorial computation.** ... Tensorial computation furnishes the bridge that allows ICS to cross the chasm separating the high ground of symbolic theories of mind from the low ground of connectionist theories of brain. On the higher side [i.e., the higher c-level], tensorial computation mirrors—is **isomorphic** to—key aspects of symbolic computation. On the lower side [i.e., the lower c-level], tensorial computation is simply parallel, distributed connectionist computation, which is plausibly isomorphic to neural computation. (2006, vol. 2, 514)

The bridge that allows ICS to cross the chasm is just the procedure for mapping binary trees to vectors, where symbols in roles are represented by tensor products. He takes that procedure to be his crucial innovation.

As I noted earlier, the procedure in question fails if "Σ" in the formula for the mapping (namely, $\mathbf{s} = \Sigma_i \, \mathbf{f}_i \otimes \mathbf{r}_i$) means superposition. This is the problem that Smolensky has in mind when he cryptically alludes to "a technical difficulty in the definition of recursive representations" on page 156 of volume 1. The technical difficulty is addressed in a section of volume 1 entitled "Recursive role vectors for binary trees." I will quote the relevant passages at length.

After a discussion of how to use the two role vectors $\mathbf{r}_0$ and $\mathbf{r}_1$ in characterizing trees that contain trees as constituents, Smolensky says:

A final question we will take up here is the meaning of '+' in such equations as ...

(44) $\mathbf{A} \otimes \mathbf{r}_o + (\mathbf{B} \otimes [\mathbf{r}_o \otimes \mathbf{r}_1] + \mathbf{C} \otimes [\mathbf{r}_1 \otimes \mathbf{r}_1])$.

In what vector space is this addition performed? This question arises because, under the simplest interpretation, the first term $\mathbf{A} \otimes \mathbf{r}_o$ and the second term $\mathbf{B} \otimes [\mathbf{r}_o \otimes \mathbf{r}_1]$ are in different vector spaces. Because the latter vector involves on additional tensor product with a fundamental role vector ($\mathbf{r}_1$), the number of numerical components in the vector is greater, by a factor equal to the number of numerical components of the fundamental role vectors (which must be at least two, so that $\mathbf{r}_o$ and $\mathbf{r}_1$ may be independent). In connectionist terms, the number of activation values (units) needed to realize the B-binding is greater than the number needed to realize the A-binding, because the former involves a deeper tree position. Indeed, just like the total number of roles, the number of units over which bindings are realized increases geometrically with tree depth; it equals AR^d, where A is the number of units used to realize atomic symbols, and R is the number used to realize the fundamental vectors $\mathbf{r}_o$ and $\mathbf{r}_1$. (2006, vol. 1, 185)

Smolensky goes on to say:

From the perspective of vector space theory, the natural interpretation of (44) involves the **direct sum** of vector spaces. To set up this sum, let $V_{(1)}$ be the vector space including the first term in (44), $\mathbf{A} \otimes \mathbf{r}_o$; the vectors composing $V_{(1)}$ are all the linear combinations of the vectors of the form $\mathbf{a} \otimes \mathbf{r}_i$, where $\mathbf{a}$ is the vector realizing an atomic symbol and $\mathbf{r}_i$ is $\mathbf{r}_0$ or $\mathbf{r}_1$. These vectors have AR numerical components and realize symbols at depth 1 in tree. Similarly, let $V_{(2)}$ be the vector space including the second (and third) term of (44), $\mathbf{B} \otimes [\mathbf{r}_o \otimes \mathbf{r}_1]$. These are all linear combinations of vectors of the form $\mathbf{a} \otimes \mathbf{r}_i \otimes \mathbf{r}_j$, with AR^2 components, realizing symbols at depth 2. In general, let $V_{(d)}$ be the AR^d-dimensional vector space formed from vectors of the form $\mathbf{a} \otimes \mathbf{r}_i \otimes \mathbf{r}_j \otimes \dots \mathbf{r}_k$ with d role vectors; these realize symbols at depth d. $V_{(0)}$ is just the vector space of linear combinations of the vectors $\mathbf{a}$ realizing symbols (or, equivalently, symbols located at depth 0, the tree root).

A vector space containing all the vectors $V_{(d)}$ for all d is the direct sum in (45);

(45) $V^\star \equiv V_{(0)} \otimes V_{(1)} \otimes V_{(2)} \otimes \dots$.

A vector s in this space is a sequence of vectors $(s_{(0)}; s_{(1)}; s_{(2)}; ...)$ where each $s_{(d)}$ is a vector in $V_{(d)}$. s is also written $s_{(0)} \otimes s_{(2)} \otimes. ...$ So (44) can be interpreted as

(46) $s = (A \otimes r_0) \otimes (B \otimes [r_0 \otimes r_1] \otimes C [r_1 \otimes r_1])$. (2006, vol. 1, 185–86)

So, we are to use direct sum rather than superposition.

Smolensky notes: "If tree depth is unbounded, then the number of units in the network realizing them, or the number of components in the vectors, must be unbounded as well. Algebraically, this is no problem; nothing in our mathematical analysis will depend on the vectors being of finite length" (2006, vol. 1, 187). He reminds us, moreover, that

> it is standard to adopt the idealization of no resource limitations in the symbolic case, it is appropriate to do the same in the connectionist case. In place of an unbounded or infinite Turing machine tape or von Neumann machine stack ... we have an unbounded or infinite set of units. (2006, vol. 1, 187)

On that point, suffice it to note that connections should be allowed their idealizations just as LOT theorists are.

Turn to the interpretation of (44). The interpretation of (44) as (46), which uses the direct sum operation, may seem complicated. But actually it is very simple. As Smolensky mentions in a tutorial on linear algebra, the direct sum of two vector spaces, $V \otimes U$, is the set of ordered pairs of vectors from V and U; that is, it is the Cartesian product $V \times U$ (2006, vol. 1, 156). So far as combining (adding in the sense of direct sum) trees at different depths is concerned, all that is going on is the formation of ordered pairs.[35]

As Smolensky goes on to note:

> In terms of connectionist networks, the direct sum approach amounts to the assumption that the network has a group of units for realizing symbols at depth 0 in the tree (simple atoms), a distinct and larger group of units for realizing symbols at depth 1, and so on down to deeper depths. The patterns of activity realizing the symbols at a given depth are all superimposed upon one another, but different tree depths are separately localized; this the **stratified realization** of binary trees. (2006, vol. 1, 187)

The stratified realization captures trees with trees as constituents.

If we use direct sum in the mapping from trees to vectors, then we lose a disanalogy with LOT representations and their constituents that Smolensky stresses as part of the reason symbols are not process relevant. Using "PSA" to stand for "Purely Symbolic Architecture," he says:

> In PSA, symbolic constituent structure is process-relevant; in ICS, it is not. This is in part because in ICS constituents are combined via **superposition** of overlapping

distributed patterns of activity, while in PSA, they are combined via some sort of nonoverlapping **concatenation**. (2006, vol. 2, 575)

But, as we just saw, in ICS, trees of different depths are combined not via superposition, but rather via direct sum. And as he himself acknowledges at one point (in 2006, vol. 1):

The recursive realization of binary trees described in Section 4.1 has been called a *stratified* representation: different levels of the tree are realized separately (over different connectionist units)—they are essentially *concatenated* by the direct sum in the construction of the vector space V, rather than truly superimposed, as in fully distributed representations. (2006, vol. 1, 341)

Given the use of direct sum in the mapping, ICS, like PSA, will contain molecular representations whose constituents are combined by concatenation.

It remains, though, that symbols at a given depth in a tree will be combined by superposition. In some places, Smolensky seems to take that to be relevant to the issue of whether symbols are cognitive process relevant. In arguing that the higher c-level is not isomorphic to the lower c-level, he says:

By definition, in a distributed representation, a single unit β will participate in the realization of multiple symbols—say, **B** and **C**. Then the mapping p_c will not respect the decomposition Σ_{cf} as an isomorphism must. At the higher sublevel [i.e., the higher c-level], **B** and **C** are separate constituents, but at the lower sublevel [i.e., the lower c-level], the decomposition into individual unit activations mixes **B** and **C** together: for example, the activation of unit β, s_β ... is a single number combining contributions from both **B** and **C**. (2006, vol. 2, 567–568)

To see what he has in mind by "mixing," let us consider one of his examples of superposition that I briefly alluded to in section 5 in my discussion of Feldman's telling question about fully distributed representations. Smolensky asks to consider "an extremely simple symbol structure," a set that has two atomic symbols as members, {A, B}. And he says:

In symbolic computation, this structure is the set {A, B}. Its connectionist realization, according to the Superpositional Principle ... is the vector sum of **A** and **B**, that **A** + **B**. ... To add the activation lists **A** = (1.0, −1.0, 0.8) and **B** = (0.0., 0.5, 0.3), we just add corresponding components, getting **A** + **B** = (1.0 + 0.0, −1.0 + 0.5, 0.8 + 0.3) = (1.0, −0.5, 1.1). (2006, vol. 1, 162)

Notice that in this example almost all of the components of the superimposed vectors get "mixed," and so do not appear in the superposition. Thus, consider the second and third components of **A** (viz., −1.0 and 0.8), and all three components of **B** (viz., 0.0, 0.5, and 0.3). None of those components is a component of the vector **A** + **B**, that is, the vector (1.0, –0.5, 1.1). This is the sort of thing that Smolensky has in mind when he says in the passage quoted in the previous paragraph (from 2006, vol. 2, 567–568) that the separate constituents **B** and **C** get "mixed" together. When **B** and **C** are superimposed, the components of **B** and components of **C** will not be components of the superposition vector.[36]

It might thus seem that even though distributed representations of trees at different depths will be concatenated, constituents at a given level in a tree will not be. They will be superimposed, and thus "mixed."

First, however, if we already use direct sum to combine trees at different depths, which just involves forming ordered pairs, we could as well simply form ordered pairs of the two constituents at a given depth, rather than using superposition. Then, molecular symbols would always be formed by concatenation. That is compatible with the architecture being an ICS architecture since, as we saw, such an architecture can contain concatenated representations. Thus, an ICS architecture can be such that all of its molecular symbols are formed by concatenation. One clear advantage of this is that Feldman's worry about fully distributed representations would be avoided. A given unit in a network, you will recall, can have only one activation level at a time; thus, a group of n-units in a network can have only one pattern of activation over those units at a time. Smolensky does not succeed in answering Feldman's question when he points out that an activity vector (1.0, –0.5, 1.1) is the vector sum of the activity vectors to which **A** and **B** are mapped, respectively, namely (1.0, –1.0, 0.8) and (0.0, 0.5, 0.3). If we have real value activation levels, there will be infinitely many pairs of activation vectors that sum to (1.0, –0.5, 1.1).

Second, a set of symbols is not a symbol structure in the sense of a syntactic structure. Binary trees are syntactic structures. Let us consider, then, the simplest sort of binary tree, one with an atomic symbol **A** as its left child and an atomic symbol **B** as its right child. And let **A** and **B** be the same vectors described above. Let us follow Smolensky in assigning r_0 and r_1 the standard or normal bases for a two-dimensional space, (1 0) and (0 1), respectively. **A** and **B** are at the same depth in the tree, namely, depth 1. Thus, we can superimpose $\mathbf{A} \otimes r_0$ and $\mathbf{B} \otimes r_1$. $\mathbf{A} \otimes r_0 = (1.0, -1.0, 0.8, 0,$ $0, 0)$.[37] $\mathbf{B} \otimes r_1 = (0, 0, 0, 0.0, 0.5, 0.3)$. $(\mathbf{A} \otimes r_0) + (\mathbf{B} \otimes r_1) = (1.0, -1.0, 0.8,$ $0.0, 0.5, 0.3)$. That vector will represent a certain six-tuple of units with the corresponding activation values in question. Here the symbols at a given depth (depth 1) are combined using superposition. Notice that both the components of the vector assigned to **A** and the components of the vector assigned to **B** are components of this vector and appear in the same order, and that the components of the vector assigned to **A** are followed by the components of the vector assigned to **B**. The components do not disappear in a "mix." That is because we used the normal bases for the role space. **A** and **B** look to be concatenated. This completely generalizes. For any tree depth d, no matter what linearly independent vectors are assigned to atomic symbols, and no matter what vector space distinct from the atomic symbol space the role space is, so long as we use the normal base vectors for the role space, we will get the same kind of result when we superimpose symbols at depth d.

Vectors (1 0) and (0 1) are, of course, by no means the only two linearly independent vectors in a two-dimensional vector space. As I noted earlier, the vectors (1 2) and (3 5) are also in the two-dimensional vector space and are also linearly independent. We could instead map r_0 to (1 2) and r_1 to (3 5), rather than mapping them to the normal bases for a two-dimensional vector space. With these roles' assignments, none of the components of either **A** or **B** will appear in the vector $(A \otimes r_0) + (B \otimes r_1)$. This different mapping of the binary tree to a vector is mathematically equivalent to the one above. From a mathematical point of view, it makes no difference which linearly independent base vectors we use.

But it can make a difference from an explanatory point of view along the epistemic dimension of explanation. Suppose, to use one of Smolensky's examples, the activity pattern (1.0, –1.0, 0.8) over a three-tuple of units is an atomic symbol, **A**. Then, that must be so in virtue of something about a three-tuple of units having that activation pattern.[38] The distributed activation state must function in the network in a way that makes it an atomic representation. It must realize an appropriate function in the functionalist sense of "realize." It must play an appropriate role in the network. Similarly for any distributed state that is an atomic symbol. The roles vectors represent not distributed activation states in the network, but rather roles that distributed representations that are atomic symbols in the network can fill. There is an explanatory value (along the epistemic dimension of explanation) to using normal base vectors for the role vectors since, then, the vectors to which trees are mapped will display the activation values of the units in the distributed activation states that are the atomic symbols in the tree. This is not, of course, to deny that we could use non-standard bases from the role vector space to represent the very same distributed activation states. But if an ICS architecture contains concatenated constituents, it does so whatever linearly independent vectors we assign to the role vectors. If we do not use the base vectors for the role space, all that follows is that the vectors that represent trees will not display the atomic symbols that are constituents of the trees, though they will of course represent them.

Smolensky makes one other proposal for how to map binary trees to vectors, a proposal that uses superposition rather than direct sum, while capturing recursion. He makes the proposal because he would like fully distributed representations, representations that involve all of the units in a network or in a certain portion of a network, on the grounds that it is then easier to get graceful degradation. Unfortunately, he does not address the issue of whether that has any neural plausibility. Surely, not every neuron is involved in every representation. Further, there is Feldman's problem. But, in any case, Smolensky says:

> The crux of the idea is to add to the fundamental role vectors $\{r_0, r_1\}$ of the stratified representation a third vector v that serves basically as a placeholder, like the digit 0 in numerals. Instead of representing an atom B at a position r_{01}, by $B \otimes r_0 \otimes r_1$, we use $B \otimes v \otimes v \otimes \ldots \otimes v \otimes r_0 \otimes r_1$, using as many vs as necessary to pad the total tensor product to produce a tensor of some selected rank $D + 1$. Now, atoms at all depths are realized by tensors of the same rank; the new vector space of representations of binary trees is just a space V' of tensors of rank $D + 1$, and the realization of all atoms can fully superimpose: this representation is *fully distributed*. (341)

The padding idea, which indeed assures tensors of the same desired rank, is quite simple. To ensure that we have tensors of the same desired rank, and so tensors that can be superimposed, we introduce a "dummy vector" such that tensor multiplications using this dummy vector, however many times is necessary, ensures that we get tensors of the rank in question. Given that they are of the same rank, they can be superimposed.

Smolensky notes:

> The stratified realization of Section 4.1 can be straightforwardly embedded as a special case of this new fully distributed representation by mapping $r_0 \rightarrow (r_0, 0)$, $r_1 \rightarrow (r_1, 0)$ and by setting $v \equiv (\mathbf{0}, 1)$, where 0 is the zero vector with the same dimensionality as r_0 and r_1. (2006, vol. 1, 341)

If r_0 is (1 0) and r_1 is (0 1), then v is (0 0 1). He further tells us that "in the *general* case, $\{r_0, r_1, v\}$ are three linearly independent vectors, each with nonzero components along all coordinate axes; in this case, every unit will take part in the realization of every atom, regardless of its depth in the tree" (2006, vol. 1, 341). So, the distributed representations will be fully distributed.

The dummy vector is of course just to be a mathematical device for getting tensors of some desired rank, so they can be superimposed. The zeros padded in by, for example, $B \otimes v \otimes v \otimes \ldots \otimes v$ act, as Smolensky notes, as place holders to ensure that the tensor has as many components as desired. The zeros that get padded in do not function to represent activation values of units. As Smolensky notes, the stratified realization, which involves concatenation, is a special case of the above strategy. With this dummy vector proposal, as with the direct sum proposal, so long as we use the normal vectors for the role space, the components of the vectors assigned to atomic symbols will appear in order, and in the order of the atomic symbols in a bracket notation for the tree, in the vector to which the tree is mapped. Thus, consider the following symbol structure: [L,[S,K]]. The vector to which this is mapped using the dummy vector strategy will have as its components the components of the vector assigned to L, followed by the components of the vector assigned to S, followed by the components of the vector assigned to K. To use Smolensky's mixing metaphor, none of those components will disappear in a mix. They will be on display in the vector to which [L,[S,K]] is mapped.

An ICS architecture can contain concatenated representations. But it is clear that Smolensky would not claim that an ICS architecture is an implementation architecture for a symbolic one if direct sum is used in the mapping from binary trees to vectors, so that trees at different depths are concatenated. That is in fact one of his two proposals (the other being the dummy vector proposal), and he emphatically denies that an ICS architecture is such an implementation architecture. Indeed, he would claim that an ICS architecture is not such an implementation architecture even if all molecular symbols in the architecture are formed by concatenation. He maintains that no ICS architecture is such an implementation architecture, because it will include no symbolic algorithms. And he maintains that an ICS architecture will include no symbolic algorithms because symbols are not cognitive process relevant in such an architecture. They will not be cognitive process relevant, he holds, because the f-level and the higher c-level are not isomorphic to the n-level.

That the f-level and higher c-level are not isomorphic to the n-level is something I have already acknowledged. But let us turn at long last to the issue of whether an ICS architecture can include symbolic algorithms as derivative algorithms, as in an implementation architecture.

11   The Combinatorial Strategy and Process Relevance

Smolensky and Legendre state:

The mental representations characterized by our first principle Rep$_{ics}$ [i.e., P$_1$] are of course only useful if they can be appropriately manipulated to support cognition.

These **mental processes** are the subject of the second ICS principle, P_2. (2006, vol. 1, 71)

They are certainly right that it will be useful to posit mental representations only if "they can be appropriately manipulated to support cognition." Representations in an ICS architecture are distributed representations, and thus they are kinds of distributed representational states. The state types in question must combine and recombine in ways that support cognition; they must recombine into different types of complex distributed activation states that are themselves distributed connectionist representations, ones whose contents are a function of the contents of the constituent representations and their mode of combination. (Smolensky holds, you will recall, that the system of distributed representations in an ICS architecture will have a compositional semantics [2006, vol. 2, 544–545].) Indeed, the combinatorial strategy would be unavailable to ICS theorists to explain systematicity and productivity unless distributed representations in an ICS architecture get manipulated in that sense. And, as Smolensky and Legendre are happy to acknowledge, they must appeal to the combinatory strategy to explain systematicity and productivity. In the final chapter of volume 2, Smolensky states:

The combinatorial strategy must explain not just *what* the mind achieves, but, to some extent, *how* the mind achieves it. Cognition is productive because at some level the mechanisms that actually generate it have parts that recombine in accord with the combinatorial description Σ_f. That is, at some sublevel of C [i.e., at some c-level of description] there must be a computational structure—call it "Σ_{cf}"—that is isomorphic to Σ_f. (2006, vol. 2, 557)

We have, however, already seen why Smolensky nevertheless maintains that symbols fail to be cognitive process relevant. At the lower c-level there is an isomorphism with the n-level. There is an isomorphism between the f-level and higher c-level; but there is only a realization from the f-level and higher c-level to the lower c-level—the mapping is not an isomorphism. So, there is no isomorphism between the f-level and the higher c-level, only a realization. Given Smolensky's technical definition of cognitive process relevance, it follows that symbols are not cognitive process relevant in an ICS architecture.

The symbols in an ICS architecture are distributed representations, and so distributed activation states, states involving n-tuples of units having an n-adic activation patterns. The connectionist processes of an ICS architecture do not consist in the manipulation of symbols, that is, the manipulation of such distributed representations. Such processes are defined over individual unit activations and patterns of connectivity.

But how, then, do distributed representations get manipulated in the ways that they must be manipulated to explain systematicity and productivity? The answer is that distributed representations get so manipulated (or are so manipulable) in an ICS architecture by means of algorithmic connectionist processes. The algorithmic connectionist processes do not consist in the manipulation of symbols (distributed representations). So, they are, in that sense, not symbol manipulation algorithms. But symbols are manipulated by means of them. There is thus a straightforward sense in which symbols, distributed representations, do not participate in the connectionist algorithmic processes in an ICS architecture, yet nevertheless get manipulated.

If, however, distributed representation manipulations go as Smolensky and Legendre maintain they will, then there will be macro-level (relative to the micro-level of individual units) patterns of manipulation of distributed representations that can explain systematicity and productivity. Indeed, it is essential that there be such macro-level processes if the combinatorial strategy is to be deployed in explaining systematicity and productivity. Without that, there will be no recombining of representational parts in accord with combinatorial descriptions.

There will thus be macro-patterns of transitions between distributed activation states that are representations that count as symbol manipulations. Those patterns will be patterns of causal dependency, macro-causal processes within the architecture. The macro-causal processes would be determined by the micro-causal connectionist processes. Such processes will thus not be fundamental processes within the connectionist network. They will be wholly implemented by micro-processes defined over individual units and their patterns of connectivity.[39] But that is exactly what would have to be the case for a connectionist architecture to implement a LOT architecture.

Smolensky's own term for what I have been calling "fundamental processes" is "primitive operations" (2006, vol. 2, 521). The primitive operations of an ICS architecture will not be symbolic operations. But a necessary condition on an architecture's being an implementation architecture for another architecture is that its primitive operations not be the primitive operations of the architecture being implemented. The issue is not whether an ICS architecture is a symbolic architecture. It is whether it would have to implement one to explain systematicity and productivity.

There seems no reason to deny that the nonfundamental algorithmic processes are cognitive processes. They are cognitive processes because distributed representations participate in them. The f-level and the higher c-level will be isomorphic to the macro-level in the architecture at which such processes occur. Smolensky's stipulative notion of cognitive process relevance in terms of isomorphism to the n-level is too strong to capture an appropriate notion of cognitive process relevance. The fact that there is no isomorphism between chemical elements and the subatomic structures that participate in quantum mechanical processes is no reason to think that chemical elements as wholes are not chemical process relevant. Indeed, an analogous point can be made about neurons and their molecular structures. J. J. Thomson, the discoverer of the electron, was mistaken when he said that all there is is physics and stamp collecting. There are higher-level patterns of causal dependence. And they are largely the business of the various special sciences.

My main claim is that if symbols in an ICS architecture get manipulated, and are manipulable, in the ways that they must be if systematicity and productivity are to be explained, then the symbols are cognitive process relevant in such an ICS architecture. To repeat, they are so because they will participate in macro-causal processes in the architecture. To be a distributed representation, a distributed activation state would have to play an appropriate functional and, so, causal role within the architecture. Moreover, the roles vectors r_1 and r_2 would have to represent certain types of functional roles, ones constitutive of syntactic roles, that distributed connectionist representations that are atomic symbols can in fact play in the architecture. To repeat, the macro-processes in which distributed representations participate will be wholly implemented by connectionist micro-processes in which units with patterns of connectivity participate. The macro-processes will thus be derivative processes, not fundamental ones in the ICS architecture. So, the symbols will not be fundamental process relevant. But that is exactly what is required of an implementation architecture for a symbolic architecture. If an ICS architecture includes such nonprimitive symbolic operations, then it would be an implementation architecture for a symbolic (LOT) architecture.

I thus do not think that Smolensky has shown how to explain systematicity and productivity in a connectionist architecture that does not implement a LOT architecture.

12 A Plug for Implementational Connectionism

Smolensky characterizes Fodor and Pylyshyn's general challenge to connectionism as follows:

Connectionism, they assert, faces a serious dilemma. On one horn, connectionist computation could be used to *eliminate* symbols from cognitive theory; but then connectionist couldn't explain central aspects of higher cognition, for which symbols seem necessary. On the other horn, connectionist computational could be used to literally *implement* symbolic computation; but then connectionism can teach us nothing new about cognition proper, only (at best) something about the neuroscience underlying the "classical" symbolic theory of cognition that we already have. (2006, vol. 2, 506)

He has misunderstood the dilemma. The misunderstanding begins with the "but then" statement. The second horn is not that "connectionism can teach us nothing new about cognition proper." It is that if connectionism cannot offer an adequate explanation of the aspects of higher cognition in question without implementing a symbolic (LOT) architecture, then it cannot offer an adequate alternative to the LOT hypothesis—the hypothesis that our cognitive architecture is a LOT architecture. It does not follow from that claim that connectionism can teach us nothing new about cognition proper. If two computer architectures include all and only the same algorithms, then they will be able to compute all and only the same functions. But two architectures can be Turing equivalent, and so compute all and only the same functions, yet include different algorithms. Given that a symbolic cognitive architecture must be realized in the brain, cognitive algorithms must be executable within the brain, in particular, in the brain structures that realize (in the functionalists sense) the cognitive architecture. The relevant brains structures will thus place lower-level constraints on what symbolic algorithms are psychologically plausible: they must be ones that could be implemented by such structures in the brain. If our cognitive architecture must be implemented by neural networks, because the relevant brain structures are neural networks, then that places constraints on what symbolic algorithms are psychologically plausible. Psychologically plausible algorithms must be implementable by neural processes. Implementational considerations are thus highly relevant to cognition proper.

Smolensky thinks Fodor and Pylyshyn hold that "a 'mere' implementation of a classical [i.e., symbolic] architecture would add nothing to its status as a cognitive architecture" (2006, vol. 2, 513). That is true in one sense, patently false in another. It is true that Fodor and Pylyshyn hold that it is the symbolic architecture that is a cognitive architecture, and so that the implementation architecture is not itself a cognitive architecture, but rather an architecture that explains how the cognitive architecture is implemented in the brain. But, as I just emphasized, facts about the brain, the organ in which that architecture is realized (in the functionalist's sense) and implemented, are of course highly relevant to whether a would-be symbolic architecture is in fact our cognitive architecture.

Smolensky writes as if he thinks that the implementational connectionism option is that the connectionist will be handed a symbolic architecture and told to implement it using a connectionist architecture. But that is not at all the idea. We do not yet know what the grammar of our LOT is or what the algorithms of our LOT are. Top-down considerations will play an essential role in answering those questions. But bottom-up considerations will play an essential role too. LOT is implemented in our brains. Thus, to repeat, the algorithms must be such that they can be implemented in our brains.

In discussing the relationship between a symbolic architecture and an ICS architecture, Smolensky draws an analogy between classical mechanics and quantum mechanics, with the symbolic architecture in the role of classical mechanics, and connectionism in the role of quantum mechanics. I think that a more apt analogy—*if* some connectionist architecture (perhaps an ICS architecture) proves to be a neural architecture that underlies our cognition (a big "if")—is the relationship quantum mechanics bears to chemistry. On this analogy, connectionism still gets to be quantum mechanics. But the symbolic approach to cognition plays the role of chemistry to its quantum mechanics. Quantum mechanics attributes structure to atoms and describes causal processes that implement bonding relationships between atoms. But chemistry remains a science in its own right with its own laws; and chemists can certainly describe chemical processes without appeal to quantum mechanics. Even in a large chemistry department, there will typically be only one quantum chemist.

In the early twentieth century, chemists were engaged in top-down work on atoms and quantum theorists were engaged in bottom-up work on atoms. In 1926, Schrödinger stated his famous equation. Several years later, it was shown that quantum mechanics can, in principle, explain chemical bonding. Quantum mechanics revolutionized chemistry. But it is also the case that the aim of explaining chemical phenomena led to quantum mechanics, which revolutionized physics. Work on mental representations should try to be like early twentieth-century work on the atom, proceeding by examining both top-down and bottom-up considerations. Hopefully, top-down work in the symbolic theory will someday be integrated with bottom-up work on neural networks to yield an account of our computational cognitive architecture. But that remains to be seen.

Smolensky and Legendre have, I claim, not shown us how to explain systematicity and productivity without implementing a LOT architecture. But they deserve praise for their truly impressive pioneering work on the relationship between symbolic computation and artificial neural networks.

Notes

1. A connectionist architecture consists of a network of units (or nodes) such that the units have activation values; there are weighted patterns of positive (excitatory) and negative (inhibitory) connectivity among units; and there are algorithms that govern the propagation of activation throughout the network (see Rumelhart and McClelland 1986).

2. Some the challenges concerns wider areas of cognition than just thought. But I will here focus exclusively on thought—more specifically, on the thought abilities of normal human beings.

3. LOT architecture includes a system of symbols. The symbols are either atomic or molecular. Molecular symbols contain other symbols as constituents, while atomic symbols do not. The system has a finite base of atomic symbols and a compositional semantics, so that the semantic value of a molecular symbol is a function of the semantic values of the atomic symbols that are its constituents together with the molecular symbol's constituent structure—a kind of syntactic structure. (The symbol system is in that sense language-like.) A LOT architecture includes, in addition, symbol construction and symbol manipulation algorithms. (LOT has further important features, including being innate, but the ones mentioned above are the only ones that will concern us here.) The grammar of LOT and its symbolic algorithms are as yet unknown. There are, however, computer architectures that share some of the features in question. For example, programming languages that are members of the family of languages, Lisp, which was developed by John McCarthy (1958) and based on Alonzo Church's (1941) lambda calculus, are Turing-equivalent symbolic architectures that include atomic symbols (atoms) and molecular symbols (lists), and symbol manipulation algorithms.

4. There are connectionist implementations of Turing machines and of production systems. (Production systems are Turing-equivalent classical systems.) But a LOT architecture is not a Turing machine architecture. Nor is it a von Neumann architecture. As I say in note 3, the grammar and algorithms of a LOT architecture are not yet known.

5. I discuss the main misunderstandings in detail in McLaughlin 2009.

6. I hasten to note, though, that Fodor and Pylyshyn (1988) never state that thought abilities are constituted by conceptual abilities; in fact, they do not even use the

term "conceptual abilities" or the term "constitution." But this is a natural interpretation of their "intrinsic connections" idea. Moreover, it is, I believe, clear that LOT theorists are committed to the view that thought abilities are constituted by conceptual abilities (McLaughlin 2009).

7. Gareth Evans (1982) defends essentially this thesis, under the name "the Generality Constraint." See also Davies 1991.

8. This point was made by Evans (1982).

9. I think that "A thinker has the ability to mentally represent in thought that *Sandy loves Kim* if and only if the thinker has the ability to mentally represent in thought that *Kim loves Sandy*" is a generic claim, and so not exceptionless, yet a law. I think special science laws are typically generics. But that is a topic for another occasion.

10. See, e.g., McLaughlin 1987; Fodor and McLaughlin 1990; Fodor and Lepore 1992; McLaughlin 1993a,b, 1997; and Aizawa 2003.

11. Appeal to this sentence schema has led to misunderstandings that none of us who have used it ever anticipated. I respond to them in McLaughlin 2009.

12. Capacities are abilities to acquire abilities.

13. Aizawa (2003), whose examination of the systematicity debate is otherwise excellent, misses this point. He is certainly right that some Turing machine, for instance, might have a symbol that represents aRb, but no symbol that represents bRa. But that is irrelevant to the debate. *By hypothesis*, the grammar of LOT will be such that it will generate a mental symbol that means that ψ, for all and only the ψs such that a normal human being has the capacity to mentally represent in thought that ψ.

14. Connectionists also talk about implicit representations, maintaining that certain matters can be implicitly represented in patterns of connectivity of units. I have no quarrel with that notion of implicit representation. The distinction between local and distributed representations is between explicit representations.

15. In some networks, units have only two activation values: on and off. But in some networks, they can have a range of activation values; indeed in some they have real value levels of activation.

16. Van Gelder points out that one representation may contain another without containing it as a proper part. He points out that holograms are such representations. A hologram of a face will in some sense contain a representation of a nose, but that representation will not be a proper part of the hologram. It is physically impossible to delete the representation of the nose from the hologram of the face. One can at best only degrade the entire hologram. This would be relevant to thought if there were evidence that the representations involved in thought are like holographic representations. But, on the contrary, there is compelling evidence that they

are not. To cite one piece of evidence, one can, for instance, forget what day Lincoln was born while pretty much not forgetting anything else one knows about him.

17. Hereafter, I will suppress reference to times.

18. See, e.g., Smolensky 1987, 1995, and Smolensky, Legendre, and Miyata 1992.

19. Some of the chapters in the volumes are coauthored and some are single authored by Smolensky. Several different authors figure as coauthors of one or more chapters. The chapters I will discuss are either single authored by Smolensky or coauthored by Smolensky and Legendre. I will either attribute claims to Smolensky or claims to Smolensky and Legendre, depending on which chapter the passage I am discussing appears in. This will require some awkward shifts from "Smolensky says" to "Smolensky and Legendre say," but it is the only fair way to proceed so as to give proper credit.

20. A discussion of whether ICS architecture is neurally plausible is well beyond the scope of this essay.

21. Smolensky and Legendre make frequent use of italics and bold font. In all of the quotes from Smolensky and Legendre 2006 in what follows, whenever either bold font or italics are used, the bold font and italics are theirs. Any insertions by me will appear in brackets.

22. Smolensky and Legendre do not address the issue of how primitive symbols get their meanings. Nor will I. Psychosemantics is far afield of our main concerns here.

23. I will hereafter use "symbolic architecture" and "LOT architecture" interchangeably. In a broader use of "symbolic architecture," Lisp is a symbolic architecture. (It should go without saying that Lisp is not LOT.) To avoid confusion, I will refer explicitly to Lisp whenever Lisp comes up.

24. The last two principles concern, respectively, harmonic grammar and optimality theory. As concerns harmony theory, nothing I say here casts any doubt on the interest and importance of the idea that it may play a central role in an implementational theory of a computational grammar. Also, nothing I say here casts any suspicion on optimality theory. As Smolensky notes, "the constraints defining optimality can be stated using symbolic devices, as can their mode of interaction" (2006, vol. 2, 564). Indeed, in actual practice, optimality modeling is done in symbolic terms. LOT theory can take optimality theory on board. Given that the LOT–connectionism debate about cognitive architecture has faded from the limelight, most of the interest generated by Smolensky and Legendre's two volumes has been on their discussion of optimality theory. From my conversations with Rutgers linguists, I have come to have a warm place in my heart for optimality theory. It is not in question here.

25. I use "realized by" here in the functional sense. See the discussion of functionalist sense of realization below.

26. I assume that "describable as" here means "correctly describable as."

27. I use "index" here in its representational sense. I am not using "index" in its technical sense from computer science.

28. This is a more accurate way of putting it than saying that a distributed activation state is represented *as* a point in vector space. It cannot be correctly represented as a point in vector space unless it is identical with a point in vector space. But a distributed activation state, a kind of biophysical state, is not identical with a point in vector space. Rather, it is represented by a point in vector space.

29. Similarly, Fodor and Pylyshyn (1988) raise no issues about vectors in a vector space as a medium of representation. Their concern is with a connectionist network as a medium of representation.

30. I believe that "or" in the first sentence should be "of" and that "or" in the second sentence should also be "of."

31. The mapping, they point out, will not be a bijection. There will be vectors in the vector space to which no symbolic structure (no binary tree structure) will be mapped. The members of the set of binary trees will be mapped into a vector space, but not onto that vector space.

32. Although Smolensky does not demonstrate that all Lisp operations can be computed by matrix multiplication operations, I will not challenge that claim here.

33. *Sic.* The text should read, "those functions."

34. Indeed, he says in one place: "The c-level generally consists of multiple sublevels; this is illustrated by the higher and lower-sublevel descriptions of tensorial nets discussed above" (2006, vol. 2, 516). But only the distinction between higher and lower-sublevels need concern us here.

35. I should mention that the number of units required to represent trees quickly becomes astronomical. But I will not press that issue here.

36. If units are neurons, what property of a neuron is "0.0," for instance, supposed to represent? At a time, a neuron is either firing or not. "0" could represent that the neuron is not spiking. There is, however, no good reason to think that cognitive information encoding is done just by neural firings, and so for taking activation values of units in ICS to be either 1 or 0. As I noted earlier, Smolensky mentions the possibility that activation values may be firing rates. A neuron, however, has a base firing rate, so it is uncertain what "0.0" would represent. Perhaps it could be used to represent that the neuron has its base firing rate (at the time in question). But a neuron's being at its base firing rate is normally taken to indicate that it is not at that time playing a role in information processing. Temporal encoding is naturally represented using sequences of 1s and 0s, where 1 means "spike" and 0 means "no spike." But the sequence of 1s and 0s represents the temporal encoding

of information by a single neuron over time, not a pattern over a group of neurons (Theunissen and Miller 1995). I cannot examine the controversial issue of neural plausibility here.

37. Here I have followed Smolensky in inserting commas between the components of the vector. That is not standardly done, except when the "< >" notation is used.

38. That fact can slip one's mind when one always operates by simply stipulating what distributed activation states in a network are distributed representations.

39. I am assuming that the macro-level causal patterns will be strongly supervenient on the connectionist processes in the network, so that it will be impossible for two networks to have all and only the same connectionist processes without having all and only the same macro-level causal patterns.

References

Aizawa, K. 2003. *The Systematicity Argument*. Dordrecht: Kluwer Academic.

Byrne, A., and N. Hall. 1999. Chalmers on consciousness and quantum mechanics. *Philosophy of Science* 66:370–390.

Church, A. 1941. *The Calculi of Lambda-Conversion*. Princeton, NJ: Princeton University Press.

Davies, M. 1991. Concepts, connectionism, and the language of thought. In *Philosophy and Connectionist Theory*, ed. W. Ramsey, S. Stich, and D. Rumelhart, 229–257. Hillsdale, NJ: Erlbaum.

Evans, G. 1982. *The Varieties of Reference*. Oxford: Clarendon.

Fodor, J., and E. Lepore. 1992. *Holism: A Shopper's Guide*. Oxford: Blackwell.

Fodor, J., and B. P. McLaughlin. 1990. Connectionism and the problem of systematicity: Why Smolensky's solution does not work. *Cognition* 35:183–204.

Fodor, J., and Z. Pylyshyn. 1988. Connectionism and cognitive architecture: A critical analysis. *Cognition* 28:3–71.

Marr, D. 1982. *Vision: A Computational Investigation into the Human Representation and Processing of Visual Information*. New York: W. H. Freeman.

McCarthy, J. 1958. Programs with common sense. Paper presented at the Symposium on the Mechanization of Thought Processes, National Physical Laboratory, Teddington, England, Nov. 24–27, 1958. (Published in *Proceedings of the Symposium* by H. M. Stationery Office.)

McLaughlin, B. P. 1987. Tye on connectionism. *Southern Journal of Philosophy*, Spindel Issue, 26:185–193.

McLaughlin, B. P. 1993a. The connectionism/classicism battle to win souls. *Philosophical Studies* 71:163–190.

McLaughlin, B. P. 1993b. Systematicity, conceptual truth, and evolution. In *Philosophy and Cognitive Science*, Royal Institute Philosophy Supplement No. 34, ed. C. Hookway and D. Peterson. Cambridge: Cambridge University Press.

McLaughlin, B. P. 1997. Classical constituents in Smolensky's ICS architecture. In *Structures and Norms in Science: Volume Two of the Tenth International Congress of Logic, Methodology and Philosophy of Science, Florence, August 1995*, ed. M. L. D. Chiara, K. Doets, D. Mundici, and J. van Benthem, 331–334. Dordrecht: Kluwer.

McLaughlin, B. P. 2009. Systematicity redux. *Synthese* 170:251–274.

Priest, G. 2006. *In Contradiction: A Study in the Transconsistent*. Oxford: Oxford University Press.

Rumelhart, D., and J. McClelland. 1986. PDP models and general issues in cognitive science. In D. Rumelhart, J. McClelland, and the PDP Research Group, *Parallel Distributed Processing: Explorations in the Microstructure of Cognition*, vol. 1: *Foundations*. Cambridge, MA: MIT Press.

Smolensky, P. 1987. The constituent structure of mental states: A reply to Fodor and Pylyshyn. *Southern Journal of Philosophy*, Spindel Issue, 26:137–160.

Smolensky, P. 1995. Reply: Constituent structure and explanation in an integrated connectionist/symbolic architecture. In *Connectionism: Debates on Psychological Explanation*, ed. C. Macdonald and G. Macdonald, 223–290. Oxford: Oxford University Press.

Smolensky, P., and G. Legendre. 2006. *The Harmonic Mind: From Neural Computation to Optimality-Theoretic Grammar*, vols. 1 and 2. Cambridge, MA: MIT Press.

Smolensky, P., G. Legendre, and Y. Miyata. 1992. Principles for an integrated connectionist/symbolic theory of higher-cognition. In *Technical Report CU-CS-600-92*. Department of Computer Science, University of Colorado at Boulder.

Theunissen, F., and J. P. Miller. 1995. Temporal encoding in nervous systems: A rigorous definition. *Journal of Computational Neuroscience* 2:149–162.

van Gelder, T. 1991. Classical questions, radical answers: Connectionism and the structure of mental representation. In *Connectionism and the Philosophy of Mind*, ed. T. Horgan and J. Tienson, 355–381. Dordrecht: Kluwer.

3 Tough Times to Be Talking Systematicity

Ken Aizawa

During the 1980s and 1990s, Jerry Fodor, Brian McLaughlin, and Zenon Pylyshyn argued that thought is in various respects systematic. Further, they argued that a so-called classical syntactically and semantically combinatorial system of mental representations provides a better explanation of the systematicity of thought than do nonclassical alternatives.[1] During the 1990s, part of what made the systematicity arguments problematic was the subtlety of the idea of providing a better explanation. In what sense is the classical account better than its rivals? During what we might call the post-connectionist era of roughly the last ten years, however, theoretical shifts have made it even more difficult to bring considerations of systematicity to bear on the nature of cognition.

This chapter will have a very simple structure. The first section, "The Systematicity Arguments Then," will describe one type of systematicity and provide some reason to think that human vision displays this type. In this section, the principal concern will be to draw attention to the challenge of explicating the notion of better explanation that was in play during the 1990s. The second section, "The Systematicity Arguments Now," will describe one of the new challenges facing the systematicity arguments. Part of this challenge stems from a shift in research emphasis away from cognition and on to behavior. Insofar as one is interested in behavior rather than cognition, one is certainly going to be less interested in putative properties of cognition, such as its systematicity. More dramatically, post-connectionist cognitive science often displays a breakdown in the earlier consensus regarding the relationship between cognition and behavior. Insofar as one denies the existence of cognition or simply equates it with behavior, one is all the more likely to reject putative properties of cognition and cognitive architecture, properties such as systematicity.

1   The Systematicity Arguments Then

In their seminal paper, Fodor and Pylyshyn provide a simple empirical argument that thought is systematic.[2] They note that language is systematic—that there are patterns among the set of grammatical sentences of natural language. Indeed, discovering and explaining these patterns is essentially the raison d'être of syntacticians, and they have traditionally explained these patterns by appeal to properties of a grammar and a lexicon. Consulting the linguistics literature will bear this out. If one adds to this observation the assumption that understanding a sentence of one's natural language involves having the thought expressed by that sentence, then the existence of patterns in the sentences of one's natural language implies that there will be corresponding patterns in the thoughts that one can entertain. It is these patterns in thought that Fodor and Pylyshyn contend deserve an explanation, and indeed an explanation in terms of a classical system of mental representation.[3]

Fodor and Pylyshyn provide one path to understanding and accepting systematicity, but McLaughlin suggests another. In "Systematicity, Conceptual Truth, and Evolution," McLaughlin invites us to consider the following cognitive capacities:

1. the capacity to believe that the dog is chasing the cat and the capacity to believe that the cat is chasing the dog,
2. the capacity to think that if the cat runs, then the dog will and the capacity to think that if the dog runs, then the cat will,
3. the capacity to see a visual stimulus as a square above a triangle and the capacity to see a visual stimulus as a triangle above a square, and
4. the capacity to prefer a green triangular object to a red square object and the capacity to prefer a red triangular object to a green square object. (McLaughlin 1993, 219)

Of these capacities, he notes that they are capacities to have intentional states in the same intentional mode (e.g., preference, belief, seeing as). Moreover, the paired capacities are semantically related. This is indicated in a rough-and-ready way by the fact that we use the same English words to describe each of the capacities in a pair.

McLaughlin's third pair of capacities, namely, the capacity to see a visual stimulus as a square above a triangle and the capacity to see a visual stimulus as a triangle above a square, provides an alternative, nonlinguistic source of evidence for systematicity. One can detect some systematicity of perception by visual inspection of images. More importantly for some, the systematicity of perception is well attested in the vision science literature, such as the literature on amodal completion.[4] Seeing is not thinking, but the case will nevertheless enable us to move beyond questions regarding the existence of, and evidence for, systematicity and on to the matter of the structure of the systematicity arguments.

What is amodal completion? In normal human environments, some objects often occlude other objects. Indeed, objects typically occlude parts of themselves. Nevertheless, in pedestrian contexts, we do not notice the lack of information from the occluded parts of objects. We, instead, perceptually complete the occluded objects. We perceive occluded objects as wholes. In an extremely simple case, consider a black square abutting a gray Pac-Man-shaped figure. This figure is not (on sustained viewing) perceived as a black square abutting a gray Pac-Man figure; it is instead perceived (on sustained viewing) as a black square occluding a gray circle. (See figure 3.1.) The Pac-Man example is useful, since it is clear and simple, it is (relatively) phenomenologically salient, and its existence is supported by a sizeable body of "objective" psychophysical evidence. In addition, the entire phenomenon of amodal completion is theoretically interesting, because it is such a pervasive feature of lived human experience. So, for example, one arguably perceives a whole tomato, even though one only detects the light coming from the front face of the tomato. One arguably perceives a cat behind a white picket fence, even though one only detects light that comes from the cat through the gaps in the fence. The phenomenon is not an artifact of contrived laboratory conditions.

So, consider now the systematicity of amodal completion on the model described by McLaughlin. Consider the following fourfold combination of capacities:

Figure 3.1
A simple example of amodal completion.

Figure 3.2
Systematicity in simple amodal completion.

i. The capacity to see a black square occluding a gray circle
ii. The capacity to see a gray square occluding a black circle
iii. The capacity to see a black circle occluding a gray square
iv. The capacity to see a gray circle occluding a black square

(See figure 3.2.) These capacities are systematic in McLaughlin's sense of being semantically related capacities in a single intentional mode. Further, this example meets McLaughlin's rough-and-ready test for systematicity, namely, that we use the same English words to describe paired capacities. So, if we let a = "black square," b = "gray circle," and R = "occludes," then we get an instance of the idea that an agent who can see aRb can also see bRa. These are cases (i) and (iv). Moreover, if we let a = "black circle," b = "gray square," and R = "occludes," then again we get an instance of the idea that an agent who can see aRb can also see bRa. These are cases (ii) and (iii). The extent of this kind of systematicity becomes clearer when one inspects shapes with various colors and patterns.[5]

So, it appears that we have a well-attested psychological phenomenon. Normal human observers who can perceptually experience that a occludes b can also perceptually experience that b occludes a.[6] This does not exhaust everything that has been included under the rubric of systematicity, but it does provide one path to a relevant explanandum. This, in turn, will be enough to help illustrate the challenge of articulating the concept of "better explanation" that is implicit in the Fodor-McLaughlin-Pylyshyn argument for classicism.

Consider two accounts that might be given of the systematicity of amodal completion: a classical account and a representational atomist account. The point of this comparison is not to attribute representational atomism to any particular psychologist or philosopher. Instead, the comparison is meant to put two simple, relatively clear accounts on the table in order to focus attention on what is supposed to make for better explanations in these sorts of arguments. So, a typical classical sort of explanation of the systematicity of amodal completion will postulate that, in coming to perceive an instance of amodal completion, the visual system will (after about 250 msec of viewing) partition the visual field into two objects, such as a square and a circle, forming a representation of each. It will also combine these two mental representations with a third mental representation, a representation of the occlusion relation, ultimately forming a mental representation having the syntactic form OBJECT$_1$ OCCLUDES OBJECT$_2$. So, when viewing a black square occluding a gray circle, the classicist proposes that the visual system constructs a representation like BLACK SQUARE OCCLUDING GRAY SQUARE. Amodal completion is, then, systematic, because there is a grammar to mental representations that enables the formation of a collection of mental representations, such as:

BLACK SQUARE OCCLUDING GRAY CIRCLE
GRAY SQUARE OCCLUDING BLACK CIRCLE
BLACK CIRCLE OCCLUDING GRAY SQUARE
GRAY CIRCLE OCCLUDING BLACK SQUARE.

By contrast, the representational atomist will maintain that, while the visual system does traffic in mental representations, it does not traffic in mental representations that have a classical combinatorial syntax and semantics. Instead, according to the representational atomist, mental representations are one and all syntactically and semantically atomic. So, the representational atomist who is coming to grips with the systematicity of amodal completion might propose that, when viewing the stimuli for amodal completion, the visual system produces mental representations with the syntactic forms ■, ◆, ✚, and ★. On this scheme, ■ might mean black square occluding a gray circle, ◆ might mean gray square occluding a black circle, ✚ might mean black circle occluding a gray square, and ★ might mean gray circle occluding a black square.

Notice that the systematicity argument is not set up in such a way as to propose that the representational atomist has absolutely nothing to say about how systematicity comes about. It is not as if the representational atomist has no story to tell. The problem, instead, is that the representational atomist account is not as good as the classical account. This is just as the situation was for connectionism and classicism. The problem was not (always) that the connectionists had nothing at all to say about systematicity. Instead, the problem was that the classical account was supposed to provide a better explanation of systematicity than did the representational atomist (or connectionist) account. Some such idea was in play in Fodor's early discussion of systematicity in the appendix to *Psychosemantics*:

For, of course, the systematicity of thought does not follow from ... [representational atomism]. If having the thought that John loves Mary is just being in one Unknown But Semantically Evaluable Neurological Condition, and having the thought that Mary loves John is just being in another Unknown But Semantically Evaluable Neurological Condition, then it is—to put it mildly—not obviously [*sic*] why God couldn't have made a creature that's capable of being in one of these Semantically Evaluable Neurological conditions but not in the other, hence a creature that's capable of thinking one of these thoughts but not the other. But if it's compatible with [representational atomism] that God could have made such a creature, then [representational atomism] doesn't explain the systematicity of thought. (Fodor 1987, 151)

Some such idea was also apparently in play in the closing pages of Fodor and McLaughlin 1990:

But this misses the point of the problem that systematicity poses for connectionists, which is not to show that systematic cognitive capacities are possible given the assumptions of a connectionist architecture, but to explain how systematicity could be necessary—how it could be a law that cognitive capacities are systematic—given those assumptions.

No doubt it is possible for Smolensky to wire a network so that it supports a vector that represents aRb if and only if it supports a vector that represents bRa. ... The trouble is that, although the architecture permits this, it equally permits Smolensky to wire a network so that it supports a vector that represents aRb if and only if it supports a vector that represents zSq; or, for that matter, if and only if it supports a vector that represents The Last of The Mohicans. The architecture would appear to be absolutely indifferent as among these options.

Whereas, as we keep saying, in the Classical architecture, if you meet the conditions for being able to represent aRb, YOU CANNOT BUT MEET THE CONDITIONS FOR BEING ABLE TO REPRESENT bRa; the architecture won't let you do so because (i) the representation of a, R and b are constituents of the representation of aRb, and (ii) you have to token the constituents of the representations. ... So then: it is built into the Classical picture that you can't think aRb unless you are able to think bRa, but the Connectionist picture is neutral on whether you can think aRb even if you can't think bRa. But it is a law of nature that you can't think aRb if you can't think bRa. So, the Classical picture explains systematicity and the Connectionist picture doesn't. So the Classical picture wins. (Fodor and McLaughlin 1990, 202–203)

In these passages, Fodor and McLaughlin are not challenging the idea that nonclassical theories of cognition, such as connectionism, can put forward some story about systematicity.[7] Classicism's rivals, such as representational atomism, or connectionism, or Paul Smolensky's tensor product theory (Smolensky 1987, 1988, 1990), are (ultimately) not completely silent on how to get a model to display systematicity. Instead, explaining the systematicity of thought is more than just having some story to tell. One of the difficulties in pressing the systematicity arguments back in the 1990s was the relatively delicate matter of developing a theory of explanation that explicates the explanatory principle or principles to which Fodor, and Fodor and McLaughlin, were appealing.

There is more than one way to make out the idea that merely having a story to tell about systematicity is not enough. One way is to recognize that hypothetico-deductive accounts of explanation do not work. In order to explain some phenomenon, it is not enough that one have some theory and set of background conditions that enable one to deduce the explanandum. Much of the story of the failure of hypothetico-deductivism is the vast array of counterexamples to the deductive-nomological (D-N) model of explanation. So, according to Carl Hempel and Paul Oppenheim, one might explain some phenomenon by constructing a proof of a sentence E that describes the phenomenon.[8] This proof would involve a set of laws, $L_1, L_2, ..., L_n$, and a set of initial and boundary conditions, $C_1, C_2, ... C_m$. Among the familiar problems with this sort of account is that it is just as easy to provide a proof of the length of the shadow of a flagpole given the height of the flagpole, the position of the sun, the laws of trigonometry, and the rectilinear propagation of light as it is to provide a proof of the height of a flagpole given the length of the shadow, the position of the sun, the laws of trigonometry, and the rectilinear propagation of light. Yet, where one might think that one can explain the length of a flagpole's shadow using this sort of deduction, no one apparently thinks that one can explain the height of the flagpole this way. As another example, consider the law of the pendulum (in a vacuum with massless support) that works for small amplitudes:

$$P = \frac{2\pi}{\sqrt{g / R}}$$

where P is the period, g is the force of gravity at the surface of the earth, and R is the length of the pendulum. It is just as easy to deduce the value of P, given 2π, the value of g, and the value of R, as it is to deduce the value of R, given 2π, the value of g, and the value of P. Yet, while one might explain the period of a pendulum using such a deduction, one cannot explain the length of the pendulum using this sort of deduction.

Another way to try to make out the idea that merely having a story to tell is not enough to make for a good explanation is to examine some examples drawn from the history of science. One might argue that the representational atomist's account of systematicity is inadequate for essentially the same reason as the Ptolemaic account of the limited elongation of Mercury and Venus is inadequate. There are other examples that might be taken from the history of science, but the Copernican/Ptolemaic example of limited elongation is probably the simplest. To see the analogy, consider the phenomenon. The ancient Greek astronomers had observed that Mercury was always to be found within about 28° of the Sun, while Venus was always to be found within about 45° of the Sun. Both Copernican heliocentric astronomy and Ptolemaic geocentric astronomy had stories to tell about why this regularity occurred. According to the Copernican account, Mercury, Venus, and the Earth orbit the Sun in increasingly larger orbits. (See figure 3.3.) Because of this, it must be the case that Mercury and Venus display only limited elongations from the Sun. Given the physics of the Copernican theory, the appearances must be as they are. Ptolemy, too, had an account. He maintained that the Sun, Mercury, and Venus each orbit the Earth on deferents. These orbits were further complicated by rotations on epicycles at the ends of the deferents. (See figure 3.4.) Because of the collinearity of the deferents of the Sun, Mercury, and Venus and the additional movements of Mercury and Venus on their epicycles, it must have been the case that, from the terrestrial perspective, Mercury and Venus never appear very far from the Sun. Given the hypotheses of Ptolemaic astronomy, the appearances must be as they are.

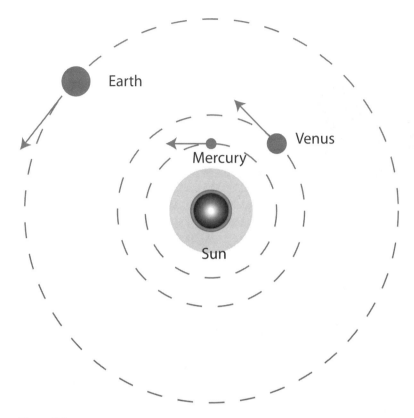

Figure 3.3
Copernican heliocentric theory.

Perhaps this analogy taps in to the explanatory standard implicitly in play in the systematicity arguments; perhaps not. Supposing, for the time being, that it does, it is nevertheless a subtle matter to articulate explicitly and precisely why the Copernican and classical explanations are superior to the Ptolemaic and nonclassical explanations. They are the kinds of subtleties that arise in making theories of explanation explicit. This is not the occasion to delve into many of the options; only a few will be presented, in order to draw attention to the nature of the challenge.

Recall Fodor and McLaughlin's claim that "it is built into the Classical picture that you can't think aRb unless you are able to think bRa, but the Connectionist picture is neutral on whether you can think aRb even if you can't think bRa." In response to this, one might wonder why it is built into the classical picture that you cannot think aRb unless you can think

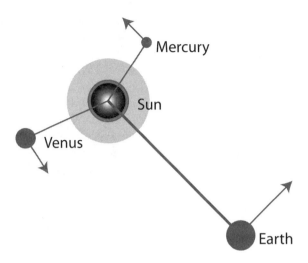

Figure 3.4
Ptolemaic geocentric system.

bRa, but it is not built into the connectionist (or representational atomist) picture. If you grant both theories their respective hypotheses, both end up postulating cognitive architectures that display systematicity. It deductively follows from the assumptions of both theories that thought is systematic, or that one can see a black square occluding a gray circle just in case one can see a gray circle occluding a black square. Similarly, if you grant both Copernicus and Ptolemy their respective assumptions, both deductively entail that Mercury and Venus will display only limited elongation from the Sun. So, what sense is there in Fodor and McLaughlin's claim that "it is built into the Classical picture that you can't think aRb unless you are able to think bRa, but the Connectionist picture is neutral on whether you can think aRb even if you can't think bRa"? How do we formulate literally the "built into" metaphor?

Maybe the idea is this. The classical picture does not need to rely on auxiliary hypotheses to secure the view that you can't think aRb unless you are able to think bRa, but the connectionist does need to rely on auxiliary hypotheses. Unfortunately, further reflection reveals two reasons that this cannot be right. First, classicism postulates a system of syntactically and semantically combinatorial representations, but must also invoke some auxiliary hypotheses about the grammar of this system of representations. For example, classicism must assume that the grammar is not just syntactically and semantically combinatorial, but also that it generates both BLACK

SQUARE OCCLUDING GRAY CIRCLE and GRAY CIRCLE OCCLUDING BLACK SQUARE. This assumption constitutes an auxiliary hypothesis. Second, rejecting explanations that invoke auxiliary hypotheses would undermine the standard view of the Copernican and Ptolemaic cases, according to which the Copernican account is superior to the Ptolemaic account. Both Copernican and Ptolemaic astronomy explain the appearances in the motions of Mercury and Venus by relying on the assumption that light propagates in straight lines. For both theories, the rectilinear propagation of light is a crucial auxiliary assumption.

Perhaps the foregoing hypothesis can be made to work through a refinement. Perhaps we should say that a classical system of representation does not rely on ad hoc auxiliary hypotheses to explain systematicity, whereas representational atomism does rely on ad hoc auxiliary hypotheses. Notice that ad hoc hypotheses enable one to give a superficially correct account of why something occurs. The problem with them stems from the fact that, roughly and intuitively speaking, they are introduced only to get a theory to work out correctly. They are hypotheses that are not independently motivated. This is just the kind of analysis that has often been applied to the Copernican and Ptolemaic case. Given the order of the planets and the rectilinear propagation of light—auxiliary hypotheses that can be independently checked—the Copernican theory guarantees that the phenomena be as they are. By contrast, Ptolemaic astronomy relies on what has long been regarded as the dubious, ad hoc assumption that the deferents of the Sun, Mercury, and Venus are collinear. This collinearity appears to be unmotivated outside of the need to explain the limited elongation of Mercury and Venus. To put the matter in a way reminiscent of what Fodor and McLaughlin wrote, no doubt it is possible for the Ptolemaic astronomer to arrange epicycles and deferents in such a way as to secure the limited elongation of Mercury and Venus. The trouble is that, although the theory permits this, it equally permits the Ptolemaic astronomer to arrange epicycles and deferents in such a way as to secure the limited elongation of Mercury and Venus from Jupiter, rather than the Sun.

An analysis of the superiority of the classical and Copernican accounts over the nonclassical and Ptolemaic accounts in terms of ad hoc auxiliary hypotheses holds considerable promise. Yet, there remains the question of whether, in fact, nonclassical theories of mental representation rely on ad hoc auxiliary hypotheses, where classicism does not. That, too, is a complicated issue that will not be pursued here.[9] The goal here is simply to draw attention to one reason it was so difficult for classicists to make the systematicity arguments thoroughly compelling during the 1990s, namely, the difficulty in articulating the implicit standards of good explanation. Unfortunately, the task of making the systematicity arguments compelling has only gotten more difficult since then.

2   The Systematicity Arguments Now

Twenty years ago, a distinction between cognition and behavior was widely embraced and relatively well understood. That was during the "connectionist era," but now things are much murkier. Strange to say, but cognition and cognitive architecture are not as central to cognitive science as they used to be. Accordingly, systematicity arguments for cognitive architecture receive less attention than they used to. It is this shift in the focus of cognitive science research away from cognition that, to a significant degree, makes it so much more difficult to make headway with the systematicity arguments. More significantly, however, many of the postconnectionist accounts that one finds of the very distinction between cognition, cognitive mechanisms, and cognitive architecture, on the one hand, and behavior, on the other, make matters much less clear than they used to be. This confusion makes it all the more difficult to make the case for any cognitive architecture, much less a syntactically and semantically combinatorial cognitive architecture.

During the 1980s and into the 1990s, cognitive scientists regularly drew a distinction between cognition and behavior. Recall that, in *Verbal Behavior*, B. F. Skinner held that behavior is a matter of physically moving about the world and manipulating it.[10] This physical bodily movement was methodologically important to many behaviorists, since it made for observable, hence "objective," scientific psychological facts. Skinner contrasted the scientifically acceptable study of behavior with the scientifically dubious study of "mental" or "cognitive" processes. Mentalism is dubious, according to Skinner, in part because it provides only vacuous pseudo-explanations of verbal behavior.

A significant component of the cognitive revolution involved a repudiation of the limited Skinnerian ontology. Noam Chomsky, for example, argued that overt, observable behavior was, in fact, the product of many distinct cognitive factors such as attention, memory, and set.[11] Most famously, he also argued that verbal behavior is also the product of what one knows about one's language. Verbal behavior depends on one's linguistic competence.

The distinction between cognitive processes and behavior was presupposed in the formulation of the systematicity arguments some twenty years ago. The very statement of some of the systematicity phenomena alludes to the idea of competence, rather than performance. The idea was not that anyone who thinks that John loves Mary also thinks that Mary loves John. It was that any who *can* think that John loves Mary *can* think that Mary loves John. Recall that McLaughlin describes systematicity in terms of capacities such as the capacity to see a visual stimulus as a square above a triangle and the capacity to see a visual stimulus as a triangle above a square. Moreover, in the text leading up to his introduction of the systematic capacities, McLaughlin writes, "To begin, cognitive capacities are not fundamental capacities: possession of a cognitive capacity consists in possession of other capacities. Cognitive capacities thus have what we will call 'constitutive bases': there are other capacities possession of which constitutes possession of the capacities in question. A theory of cognition should explain what possession of cognitive capacities consists in: it should describe constitutive bases for such capacities" (McLaughlin 1993, 219). So, the advocates of the systematicity arguments maintained a distinction between behavior and cognition. Additionally, those connectionists, such as Paul Smolensky and Robert Hadley, who accepted the challenge to explain the systematicity of thought implicitly accepted a distinction between behavior and the cognitive mechanisms that give rise to behavior.[12] Steven Phillips and William Wilson also seem to recognize the cognition–behavior distinction when they postulate a cognitive architecture based in category theory.[13] Like classicists, they invoked information-processing capacities that they supposed would generate systematic sets of cognitive capacities.

This long-standing consensus notwithstanding, there are important threads of post-connectionist cognitive science that wish to deemphasize cognition and, in fact, obliterate the former distinction between cognition and behavior. This shift in interest and emphasis takes many forms, not always explicitly recognized, but it emerges clearly in numerous passages from the literature on ecological psychology, enactivism, adaptive behavior, and extended cognition.

2.1 Ecological Psychology

One important source of inspiration for certain strains of post-connectionist cognitive science is the ecological psychology of J. J. Gibson, which is typically hostile to classical approaches to vision of the sort favored by Fodor, Pylyshyn, and McLaughlin. As just a brief sample, Gibson himself wrote

Not even the current theory that the inputs of the sensory channels are subject to "cognitive processing" will do. The inputs are described in terms of information theory, but the processes are described in terms of old-fashioned mental acts: recognition, interpretation, inference, concepts, ideas, and storage and retrieval of ideas. These are still the operations of the mind upon the deliverances of the senses, and there are too many perplexities entailed in this theory. It will not do, and the approach should be abandoned. (Gibson 1980, 238)

In this section of his book, Gibson dismisses all previous theories of the mechanisms of visual processing. Rather than hypothesizing brain mechanisms that might realize a cognitive architecture for cognitive processes, Gibson famously proposed that we ask not what's inside your head, but what your head's inside of.[14] Gibson would, thus, dismiss attempts to explain amodal completion by appeal to such things as mechanisms for computing the relatability of edges. If, like Gibson, one is uninterested in cognitive mechanisms, one will not be interested in the putative properties of cognitive mechanisms, properties such as systematicity.

2.2 Enactivism

Like many terms in contemporary cognitive science, "enactivism" means different things to different cognitive scientists, but a common thread appears to be either the diminution or even erasure of a distinction between behavior and cognition. To see this, consider first a well-known passage from Humberto Maturana and Francisco Varela's *Autopoesis and Cognition*: "A cognitive system is a system whose organization defines a domain of interactions in which it can act with relevance to the maintenance of itself, and the process of cognition is the actual (inductive) acting or behaving in this domain. Living systems are cognitive systems, and living as a process is a process of cognition. This statement is valid for all organisms, with and without a nervous system" (Maturana and Varela 1980, 13). The most striking feature of this passage, for a classicist at least, is that thinking is supposed to be a property of all organisms. Trees cognize, mushrooms cognize, slime molds cognize. A classicist might well be tempted to claim that trees, mushrooms, and slime molds are obvious counterexamples to Maturana and Varela's theory of cognition. For the classicist, it is something like a home truth that trees, mushrooms, and slime molds do not cognize; they don't think. A more generous interpretation, perhaps, is that Maturana and Varela are simply not trying to characterize the same thing as are classicists. Maturana and Varela are not theorizing about the same targets as are classicists; they do not mean by a "cognitive system" what classicists mean by a "cognitive system." Set this issue aside. What might escape notice is Maturana and Varela's equation of cognition and behavior: the process of cognition is behaving in a domain. If one equates cognition and behavior, there is hardly room for cognitive mechanisms as a causal factor underlying behavior.[15]

Evan Thompson embraces a similar approach to cognition. In the preface to his 2007 book, *Mind in Life*, he writes, "Where there is life there is mind." This appears to embrace the Maturana and Varela view, yet in the very next sentence, we get a different view: "Life and mind share a core set of formal or organizational properties, and the formal or organizational properties distinctive of mind are an enriched version of those fundamental to life" (Thompson 2007, ix). Indeed, not only does Thompson's second sentence offer a different view than does the first, the second apparently contradicts the first. If the properties distinctive of mind are an enriched version of those fundamental to life, this strongly suggests that minds have properties that not all forms of life share. So, it would not be the case that where there is life there is mind. That aside, if living and cognizing are the same thing, how does cognizing differ from behaving, if at all?

Chemero (2009) offers another similar account of cognition. He writes, "I take it that cognition is the ongoing, active maintenance of a robust animal-environment system, achieved by closely coordinated perception and action" (Chemero 2009, 212, n. 8). Notice that this commits Chemero to a more restrictive conception of cognition than Maturana and Varela's. Chemero's theory limits cognition to animals (for no evident reason). Moreover, it is similarly more restrictive than is Thompson's contention that "where there is life there is mind." So, for Chemero, sponges might be cognitive agents, but not trees, mushrooms, or slime molds. Nevertheless, Chemero's account also seems to conflate cognition and behavior. Cognition is proposed to be the ongoing, active maintenance of a robust animal-environment system, but then where does that leave behavior? The same issue arises when Chemero writes about "how radical embodied cognitive science can explain cognition as the unfolding of a brain-body-environment system, and not as mental gymnastics" (Chemero 2009, 43). If cognition is the unfolding of a brain-body-world system, then what is behavior? Are behavior and cognition to be taken to be the same?[16]

Notice that while Chemero proposes that cognition is a feature of animals only, Paco Calvo and Fred Keijzer, who also draw on enactivism, argue for plant cognition:

Embodied Cognition takes perception-action as its major focus, and within this embodied perspective, most animal and even bacterial behavior can be considered cognitive in a limited form. *Embodied Cognition* stresses the fact that free-moving creatures are not simple, hardwired reflex automatons but incorporate flexible and adaptive means to organize their behavior in coherent ways. Even when it may go too far to ascribe a mind to such systems, they deserve an acknowledgement of the intelligence involved in the things they do, and for this reason, the notion of cognition seems appropriate. So far, the notion of cognition has not been extended to plants. One reason is simply that most cognitive scientists, even those involved in *Embodied Cognition*, are simply unaware of what plants can do. (Calvo and Keijzer 2008, 248; see also Calvo and Keijzer 2011)

Calvo and Keijzer also apparently disagree with Chemero's apparent view that *all* animals can be considered cognitive agents, proposing instead that *most* animal behavior can be considered cognitive in at least a limited form. Despite these differences, Calvo, Keijzer, and Chemero agree in effacing the difference between cognition and behavior.[17] Notice that while it is true that plants incorporate flexible and adaptive means to organize their behavior, one might well maintain that there is more to cognition than mere adaptive and flexible behavior. Cognition and behavior are different things. For classicists, for example, plants and single-celled organisms might be flexible and adapt to their environment by moving along concentration gradients for food, but that is not necessarily the same thing as being flexible and adapting to their environment by, say, thinking about how to find food.

2.3 Adaptive Behavior

The foregoing examples suggest that the study of behavior should not necessarily be equated with the study of cognition. Yet, in an influential 2003 paper, "The Dynamics of Active Categorical Perception in an Evolved Model Agent," Randall Beer offers the following perspective:

For a situated, embodied agent, taking action appropriate to both its immediate circumstances and its long-term goals is the primary concern, and cognition becomes only one resource among many in service of this objective. An agent's physical body, the structure of its environment, and its social context can play as important a role in the generation of its behavior as its brain. Indeed, in a very real sense, cognition can no longer be seen as limited to an agent's head, but can be distributed across a group of agents and artifacts. (Beer 2003, 209)

Notice, in the first sentence, that Beer hints that work with situated agents involves a shift in research emphasis. He implicitly proposes that cognitive scientists should pay greater attention to action or behavior rather than cognition. Here, the traditional distinction between cognition and behavior is preserved. Moreover, we have the view, consonant with cognitivism, that an agent's physical body, the structure of the environment, and its social context can play an important role in the determination of behavior. (How one measures the relative importance of these factors is another matter.) The final sentence, by contrast, proposes something contrary to what has gone before. In the last sentence, cognition is apparently *not* one factor among many shaping behavior. Instead, the cognitive is apparently equated with the behavioral. At the very least, a lot more of what goes on in the world counts as cognitive processing; cognitive processing is spread over a group of agents and artifacts.

Why is there such a dramatic shift in Beer's perspective? What underlies the shift from the first part of the quoted text to the last? One conjecture is that this has to do with an implicit commitment to some form of operationism. The operationist method is to consider some task that is taken to constitute a cognitive task or a cognitive behavior. That is, assume that anything that can accomplish a specific task or behavior must be doing cognitive processing. Consider the task Beer proposes in his 2003 paper, namely, classifying simulated diamond-shaped and circular-shaped objects in an environment. Beer apparently supposes that any device that can accomplish this task is performing "categorical perception." When Beer then develops a device that can perform this task, he concludes that he has a device that performs categorical perception. But, then, since the cognitive is whatever contributes to the accomplishment of this task, the artifacts in the external world will count as part of the cognitive system. This sort of operationism is quite familiar. Consider the putative cognitive task of playing chess. One might suppose that any device that can play chess is thinking or cognizing, so that if one produces a device that can do this, then that device is thinking. So, one's computer running a chess-playing program is thinking. Or, consider the putative cognitive task of carrying on a conversation. It turns out that, in online chat rooms, chatbots can do a reasonably good job of fooling humans into thinking that they are communicating with another human. So, one might think that chatbots can carry on a conversation.

2.4 Extended Cognition

The kind of operationism one finds in Beer's work also seems to color the thinking of some advocates of extended cognition, at least at times. In a recent paper, Clark writes, "What makes a process cognitive, it seems to me, is that it supports genuinely intelligent behavior. ... To identify cognitive processes as those processes, however many and varied, that support intelligent behavior may be the best we can do" (Clark 2010, 292–293). Classicism provides one way of understanding the idea that cognitive processes support intelligent behavior, namely, that they are one factor that shapes behavior. Operationism provides another. This is the idea Clark articulates in the second sentence—that whatever processes support intelligent behavior are cognitive processes. Relying on such an assumption would make sense of many of his proposals to treat various activities as instances of extended cognition. One example is his analysis of writing a philosophy paper. Perhaps Clark assumes that writing a philosophy paper is a cognitive task and that whatever supports this task counts as cognitive processing. Given this assumption, Clark's observation that environmental props causally contribute to the writing of a paper naturally leads to the conclusion that the use of these environmental props is cognitive processing. That is, it is natural to suppose that one has extended cognitive processing. This is how Clark might arrive at the conclusion that, in writing a paper, "the intelligent process just *is* the spatially and temporally extended one which zig-zags between brain, body, and world" (Clark 2001, 132). Here again, the traditional distinction between cognition and behavior threatens to break down.

This kind of equation of cognition and behavior might be the proper way to read Mark Rowlands's manipulation thesis, according to which "cognitive processes are not located exclusively in the skin of cognising organisms because such processes are, in part, made up of physical or bodily manipulation of structures in the environments of such organisms" (Rowlands 1999, 23). Maybe this is Rowlands's way of saying that cognition is behavior. Richard Menary, at times, also appears to equate behavior and cognition: "The real disagreement between internalists [like Adams and Aizawa] and integrationists [like Menary] is whether the manipulation of external vehicles *constitutes* a cognitive process. Integrationists think that they do, typically for reasons to do with the close coordination and causal interplay between internal and external processes" (Menary 2006, 331). Clearly, identifying cognition and behavior provides an easy path to extended cognition. Behavior is clearly extended into brain, body, and world. Returning to the topic of systematicity, anyone who identifies cognition with behavior will not be looking for internal mechanisms, such as brain mechanisms, that realize cognitive mechanisms. Moreover, anyone who identifies cognition with behavior will not be looking for cognitive mechanisms, such as the manipulation of mental representations, that would underlie behavior. Nor would one be looking for systematic properties of such mechanisms. So, many of those working in the post-connectionist era will simply not be theoretically oriented to the project of explaining the systematicity of thought.

3 Conclusion

It has always been difficult for classicists to make the systematicity arguments stick. A significant contributor to this difficulty has been the background assumptions regarding what makes one explanation better than another. In the absence of clearly and explicitly stated and well-defended principles regarding what makes for superior explanations, it is easy to simply stand by classicism and maintain that it offers a better explanation of systematicity than does connectionism or to simply stand by connectionism and maintain that it offers just as good an explanation of systematicity as classicism. In this context, historical cases provide a useful resource in the development of well-motivated principles of good explanation. That said, even if one is willing to try to understand good scientific explanation in terms of a multiplicity of episodes in the history of science, there remains the familiar challenge of correctly explicating what is involved in actual scientific explanatory practice. Scientists make claims about particular instances of explanatory superiority, but it is often difficult to articulate these views in a way that is simultaneously precise and true. Philosophy of science has its challenges.

Unfortunately, much of the post-connectionist reorientation of the last twenty years has made it even more difficult for many cognitive scientists to see the relevance of the systematicity arguments. Part of this reorientation has been to focus more attention on behavior than on cognition. For those interested in behavior, the systematicity of cognition will be, at best, a topic of secondary interest. A more radical shift in post-connectionist thinking, however, has gone farther and apparently identified cognition and behavior. One can see intimations of this view in comments by Gibson, Maturana and Varela, Thompson, Chemero, Calvo and Keijzer, Beer, Clark, Rowlands, Menary, and many others. If cognition just is behavior, then there is no need to search for putative cognitive mechanisms underlying behavior that might, or might not, be systematic. Either type of post-connectionist theorizing, thus, makes for tough times to be talking systematicity.

Acknowledgments

Thanks to Paco Calvo and John Symons for organizing the "Systematicity and the Post-Connectionist Era" workshop in Almería, Spain. Thanks also to members of the audience for comments on an earlier version of this essay. Special thanks to David Travieso, Antoni Gomila, and Lorena Lobo for their discussion of amodal completion and its bearing on the systematicity of perception.

Notes

1. There are some nuances regarding what counts as a "classical" versus a "nonclassical" system of syntactically and semantically combinatorial representations (see, e.g., Aizawa 2003), but those nuances will be set aside here.

2. See Fodor and Pylyshyn 1988, 37–39.

3. Chemero (2009, 8–9) discusses the systematicity arguments but ignores the empirical argument for the systematicity of thought given above. Instead, he observes that Fodor and Pylyshyn do not provide references to the literature, from which he concludes that no prior empirical study provides evidence for systematicity. To turn the tables on Chemero, one might observe that Chemero provides little evidence that no prior empirical study supported the systematicity of thought. That little bit of evidence is the lack of references in Fodor and Pylyshyn's article.

4. There is a vast vision science literature and an extensive literature dedicated to amodal completion. Kellman and Shipley (1991) provide a useful point from which to begin to explore amodal completion.

5. The cases of white on white, gray on gray, and black on black have an additional twist to them insofar as sustained viewing leads them to switch between being perceived as a square occluding a circle to a circle occluding a square. So, it could turn out that the systematic pattern is that a normal human who can see shape A occlude shape B can also see shape B occlude shape A.

6. Travieso, Gomila, and Lobo (this volume) draw attention to the oversimplification in this claim. These comments mirror arguments they give in Gomila, Travieso, and Lobo 2012. (Fodor and Pylyshyn [1988], 40f, noted the oversimplification.) See, e.g., Fodor and Pylyshyn 1988, 42: "It's uncertain exactly how compositional natural languages actually are (just as it's uncertain exactly how systematic they are)." To be more explicit, one might say that what is to be explained in the systematicity arguments is why amodal completion (or thought) is as systematic as it is.

7. There are, of course, other points at which they do object that, for example, Smolensky is not actually providing any account of systematicity: "So then, after

all this, what *is* Smolensky's solution to the systematicity problem? Remarkably enough, *Smolensky doesn't* say" (Fodor and McLaughlin 1990, 201).

8. Hempel and Oppenheim 1948.

9. Again, further discussion is available in Aizawa 2003.

10. Skinner 1957.

11. See Chomsky 1959.

12. See Smolensky 1987, 1988, 1990; Hadley 1994; Hadley and Hayward 1997.

13. See Phillips and Wilson 2010, 2011, 2012.

14. See Mace 1977.

15. Incidentally, enactivism might also support skepticism of the view that cognition is in fact systematic. If one thinks that mushrooms are cognitive systems, one might think that they are such simple cognitive systems that they do not display systematicity. So, maybe not all cognition is systematic.

16. In discussion at the "Systematicity and the Post-Connectionist Era" workshop, Chemero indicated that he does believe that cognition and behavior are the same.

17. Although, in correspondence, Calvo has suggested he embraces the elimination of a cognition–behavior distinction, Calvo and Keijzer (2011) repeatedly write as if cognition and behavior were distinct.

References

Aizawa, K. 2003. *The Systematicity Arguments*. Boston: Kluwer Academic.

Beer, R. 2003. The dynamics of active categorical perception in an evolved model agent. *Adaptive Behavior* 11 (4):209–243.

Calvo, P., and F. Keijzer. 2008. Cognition in plants. In *Plant-Environment Interactions: From Sensory Plant Biology to Active Plant Behavior*, ed. F. Baluska, 247–266. Berlin: Springer Verlag.

Calvo, P., and F. Keijzer. 2011. Plants: Adaptive behavior, root-brains, and minimal cognition. *Adaptive Behavior* 19 (3):155–171.

Chemero, A. 2009. *Radical Embodied Cognitive Science*. Cambridge, MA: MIT Press.

Chomsky, N. 1959. A review of *Verbal Behavior*. *Language* 35 (1):26–58.

Clark, A. 2001. Reasons, robots, and the extended mind. *Mind and Language* 16 (2):121–145.

Clark, A. 2010. Coupling, constitution and the cognitive kind: A reply to Adams and Aizawa. In *The Extended Mind*, ed. R. Menary, 81–99. Cambridge, MA: MIT Press.

Fodor, J. A. 1987. *Psychosemantics: The Problem of Meaning in the Philosophy of Mind*. Cambridge, MA: MIT Press.

Fodor, J. A., and B. P. McLaughlin. 1990. Connectionism and the problem of systematicity: Why Smolensky's solution doesn't work. *Cognition* 35 (2):183–204.

Fodor, J., and Z. W. Pylyshyn. 1988. Connectionism and cognitive architecture: A critical analysis. *Cognition* 28:3–71.

Gibson, J. J. 1980. *The Ecological Approach to Visual Perception*. Hillsdale, NJ: Erlbaum.

Gomila, A., D. Travieso, and L. Lobo. 2012. Wherein is human cognition systematic? *Minds and Machines* 22:1–15.

Hadley, R., and M. Hayward. 1997. Strong semantic systematicity from Hebbian connectionist learning. *Minds and Machines* 7 (1):1–37.

Hadley, R. F. 1994. Systematicity in connectionist language learning. *Mind and Language* 9 (3):247–272.

Hempel, C. G., and P. Oppenheim. 1948. Studies in the logic of explanation. *Philosophy of Science* 15 (2):135–175.

Kellman, P. J., and T. F. Shipley. 1991. A theory of visual interpolation in object perception. *Cognitive Psychology* 23 (2):141–221.

McLaughlin, B. P. 1993. Systematicity, conceptual truth, and evolution. *Royal Institute of Philosophy* (Supplement 34):217–234.

Mace, W. M. 1977. James J. Gibson's strategy for perceiving: Ask not what's inside your head, but what your head's inside of. In *Perceiving, Acting, and Knowing*, ed. R. E. Shaw and J. Bransford, 43–65. Hillsdale, NJ: Erlbaum.

Maturana, H., and F. Varela. 1980. *Autopoiesis and Cognition: The Realization of the Living*. Dordrecht: Springer.

Menary, R. 2006. Attacking the bounds of cognition. *Philosophical Psychology* 19 (3):329–344.

Phillips, S., and W. H. Wilson. 2010. Categorial compositionality: A category theory explanation for the systematicity of human cognition. *PLoS Computational Biology* 6 (7):1–14.

Phillips, S., and W. H. Wilson. 2011. Categorial compositionality II: Universal constructions and a general theory of (quasi-) systematicity in human cognition. *PLoS Computational Biology* 7 (8):e1002102.

Phillips, S., and W. H. Wilson. 2012. Categorial compositionality III: F-(co)algebras and the systematicity of recursive capacities in human cognition. *PLoS Computational Biology* 7 (4):e35028.

Rowlands, Mark. 1999. *The Body in Mind: Understanding Cognitive Processes*. New York: Cambridge University Press.

Skinner, B. F. 1957. *Verbal Behavior*. New York: Appleton-Century-Crofts.

Smolensky, P. 1987. The constituent structure of connectionist mental states: A reply to Fodor and Pylyshyn. *Southern Journal of Philosophy* 26:137–160.

Smolensky, P. 1988. Connectionism, constituency, and the language of thought. In *Meaning in Mind: Fodor and His Critics*, ed. B. Loewer and G. Rey. Oxford: Blackwell.

Smolensky, P. 1990. Tensor product variable binding and the representation of symbolic structures in connectionist systems. *Artificial Intelligence* 46 (1–2):159–216.

Thompson, E. 2007. *Mind in Life: Biology, Phenomenology, and the Sciences of Mind*. Cambridge, MA: Harvard University Press.

II

4 PDP and Symbol Manipulation: What's Been Learned Since 1986?

Gary Marcus

Nobody could doubt that the brain is made up of neurons and connections between them. But how are they organized?

In cognitive science, much of the excitement of mid-1980s connectionism came from a specific hypothesis: that the mind did its work without relying on the traditional machinery of symbol manipulation. Rumelhart and McClelland (1986, 119), for instance, clearly distanced themselves from those that would explore connectionist implementations of symbol-manipulation when they wrote,

We have not dwelt on PDP implementations of Turing machines and recursive processing engines [canonical machines for symbol-manipulation] because we do not agree with those who would argue that such capabilities are of the essence of human computation.

Up until that point, most (though not certainly all) cognitive scientists took it for granted that symbols were the primary currency of mental computation. Newell and Simon (1975), for example, wrote about the human mind as a "physical symbol system," in which much of cognition was built on the storage, comparison, and manipulation of symbols.

Rumelhart and McClelland (1986) challenged this widespread presumption by showing that a system that ostensibly lacked rules could apparently capture a phenomenon—children's overregularization errors—that heretofore had been the signal example of rule learning in language development. On traditional accounts, overregularizations (e.g., *singed* rather *sang*) were seen as the product of mentally represented rule (e.g., past tense = stem + -ed).

In Rumelhart and McClelland's model, overregularizations emerged not through the application of an explicit rule, but through the collaborative efforts of hundreds of individual units that represented individual

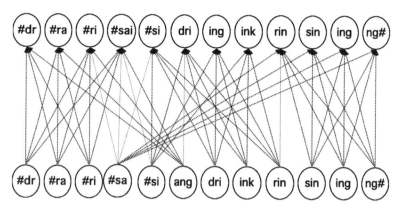

Figure 4.1

sequences of phonetic features that were distributed across a large network, with a structure akin to that in figure 4.1.

A flurry of critiques soon followed (Fodor and Pylyshyn 1988; Lachter and Bever 1988; Pinker and Prince 1988), and the subsequent years were characterized by literally dozens of papers on the development of the English past tense, both empirical (e.g., Kim, Marcus, Pinker, Hollander, and Coppola 1994; Kim, Pinker, Prince, and Prasada 1991; Marcus, Brinkmann, Clahsen, Wiese, and Pinker 1995; Marcus et al. 1992; Pinker 1991; Prasada and Pinker 1993) and computational (e.g., Ling and Marinov 1993; Plunkett and Marchman 1991, 1993; Taatgen and Anderson 2002)

In the late 1990s, I began to take a step back from the empirical details of particular models—which were highly malleable—to try to understand something general about how the models worked, and what their strengths and limitations were (Marcus 1998a,b, 2001). In rough outline, the argument was that the connectionist models that were then popular were inadequate, and that without significant modification they would never be able to capture a broad range of empirical phenomena.

In 2001, in a full-length monograph on the topic (Marcus 2001), I defended the view that the mind did indeed have very much the same symbolic capacities as the pioneering computer programming language Lisp, articulated in terms of the following seven claims.

1. The mind has a neurally realized way of representing *symbols*.
2. The mind has a neurally realized way of representing *variables*.

3. The mind has a neurally realized way of representing *operations over variables*, to form the progressive form of a verb, take the stem and add *-ing*.

4. The mind has a neurally realized way of representing *distinguishing types from tokens*, such as one particular coffee mug as opposed to mugs in general.

5. The mind has a neurally realized way of representing *ordered pairs (AB ≠ BA)*; *man bites dog* is not equivalent to *dog bites man*.

6. The mind has a neurally realized way of representing *structured units* (element C is composed of elements A and B, and distinct from A and B on their own).

7. The mind has a neurally realized way of representing *arbitrary trees*, such as the syntactic trees commonly found in linguistics.

A decade later, I see no reason to doubt any of the first six claims; PDP efforts at modeling higher-level cognition have become far less common than they once were, no major new architecture for modeling cognition has been proposed (though see below for discussion of Hinton's approach to deep learning and its application to AI), no serious critique of *The Algebraic Mind* (Marcus 2001) was ever published, and to my knowledge there has been no serious PDP attempt in recent years to capture the phenomena highlighted therein (e.g., the human facility with distinguishing types from tokens). Instead, recent theoretical works such as Rogers and McClelland (2004) continue to rely on architectures that were introduced over a decade ago, such as the PDP model of Rumelhart and Todd (1993) and remain vulnerable to the same criticisms as their predecessors (Marcus and Keil 2008).

Yet if I remain confident in the accuracy of the first six conjectures, I now believe I was quite wrong about the seventh claim—in a way that may cast considerable light on the whole debate.

The problem with the seventh claim is, to put it bluntly, people don't behave as if they really *can* represent full trees (Marcus 2013). We humans have trouble remembering sentences verbatim (Jarvella 1971; Lombardi and Potter 1992); we have enormous difficulty in properly parsing center-embedded sentences (*the cat the rat the mouse chased bit died*) (Miller and Chomsky 1963) and we can easily be seduced by sentences that are globally incoherent (Tabor, Galantucci, and Richardson 2004), provided that individual chunks of the sentence seem sufficiently coherent at a local level (e.g., *More people have been to Russia than I have*). Although tree representations are feasible in principle—computers use them routinely—

there is no direct evidence that *humans* can actually use them as a form of mental representation. Although humans may have the abstract competence to use (or at least discuss) trees, actual performance seems to consistently fall short.

In a proximal sense, the rate-limiting step may be the human mind's difficulties with rapidly creating large numbers of bindings. For example, by rough count, the sentence *the man bit the dog* demands the stable encoding of at least a dozen bindings, on the reasonable assumption that each connection between a node and its daughters (e.g., S & NP) requires at least one distinct binding; on some accounts, that sentence alone might require as many as 42 (if each node bore three pointers, one for its own identity, and one for each of two daughters).

Although numbers of between 12 and 42 (more in more complex sentences) might at first blush seem feasible, they far exceed the amount of short-term information-binding bandwidth seen in other domains of cognition (Treisman and Gelade 1980). George Miller famously put the number of elements a person could remember at 7 +/– 2 (Miller 1956), but more recent work suggested that Miller significantly overestimated; realistic estimates are closer to 4 or even fewer (Cowan 2001; McElree 2001). Similarly low limits on binding seem to hold in the domain of visual object tracking (Pylyshyn and Storm 1988). Although it is certainly possible that language affords a far greater degree of binding than in other domains, the general facts about human memory capacity clearly raise questions.

In a more distal sense, the rate-limiting step may have been the conservative nature evolution, and in particular the very mechanism of context-dependent memory that was brought up earlier. As mentioned, computers succeed in readily representing trees because their underlying memory structures are organized by location ("or address"); someone's cell phone number might be stored in location 43,212, their work number in location 43,213, and so forth. Human memory, in contrast—and indeed probably all of biological memory—appears to be organized around a different principle, known as content-addressability, meaning that memories are retrieved by content or context, rather than location (Anderson 1983). Given the short evolutionary history of language (Marcus 2008), and the fundamentally conservative nature of evolution (Darwin 1859; Marcus 2008), context-dependent memory seems likely to be the *only* memory substrate that is available to the neural systems that support language. Although content-addressability affords rapid retrieval, by itself it does not suffice to yield tree-geometric traversability. With content-addressability, one can retrieve elements from memory based on their properties (e.g., animate, nominative, plural, etc.), but not (absent location-addressable memory) their location.

Constrained in this way, we may thus be forced to rely on a sort of cobbled-together substitute for trees, in which linguistic structure can only be represented in approximate fashion, by means of sets of subtrees ("treelets") that are bound together in transitory and incomplete ways (Marcus 2013). Our brains may thus be able to afford some sort of approximate reconstruction but not with the degree of reliability and precision that veridically represented trees would demand.

As a rough sketch, imagine that a binding-limited human listener hears a sentence that begins *It was the dancer ...* As the sentence proceeds, the listener might place a set of small chunks in working memory, which we might suppose consist of elements like those in figure 4.2. The trick, of course, is to figure out the relations between those elements, presumably by drawing bindings between them, as in figure 4.3.

The trouble is that binding is an expensive operation; eventually the human parser appears not to be able to keep track of all the bindings. Even if each individual treelet (e.g., a noun phrase consisting of determiner, *the*, preceding a noun, *dancer*) were relatively automatized or chunked such that it is individually memorable, the parser soon runs out of room when trying to bind the treelets to one another. In a more complex sentence such as *It was the dancer that the fireman liked*, the binding capacity is

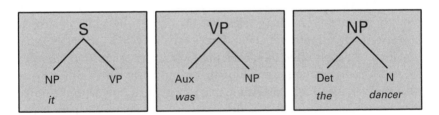

Figure 4.2

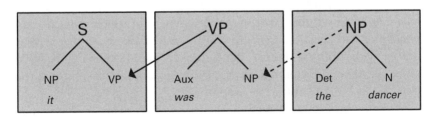

Figure 4.3

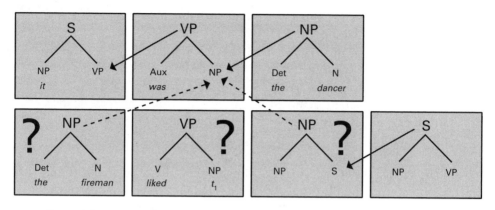

Figure 4.4

exceeded, and the system can no longer keep veridical track of all the elements and the relations between them. Instead, one might wind up with the sort of chaos depicted in figure 4.4.

At this point we would expect confusion. Who liked whom? Empirical data suggest that clefted sentences like this are quite difficult to parse, with substantial interference between elements (e.g., Gordon, Hendrick, and Johnson 2001). The resulting confusion fits naturally with a system in which the representation of trees is fragmentary rather than complete, but would make little sense in a system in which trees were veridically represented. Clefted sentences pose no special problem for machine parsers with robustly implemented binary trees, but they are challenging for people. Likewise, a set of imperfectly bound treelets could give a natural explanation for why people often infer from the garden-path sentence *While Anna dressed the baby slept* both something that *is* consistent with a proper parse (that the baby slept) and something that is consistent only with the initial parse, but *inconsistent* with the final parse (that Anna dressed the baby; example from experiments of Christianson et al. 2001). Likewise, the treelet account helps make sense of why we are vulnerable to linguistic illusions, such as the incorrect sense that *More people have been to Russia than I have* is a well-formed sentence. An account in which syntactic representations are fragmentary helps makes sense of a host of other facts as well, ranging from the human difficulty with parsing center-embedding sentences to our difficulties in remembering verbatim syntactic structures. (See Marcus 2013 for further discussion.)

While most of the particular details (e.g., of how many treelets can be represented and what their precise nature might be) are up for grabs, overall this sort of approach seems more consistent with the empirically observed facts of psycholinguistics than any framework that presumes that sentences are robustly represented on full-scale trees.

Should this conjecture that sentences are represented imperfectly (via a weakly bound sound of treelets, rather than full trees) prove correct, the upshot would be that human beings possess most—but critically, *not all*—of the symbol-manipulating apparatus that we associate with digital computers, namely, the first six elements listed above, but not the seventh, leaving us with vastly more symbolic power than simple neural networks, but still notably less than machines.

This seems, in hindsight, to be quite plausible: we are not nearly as good as machines at representing arbitrary taxonomies (what normal human, for example, can instantaneously acquire the complete family tree of the British monarchy?), yet we are fully capable of symbolic, rule-based generalizations, and still far outstrip PDP networks in our capacity to think about individuals versus kinds (Marcus 2001; Marcus and Keil 2008). The conjecture that humans possess most but not quite all of the classic tools of symbol manipulation would capture some of the spirit (though not the letter) of Smolensky's (1988) suggestion that human representations might be approximations to their digital counterparts. In retrospect, advocates of PDP probably underestimated our symbolic capacities, but may have been right that humans lack the full complement of symbol-manipulating faculties seen in machines. Connectionism's common denial of symbols still seems excessive and undermotivated, but there may well be plenty of important work to be done in finding instances in which human capacities deviate from full-scale symbol-manipulators.

Epilogue

A popular myth (which the author recently heard recounted at a public lecture by a prominent advocate of Bayesian models of cognition) is that AI in the old days was mostly symbolic, and mostly ineffective. In recent years, statistics (according to the myth, which was presented as if it were truth) had replaced symbols, and in that transition AI had finally begun to work. Watson won at Jeopardy, and Siri could kind of, sort of, understand speech. Little attention was paid to the obvious confound—machines have gotten millions of times more powerful since the early days of AI—but the more serious problem with the myth is that it is not true: statistics haven't replaced symbols, they have supplemented them. If one inspects the details of how Watson actually works, one finds a huge array of symbolic systems (e.g., temporal reasoners), working right alongside statistical systems (Ferrucci et al. 2010). Likewise, most of the leading machine parsers make heavy use of statistical information (typically in the form of Bayesian inference), but the categories that those parsers work over (e.g., probability context free grammars) are all symbolic representations, filled with variables, categories, rules, and so forth.

Champions of statistics are right to claim that statistics have revolutionized AI, but those who would eschew symbols would find no solace if they looked carefully at the mechanics of how the best AI systems work. Effective AI systems, like the human mind, use symbolic representations as part of their fundamental currency.[1]

Eliminative connectionism was a fascinating idea, but a quarter-century later, it still seems unlikely to work. A more profitable endeavor might be to figure out how to use networks of neurons in systems that unify symbols and statistics, rather than needlessly treating them as antithetical.

This chapter is adapted and expanded from an earlier article by the author that appeared in *Topics in Cognitive Sciences*, volume 1, 2009.

Note

1. A notable exception might be Google's recent "cat detector" system for recognizing images (Le et al. 2011), inspired in part by Hinton's work on deep learning (Hinton, Osindero, and Teh 2006), a 16,000-processor computer system that far exceeds its predecessors in the unsupervised learning of images—but still achieves an overall correct performance of just 15.8 percent, enough to qualify as an advance over neural networks, but not enough to qualify as a bona fide solution to the problem of vision.

References

Anderson, J. R. 1983. *The Architecture of Cognition*. Cambridge, MA: Harvard University Press.

Christianson, K., A. Hollingworth, J. F. Halliwell, and F. Ferreira. 2001. Thematic roles assigned along the garden path linger. *Cognitive Psychology* 42 (4):368–407.

Cowan, N. 2001. The magical number 4 in short-term memory: A reconsideration of mental storage capacity. *Behavioral and Brain Sciences* 24 (1):87–114, discussion 114–185.

Darwin, C. 1859. *On the Origin of Species*. London: Murray.

Ferrucci, D., E. Brown, J. Chu-Carroll, J. Fan, D. Gondek, A. A. Kalyanpur, et al. 2010. Building Watson: An overview of the DeepQA project. *AI Magazine* 31 (3): 59–79.

Fodor, J. A., and Z. Pylyshyn. 1988. Connectionism and cognitive architecture: A critical analysis. *Cognition* 28:3–71.

Gordon, P. C., R. Hendrick, and M. Johnson. 2001. Memory interference during language processing. *Journal of Experimental Psychology: Learning, Memory, and Cognition* 27 (6):1411–1423.

Hinton, G. E., S. Osindero, and Y.-W. Teh. 2006. A fast learning algorithm for deep belief nets. *Neural Computation* 18 (7):1527–1554. doi:10.1162/neco.2006.18.7.1527.

Jarvella, R. J. 1971. Syntactic processing of connected speech. *Journal of Verbal Learning and Verbal Behavior* 10:409–416.

Kim, J. J., G. F. Marcus, S. Pinker, M. Hollander, and M. Coppola. 1994. Sensitivity of children's inflection to grammatical structure. *Journal of Child Language* 21: 173–209.

Kim, J. J., S. Pinker, A. Prince, and S. Prasada. 1991. Why no mere mortal has ever flown out to center field. *Cognitive Science* 15:173–218.

Lachter, J., and T. G. Bever. 1988. The relation between linguistic structure and associative theories of language learning: A constructive critique of some connectionist learning models. *Cognition* 28:195–247.

Le, Q. V., M. A. Ranzato, R. Monga, M. Devin, K. Chen, G. S. Corrado, J. Dean, and A. Y. Ng. 2011. Building high-level features using large scale unsupervised learning. http://arxiv.org/pdf/1112.6209.

Ling, C. X., and M. Marinov. 1993. Answering the connectionist challenge: A symbolic model of learning the past tense of English verbs. *Cognition* 49:235–290.

Lombardi, L., and M. Potter. 1992. The regeneration of syntax in short term memory. *Journal of Memory and Language* 31:713–733.

Marcus, G. F. 1998a. Can connectionism save constructivism? *Cognition* 66: 153–182.

Marcus, G. F. 1998b. Rethinking eliminative connectionism. *Cognitive Psychology* 37 (3):243–282.

Marcus, G. F. 2001. *The Algebraic Mind: Integrating Connectionism and Cognitive Science.* Cambridge, MA: MIT Press.

Marcus, G. F. 2008. *Kluge: The Haphazard Construction of the Human Mind.* Boston: Houghton-Mifflin.

Marcus, G. F. 2013. Evolution, memory, and the nature of syntactic representation. In *Birdsong, Speech, and Language: Exploring the Evolution of Mind and Brain*, ed. J. Bolhuis and M. Everaert. Cambridge, MA: MIT Press.

Marcus, G. F., U. Brinkmann, H. Clahsen, R. Wiese, and S. Pinker. 1995. German inflection: The exception that proves the rule. *Cognitive Psychology* 29:186–256.

Marcus, G. F., and F. C. Keil. 2008. Concepts, correlations, and some challenges for connectionist cognition. *Behavioral and Brain Sciences* 31:722–723.

Marcus, G. F., S. Pinker, M. Ullman, J. M. Hollander, T. J. Rosen, and F. Xu. 1992. Overregularization in language acquisition. *Monographs of the Society for Research in Child Development* 57 (4, Serial No. 228).

McElree, B. 2001. Working memory and focal attention. *Journal of Experimental Psychology: Learning, Memory, and Cognition* 27 (3):817–835.

Miller, G. 1956. The magical number seven plus or minus two: Some limits on our capacity for processing information. *Psychological Review* 63 (2):81–97.

Miller, G., and N. A. Chomsky. 1963. Finitary models of language users. In *Handbook of Mathematical Psychology*, vol. II, ed. R. D. Luce, R. R. Bush, and E. Galanter, 419–492. New York: Wiley.

Newell, A., and H. A. Simon. 1975. Computer science as empirical inquiry: Symbols and search. *Communications of the Association for Computing Machinery* 19:113–136.

Pinker, S. 1991. Rules of language. *Science* 253:530–535.

Pinker, S., and A. Prince. 1988. On language and connectionism: Analysis of a parallel distributed processing model of language acquisition. *Cognition* 28:73–193.

Plunkett, K., and V. Marchman. 1991. U-shaped learning and frequency effects in a multi-layered perceptron: Implications for child language acquisition. *Cognition* 38 (1):43–102.

Plunkett, K., and V. Marchman. 1993. From rote learning to system building: Acquiring verb morphology in children and connectionist nets. *Cognition* 48 (1):21–69.

Prasada, S., and S. Pinker. 1993. Similarity-based and rule-based generalizations in inflectional morphology. *Language and Cognitive Processes* 8:1–56.

Pylyshyn, Z. W., and R. W. Storm. 1988. Tracking multiple independent targets: Evidence for a parallel tracking system. *Spatial Vision* 3:179–197.

Rogers, T. T., and J. L. McClelland. 2004. *Semantic Cognition: A Parallel Distributed Processing Approach*. Cambridge, MA: MIT Press.

Rumelhart, D. E., and J. L. McClelland. 1986. On learning the past tenses of English verbs. In J. L. McClelland, D. E. Rumelhart, and the PDP Research Group, *Parallel*

Distributed Processing: Explorations in the Microstructure of Cognition, vol. 2: *Psychological and Biological Models*, 216–271. Cambridge, MA: MIT Press.

Rumelhart, D. E., and P. M. Todd. 1993. Learning and connectionist representations. In *Attention and Performance XIV*, ed. D. E. Meyer and S. Kornblum, 3–30. Cambridge, MA: MIT Press.

Smolensky, P. 1988. On the proper treatment of connectionism. *Behavioral and Brain Sciences* 11:1–74.

Taatgen, N. A., and J. R. Anderson. 2002. Why do children learn to say "broke"? A model of learning the past tense without feedback. *Cognition* 86 (2):123–155.

Tabor, W., B. Galantucci, and D. Richardson. 2004. Effects of merely local syntactic coherence on sentence processing. *Journal of Memory and Language* 50 (4):355–370.

Treisman, A. M., and G. Gelade. 1980. A feature-integration theory of attention. *Cognitive Psychology* 12 (1):97–136.

5 Systematicity in the Lexicon: On Having Your Cake and Eating It Too

Jeffrey L. Elman

1 Introduction

Fodor and Pylyshyn's (1988, henceforth F&P) seminal work marked an important step forward in the debate about what kind of architecture is required for human cognition. By focusing on core concepts of "systematicity," "symbolic," and "compositionality," F&P suggested new ways to consider the competing claims about connectionist versus symbolic systems in an explicit and precise way. Whether or not one agrees with their definitions (and there has been considerable debate both about the correctness as well as clarity and coherence of these definitions), the narrowing of the debate has led to stimulating discussion.

Absent such discussion, for example, terms such as 'systematic' may be taken to refer to significantly different things. The equation $y(t) = f(x, y_{t-1})$, for example, describes a dynamical system that is, in general terms, systematic. The problem is that such equations can be used to characterize both connectionist and (with some stretching) symbolic systems. F&P usefully narrow the definition of systematic in a way that allows us to ask important questions: Are connectionist models capable of systematicity? Are only symbolic models systematic? Is human cognition itself systematic? The main point of F&P, of course, is that a cognitive architecture must be capable of systematicity in order to explain human cognition. A related formulation, which captures much of what F&P argued for, is that human cognition is algebraic, where that algebra requires symbols that are abstract and context free (see Marcus 2001).

Although F&P's definitions characterize both representations as well as rules, much of the ensuing debate has tended to focus on their claims regarding systematicity of rules. Representations, and in particular, lexical representations, have not figured prominently in the debate. This emphasis on rules is not surprising, given the zeitgeist of the time in which the paper was written. In that period, rules were seen as the major source of linguistic generativity. Because the lexicon was understood as involving a relatively stable inventory of entities with fixed meanings, pronunciations, and so on, the contribution of the lexicon to productive language use was felt to not be particularly interesting.

In intervening years, however, the lexicon has come into its own. Lexical knowledge is now acknowledged to be quite rich. Given proposals that the lexicon includes abstract patterns (e.g., Goldberg 2003), it is now seen as a source of linguistic productivity. One consequence of this is that boundary between the lexicon and rule systems has become somewhat blurred.

In this chapter, I consider ways in which systematicity, as F&P have defined it, might apply to the lexicon. Their treatment of the lexicon is light, which requires me to make some assumptions about how systematicity might apply to the lexicon. Whether or not these assumptions are reasonable is for the reader (and, presumably, F&P) to judge.

A number of chapters in this book challenge the claims made by F&P with regard to rule systems. I will argue that similar problems arise in considering the lexicon. So I begin by discussing empirical data that are problematic for the view of the lexicon as an enumerative database of the sort that would be required by an F&P approach. That is, the option of significantly enlarging the format of lexical entries to accommodate these data—the obvious move—not only leads to an unwieldy combinatoric explosion of information, but more seriously compromises the theoretical assumptions that motivated placing some, but not all, information in the lexicon. I then go on to argue that a very different mechanism than the lexicon is required in order to capture the richness of lexical knowledge. To lend some concreteness to this conclusion, I conclude with a preliminary model that I believe possesses some of the characteristics that are desirable in such a mechanism. This model eliminates the lexicon, *qua* lexicon, while providing an alternative way for lexical knowledge to play a role in language comprehension.

2 The Lexicon as Dictionary

The metaphor of the mental lexicon as a dictionary is pervasive and compelling. However, metaphors bring a lot of baggage with them, sometimes hidden from view. If the lexicon is to do real work for us, we need to go beyond metaphor and specify what the properties of the lexicon are.

There is considerable and important divergence on this issue, although commonalities exist across many theories. These are well captured in Jackendoff's (2002) description of the lexicon:

For a first approximation, the lexicon is the store of words in long-term memory from which the grammar constructs phrases and sentences. (130)
[A lexical entry] lists a small chunk of phonology, a small chunk of syntax, and a small chunk of semantics. (131)

A number of important questions immediately arise: (1) Just how small is "small"? (2) Why is there a limit on the quantity of this information? (3) What tests do we use to determine what information is included in the lexical representation, and what information resides elsewhere? There is no agreement on the answers to these questions. Indeed, these questions are rarely posed in an explicit form. However, there seems to be some implicit consensus that the limit on information is driven by notions of parsimony, and that only core information is included in lexical entries. "Core" may be defined operationally as information that is reliable and stable across context and is minimally sufficient to distinguish otherwise similar lexical entries.

Some awkwardness arises when dealing with polysemy. Many lexical items have multiple senses. In some cases, these senses share a common underlying meaning. The verb *admit*, for instance, can mean "to let in" as well as "to acknowledge." Because these meanings are highly associated with different syntactic frames (the first prefers a direct object, the second a sentential complement), but because there is overlap in their meaning, they might appear as distinct senses in the same lexical entry. The very different meanings of the verb "bank," on the other hand, suggest separate lexical entries that share the same phonological form. However, there are also many cases in which the distinctions in meaning may be quite subtle (cf. *The journalist checked the facts* vs. *The mechanic checked the brakes*; see Hare, Elman, and McRae 2000, for a fuller discussion of such facts and their effects on processing). The different uses of this same verb are associated with different preferences as to fillers of the patient role, but the senses are quite similar. Thus, although it is easy to come up with clear examples of clear polysemy versus homophony, there are many cases for which the vague notion of core provides no clear guide as to when two lexemes are polysemes or homophones.

In the psycholinguistics literature, an important additional processing assumption has emerged, often implicit but nonetheless important because it suggests a test for what information might be included or excluded from a lexical representation: access to lexical information is faster and precedes the access of supralexical (e.g., syntactic, pragmatic) information. For example, it has been argued that

lexical access is an autonomous subsystem of the sentence comprehension routine in which all meanings of a word are momentarily accessed, regardless of the factors of contextual bias or bias associated with frequency of use. (Onifer and Swinney 1981, 225)

Under this hypothesis, disambiguation occurs subsequent to the initial retrieval of all meanings, and extralexical information is allowed to determine the contextually appropriate meaning at a later stage in processing. Despite the fact that subsequent work has challenged this claim (e.g., Van Petten and Kutas 1987, among many others), it remains a common assumption for many in the field that lexical information is privileged by rapid access. An important corollary of this is that early access to information is a good test of whether or not that information is contained in the lexicon. Factors that affect processing later are assumed to be extralexical.

A final important assumption is that the lexicon is the bottom line as far as meaning is concerned. This is consistent with F&P's definition of compositionality of meaning as being the molecular composition of atomic meanings that are indexed in the lexicon. Although the mapping between concepts and lexical semantics is not one to one, the relationship is in practice fairly tight.

3   Problems for This View of the Lexicon

I have already noted some challenges that are problematic for the lexicon as dictionary. These include the lack of specific tests for deciding what information is contained in lexical entries, the vagueness of the definition of "core," and the gradations in meanings that often make it difficult to distinguish polysemy from homophony. I turn now to a broader set of facts that stretches the notion of what information should be contained in the lexicon, assuming that processing facts can be used to infer lexical versus extralexical information. Because this information is often lexically idiosyncratic, it creates tension between information that is general and abstract and information that is lexically specific. I then move to a second set of facts that creates more significant challenges for the traditional lexicon (see Elman 2009 for a more extensive presentation), and which leads to my final proposal for an alternative way to encode lexical knowledge.

Although many theories of the lexicon begin with nouns, I will focus on verbs. Verbs are critical for binding other elements in a sentence together, and most of us have the intuition that, absent verbs, language is little more than a primitive indexical communication system.

3.1   Lexically Specific Knowledge

3.1.1   The Relationship between Meaning and Complement Structure Preferences

One workhorse of the psycholinguistic literature has been the study of how comprehenders process sentences that contain a temporary ambiguity. In cases where competing theories predict how comprehenders will resolve this ambiguity, the comprehenders' response when the ambiguity is first encountered, or subsequently disambiguated, provide useful clues as to what information was available and what processing strategy was used.

A much-studied structural ambiguity is that which arises at the postverbal noun phrase (NP) in sentences such as *The boy heard the story was interesting.* At the point where *the story* is encountered, it could either be the direct object (DO) of *heard*, or it could be the subject noun of a sentential complement (SC, as it ends up being in this example). An early influential theory of syntactic processing (the "two-stage model"; Frazier and Rayner 1982) predicts that the DO interpretation will be favored initially. Early data supported this claim. An alternative possibility (following constraint-based approaches) is that that result might arise for other reasons: (1) the relative frequency that a given verb occurs with either a DO or SC might favor the DO bias in this case (Garnsey et al. 1997; Holmes 1987; Mitchell and Holmes 1985); (2) the relative frequency that a given verb takes an SC with or without the disambiguating but optional complementizer *that* could lead to the DO bias (Trueswell, Tanenhaus, and Kello 1993); and (3) the plausibility of the postverbal NP as a DO for that particular verb (Garnsey et al. 1997; Pickering and Traxler 1998; Schmauder and Egan 1998).

The first of these factors—the statistical likelihood that a verb appears with either a DO or SC structure—has been particularly perplexing. The prediction is that if comprehenders are sensitive to the usage statistics of different verbs, then, when confronted with a DO/SC ambiguity, comprehenders will prefer the interpretation that is consistent with that verb's bias. Some studies report either late or no effects of verb bias (e.g., Mitchell 1987; Ferreira and Henderson 1990). More recent studies, on the other hand, have shown that verb bias does affect comprehenders' interpretation of such temporarily ambiguous sequences (Garnsey et al. 1997; Trueswell et al. 1993; but see Kennison 1999). Whether or not such information is used at early stages of processing is important not only because of its processing implications but because, if it is, this implies that the detailed statistical patterns of subcategorization usage will need to be part of a verb's lexical representation.

One possible explanation for the discrepant experimental data is that many of the verbs that show such DO/SC alternations have multiple senses, and these senses may have different subcategorization preferences (Roland and Jurafsky 1998, 2002). This raises the possibility that a comprehender might disambiguate the same temporarily ambiguous sentence fragment in different ways, depending on the inferred meaning of the verb. That meaning might in turn be implied by the context that precedes the sentence. A context that primes the sense of the verb that more frequently occurs with DOs should generate a different expectation than a context that primes a sense that has an SC bias.

Hare, McRae, and Elman (2003, 2004) tested this possibility. Several large text corpora were analyzed to establish the statistical patterns of usage that were associated with verbs (DO vs. SC) and in which different preferences were found for different verb senses. The corpus analyses were used to construct pairs of two sentence stories. In each pair, the second target sentence contained the same verb in a sequence that was temporarily (up to the postverbal NP) ambiguous between a DO or SC reading. The first sentence provided a meaning-biasing context. In one case, the context suggested a meaning for the verb in the target sentence that was highly correlated with a DO structure. In the other case, the context primed another meaning of the verb that occurred more frequently with an SC structure. Both target sentences were in fact identical until nearly the end. Thus, sometimes the ambiguity was resolved in a way that did not match participants' predicted expectations. The data (reviewed in more detail in Hare et al. 2009) suggest that comprehenders' expectancies regarding the subcategorization frame in which a verb occurs is indeed sensitive to statistical patterns of usage that are associated not with the verb in general, but with the sense-specific usage of the verb. A computational model of these effects is described by Elman, Hare, and McRae (2005).

A similar demonstration of the use of meaning to predict structure is reported by Hare, Elman, Tabaczynski, and McRae (2009). That study examined expectancies that arise during incremental processing of sentences that involve verbs such as *collect*, which can occur in either a transitive construction (e.g., *The children collected dead leaves,* in which the verb has a causative meaning) or an intransitive construction (e.g., *The rainwater collected in the damp playground*, in which the verb is inchoative). Here again, at the point where the syntactic frame is ambiguous (at the verb, *The children collected...* or *The dead leaves collected...*), comprehenders appeared to expect the construction that was appropriate given the likely meaning of the verb (causative vs. inchoative). In this case, the meaning was biased by having subjects that were either good causal agents (e.g., *children* in the first example above) or good themes (*rainwater* in the second example).

These experiments suggest that the lexical representation of verbs must include not just information regarding the verb's overall structural usage patterns, but that this information regarding the syntactic structures associated with a verb is sense-specific, and a comprehender's structural expectations are modulated by the meaning of the verb that is inferred from the context. This implies a richer lexical representation for verbs than might have been assumed, though this can be easily accommodated within the traditional lexicon.

3.1.2 Verb-Specific Thematic Role Filler Preferences

Another well-studied ambiguity arises with verbs such as *arrest*. These are verbs that can occur in both the active voice (as in *The man arrested the burglar*) and in the passive (as in *The man was arrested by the policeman*). The potential for ambiguity arises because relative clauses in English (*The man who was arrested...*) may occur in a reduced form in which *who was* is omitted. This gives rise to *The man arrested...*, which is ambiguous at the verb. Until the remainder of the sentence is provided, it is temporarily unclear whether the verb is in the active voice (and the sentence might continue as in the first example) or whether this is the start of a reduced relative construction, in which the verb is in the passive (as in *The man arrested by the policeman was innocent*).

In an earlier study, Taraban and McClelland (1988) found that when participants read sentences involving ambiguous prepositional attachments, for example, *The janitor cleaned the storage area with the broom...* or *The janitor cleaned the storage area with the solvent...*, reading times were faster in sentences involving more typical fillers of the instrument role (in these examples, *broom* rather than *solvent*). McRae, Spivey-Knowlton, and Tanenhaus (1998) noted that in many cases, similar preferences appear to exist for verbs that can appear in either the active or passive voice. For many verbs, there are nominals that are better fillers of the agent role than the passive role, and vice versa.

This led McRae et al. (1998) to hypothesize that when confronted with a sentence fragment that is ambiguous between a main verb and reduced relative reading, comprehenders might be influenced by the initial subject NP and whether it is a more likely agent or patient. In the first case, this should encourage a main verb interpretation; in the latter case, a reduced relative should be favored. This is precisely what McRae et al. found to be the case. *The cop arrested...* promoted a main verb reading over a reduced relative interpretation, whereas *The criminal arrested...* increased the likelihood of the reduced relative reading. McRae et al. concluded that the thematic role specifications for verbs must go beyond simple categorical information, such as *agent, patient, instrument, beneficiary*, and so on. The experimental data suggest that the roles contain very detailed information about the preferred fillers of these roles, and that the preferences are verb specific. The preferences are expressed not only over the nominal fillers of roles, but their attributes as well. Thus, a *shrewd, heartless gambler* is a better agent of *manipulate* than a *young, naive gambler*; conversely, the latter is a better filler of the same verb's patient role (McRae, Ferretti, and Amyote 1997).

This account of thematic roles resembles that of Dowty (1991) in that both accounts suggest that thematic roles have internal structure. But the McRae et al. (1997; McRae, Spivey-Knowlton, and Tanenhaus 1998) results further suggest a level of information that goes considerably beyond the limited set of proto-role features envisioned by Dowty. McRae et al. interpreted these role-filler preferences as reflecting comprehenders' specific knowledge of the event structure associated with different verbs. This appeal to event structure, as we shall see below, will figure significantly in phenomena that are not as easily accommodated by the lexicon.

We have seen that verb-specific preferences for their thematic role fillers arise in the course of sentence processing. Might such preferences also be revealed in word-word priming? The question is important because this sort of priming has often been assumed to occur at the lexical level, that is, to reflect the ability of one word to activate another word, and thus to be a test of the context of a word's lexical entry.

The answer is yes, such priming does occur. Ferretti, McRae, and Hatherell (2001) found that verbs primed nouns that were good fillers for their agent, patient, or instrument roles. The priming also goes in the opposite direction, such that when a comprehender encounters a noun, the noun serves as a cue for the event in which it typically participates, thereby priming verbs that describe that event activity (McRae et al. 2005).

The above results are among a much larger empirical literature that significantly extend the nature of the information that must be encoded in a verb's lexical representation. In addition to sense-specific structural usage patterns, the verb's lexical entry must also encode verb-specific information regarding the characteristics of the nominals that best fit that verb's thematic roles. The lexical representation for verbs must include subentries about all the verb's senses. For each sense, all possible subcategorization frames would be shown. For each verb-sense-subcategorization combination, additional information would be indicating the probability of each combination. Finally, similar information would be needed for every verb-sense-thematic role possibility. The experimental evidence indicates that in many cases, this latter information will be detailed, highly idiosyncratic of the verb, and represented at the featural level (e.g., Ferretti, McRae, and Hatherell 2001; McRae, Ferretti, and Amyote 1997).

3.2 Flies in the Ointment

These findings, among many others in recent years, expand the contents of the verb's lexical representation. But even though these data suggest very detailed and often idiosyncratic lexical representations, they could still be accommodated by an enumerative data structure of the sort implemented by the lexicon. We now turn to additional phenomena that are problematic for the traditional view of the mental lexicon *qua* dictionary.

3.2.1 Aspect and Event Knowledge

As noted above, Ferretti et al. (2001) found that verbs were able to prime their preferred agents, patients, and instruments. However, no priming was found from verbs to the locations in which their associated actions take place. Why might this be? Ferretti, Kutas, and McRae (2007) noted that in that experiment the verb primes for locations were in the past tense (e.g., *skated—arena*), and possibly interpreted by participants as having perfective aspect. Because the perfective signals that the event has concluded, it is often used to provide background information prefatory to the time period under focus (as in *Dorothy had skated for many years and was now looking forward to her retirement*). Imperfective aspect, on the other hand, is used to describe events that are either habitual or ongoing; this is particularly true of the progressive. Ferretti et al. hypothesized that although a past perfect verb did not prime its associated location, the same verb in the progressive might do so because of the location's greater salience to the unfolding event.

This prediction was borne out. The two-word prime *had skated* failed to yield significant priming for *arena* in a short SOA naming task, relative to an unrelated prime; but the two-word prime *was skating* did significantly facilitate naming. In an ERP version of the experiment, the typicality of the location was found to affect expectations. Sentences such as *The diver was snorkeling in the ocean* (typical location) elicited lower amplitude N400 responses at *ocean*, compared to *The diver was snorkeling in the pond* at *pond*. The N400 is interpreted as an index of semantic expectancy, and the fact that typicality of agent-verb-location combinations affected processing at the location indicates that this information must be available early in processing.

The ability of verbal aspect to manipulate sentence processing by changing the focus on an event description, with implications for processing, has been noted elsewhere (e.g., Kehler 2002; Kehler et al. 2008). The results in this case, however, present a specific challenge for how to represent verb argument preferences. Critically, the effect seems to occur on the same time scale as other information that affects verb argument expectations (this was demonstrated by Experiment 3 in Ferretti et al. 2007, in which ERP data indicated aspectual differences within 400 ms of the expected word's presentation). This is a time frame that has often been seen as indicating that intralexical information is operant, and prior to adjustments that depend on extralexical (e.g., semantic, discourse, pragmatic) factors. But logically, it is difficult to see how one would encode the dynamic and context-specificity contingency on thematic role requirements that arises when aspect is manipulated. That is, although the patterns of ambiguity resolution described in earlier sections, along with parallel findings using priming (Ferretti, McRae, and Hatherell 2001; McRae et al. 2005), might be accommodated by enriching the information in the lexical representations of verbs, the very similar effects of aspect do not seem amenable to such an account.

Setting this important question aside for the moment (we return to it later), we might ask, If verb aspect can alter the expected arguments for a verb, what else might do so? The concept of event representation has emerged as a useful way to understand other results in which aspect plays a role (Kehler 2002; Kehler et al. 2008; Kertz, Kehler, and Elman 2006; Moens and Steedman 1988; Rohde, Kehler, and Elman 2006). If we consider the question from the perspective of event representation, viewing the verb as providing merely some of the cues (albeit very potent ones) that tap into event knowledge, then several other candidates suggest themselves.

3.2.2 Dynamic Alterations in Verb Argument Expectations

If we think in terms of verbs as cues and events as the knowledge they target, then it should be clear that although the verb is obviously a very powerful cue, and its aspect may alter the way the event is construed, there are other cues that change the nature of the event or activity associated with the verb. For example, the choice of agent of the verb may signal different activities. A sentence-initial noun phrase such as *The surgeon...* is enough to generate expectancies that constrain the range of likely events. In isolation, this cue is typically fairly weak and unreliable, but different agents may combine with the same verb to describe quite different events.

Consider the verb *cut*. Our expectations regarding what will be cut, given a sentence that begins *The surgeon cuts...* are quite different than for the fragment *The lumberjack cuts....* These differences in expectation clearly reflect our knowledge of the world. This is not remarkable. The critical questions are: What is the status of such knowledge, and where does it reside? No one doubts that a comprehender's knowledge of how and what a surgeon cuts, versus what a lumberjack cuts, plays an important role in comprehension at some point.

The crucial issue, for the purposes of deciding what information is included in a lexical entry and what information arises from other knowledge sources, is when this knowledge enters into the unfolding process of comprehension. This is because, as pointed out above, timing has been an important adjudicator for models of processing and representation. If the knowledge is available very early—perhaps even immediately on encountering the relevant cues—then it is a candidate for being present in the lexical representation.

Bicknell, Elman, Hare, McRae, and Kutas (2010) hypothesized that if different agent-verb combinations imply different types of events, this might lead comprehenders to expect different patients for the different events. This prediction follows from a study by Kamide, Altmann, and Haywood (2003). Kamide et al. employed a paradigm in which participants' eye movements toward various pictures were monitored as they heard sentences such as *The man will ride the motorbike* or *The girl will ride the carousel* (all combinations of agent and patient were crossed) while viewing a visual scene containing a man, a girl, a motorbike, a carousel, and candy. At the point when participants heard *The man will ride...*, Kamide et al. found that there were more looks toward the motorbike than to the carousel, and the converse was true for *The girl will ride....* The Bicknell et al. study was designed to look specifically at agent-verb interactions to see whether such effects also occurred during self-paced reading, and if so, how early in processing.

A set of verbs such as *cut*, *save*, and *check* were first identified as potentially describing different events depending on the agent of the activity, and in which the event described by the agent-verb combination would entail different patients. These verbs were then placed in sentences in which the agent-verb combination was followed either by the congruent patient, as in *The journalist checked the spelling of his latest report* or in which the agent-verb was followed by an incongruent patient, as in *The mechanic checked the spelling of his latest report* (all agents of the same verb appeared with all patients, and a continuation sentence followed that increased the plausibility of the incongruent events). Participants read the sentences a word at a time, using a self-paced moving-window paradigm.

As predicted, reading times increased for sentences in which an agent-verb combination was followed by an incongruent (though plausible) patient. The slowdown occurred at one word following the patient, leaving open the possibility that the expectation reflected delayed use of world knowledge. Bicknell et al. therefore carried out a second experiment using the same materials, but recording ERPs as participants read the sentences. The rationale for this was that ERPs provide a more precise and sensitive index of processing than reading times. Of particular interest was the N400 component, since this provides a good measure of the degree to which a given word is expected and/or integrated into the prior context. As predicted, an elevated N400 was found for incongruent patients.

The fact that what patient is expected may vary as a function of specific particular agent-verb combinations is not in itself surprising. What is significant is that the effect occurs at the earliest possible moment, at the patient that immediately follows the verb. The timing of such effects has in the past often been taken as indicative of an effect's source. A common assumption has been that immediate effects reflect lexical or "first-pass" processing, and later effects reflect the use of semantic or pragmatic information. In this study, the agent-verb combinations draw on comprehenders' world knowledge. The immediacy of the effect would seem to require either that this information must be embedded in the lexicon or that world knowledge must be able to interact with lexical knowledge more quickly than has often typically been assumed.

Can other elements in a sentence affect the event type that is implied by the verb? Consider again the verb *cut*. The *Oxford English Dictionary* shows the transitive form of this verb as having a single sense. *WordNet* gives 41 senses. The difference is that *WordNet*'s senses more closely correspond to what one might call event types, whereas the *OED* adheres to a more traditional notion of sense that is defined by an abstract core meaning that does not depend on context. Yet cutting activities in different contexts may involve quite different sets of agents, patients, instruments, and even locations. The instrument is likely to be a particularly potent constraint on the event type.

Matsuki, Chow, Hare, Elman, Scheepers, and McRae (2011) tested the possibility that the instrument used with a verb would cue different event schemas, leading to different expectations regarding the most likely patient. Using eye-tracking to monitor processing during reading, participants were presented with sentences such as *Susan used the scissors to cut the expensive paper that she needed for her project*, or *Susan used the saw to cut the expensive wood.*... Performance on these sentences was contrasted with that on the less expected *Susan used the scissors to cut the expensive wood...* or *Susan used the saw to cut the expensive paper.*... As in the Bicknell et al. study, materials were normalized to ensure that there were no direct lexical associations between instrument and patient. An additional priming study was carried out in which instruments and patients served as prime-target pairs; no significant priming was found between typical instruments and patients (e.g., *scissors-paper*) versus atypical instruments and patients (e.g., *saw-paper*; but priming did occur for a set of additional items that were included as a comparison set). As predicted, readers showed increased reading times for the atypical patient relative to the typical patient. In this study, the effect occurred right at the patient, demonstrating that the filler of the instrument role for a specific verb alters the restrictions on the filler of the patient role.

4 Lexical Knowledge without a Lexicon

4.1 Where Does Lexical Knowledge Reside?

The findings reviewed in section 3.1 strongly support the position that lexical knowledge is quite detailed, often idiosyncratic and verb specific, and brought to bear at the earliest possible stage in incremental sentence processing. The examples above focused on verbs and the need to encode restrictions (or preferences) over the various arguments with which they may occur. Taken alone, those results might be accommodated by simply providing greater detail in lexical entries in the mental lexicon, as standardly conceived.

Where things get tricky is when one also considers what seems to be the ability of dynamic factors to significantly modulate such expectations (section 3.2). These include the verb's grammatical aspect, the agent and instrument that are involved in the activity, and the overall discourse context. To be clear: that these factors play a role in sentence processing is not itself surprising. However, the common assumption has been that such dynamic factors lie outside the lexicon. This is, for example, essentially the position outlined by J. D. Fodor (1995): "We may assume that there is a syntactic processing module, which feeds into, but is not fed by, the semantic and pragmatic processing routines ... syntactic analysis is serial, with back-up and revision if the processor's first hypothesis about the structure turns out later to have been wrong" (435).

More pithily, the data do not accord with the "syntax proposes, semantics disposes" hypothesis (Crain and Steedman 1985). Thus, what is significant about the findings above is that the influence of aspect, agent, instrument, and discourse all occur within the same time frame that has been used operationally to identify information that resides in the lexicon. This is important if we are to have some empirical basis for deciding what goes in the lexicon and what does not.

All of this places us in the uncomfortable position of having to make some difficult decisions.

One option would be to abandon any hope of finding any empirical basis for determining the contents of the mental lexicon. One might simply stipulate that some classes of information reside in the lexicon and others do not. This is not a desirable solution. Note that even within the domain of theoretical linguistics, considerable controversy has emerged regarding what sort of information belongs in the lexicon, with different theories taking different and often mutually incompatible positions (cf., among many other examples, Haiman 1980; Lakoff 1971; Weinreich 1962; Jackendoff 1983, 2002; Katz and Fodor 1963; Langacker 1987; Chomsky 1965; Levin and Hovav 2005; Fodor 2002). If we insist that the form of the mental lexicon has no consequences for processing, and exclude data of this type, then we have no behavioral way to evaluate different proposals. This essentially accepts that performance tells us little about competence (Chomsky 1965).

A second option would be to significantly enlarge the format of lexical entries so that they accommodate all the above information. This would be a logical conclusion to the trend that has appeared not only in the processing literature (e.g., in addition to the studies cited above, van Berkum et al. 2003; van Berkum et al. 2005; Kamide, Altmann, and Haywood 2003; Kamide, Scheepers, and Altmann 2003; Altmann and Kamide 2007) but also many recent linguistic theories (e.g., Bresnan 2006;

Fauconnier and Turner 2002; Goldberg 2003; Lakoff 1987; Langacker 1987; though many or perhaps all of these authors might not agree with such a conclusion). The lexicon has become increasingly rich and detailed in recent years. Why impose arbitrary limits on its contents?

One problem is that the combinatoric explosion this entails, especially given the unbounded nature of discourse contexts, may render the proposal infeasible. But it also presents us with a logical conundrum: if all this information resides in the lexicon, is there then any meaningful distinction between the lexicon and other linguistic modules?

The third option is the most radical: is it possible that lexical knowledge of the sort discussed here might be instantiated in a very different way than through an enumerative dictionary?

4.2 An Alternative to the Mental Lexicon as Dictionary

The common factor in the studies described above was the ability of sentential elements to interact in real time to produce an incremental interpretation that guided expectancies about upcoming elements. These can be thought of as very powerful context effects that modulate the meaning that words have.

But suppose that one views words not as elements in a data structure that must be retrieved from memory, but rather as stimuli that alter mental states (which arise from processing prior words) in lawful ways. In this view, words are not mental objects that reside in a mental lexicon. They are operators on mental states. From this perspective, words do not *have* meaning; rather, they are *cues* to meaning (Elman 2009; Rumelhart 1979).

This scheme of things can be captured by a model that instantiates a dynamical system. The system receives inputs (words, in this case) over time. The words perturb the internal state of the system (we can call it the "mental state") as they are processed, with each new word altering the mental state in some way.

Over the years, a number of connectionist models have been developed that illustrate ways in which context can influence processing in complicated but significant ways (e.g., among many others, McClelland and Rumelhart 1981; McRae, Spivey-Knowlton, and Tanenhaus 1998; Rumelhart et al. 1988; Taraban and McClelland 1988). There is also a rich literature in the use of dynamical systems to model cognitive phenomena (e.g., Smith and Thelen 1993; Spencer and Schöner 2003; Tabor and Tanenhaus 2001; Thelen and Smith 1994).

A particularly fruitful architecture has been one that involves recurrence between processing units (e.g., Botvinick and Plaut 2004; Elman 1990;

Rogers et al. 2004; St. John and McClelland 1990). In recurrent networks, information flow is multidirectional, and feedback loops allow the current state of the system to be affected by prior states. Learning in such systems can be thought of as encoding the grammar over sequences. The grammar constrains the lawful effects that inputs have on moving the system through the network's "mental state space." If those inputs are words, then what we think of as lexical knowledge is the knowledge encoded in the connections between processing units that allows each word to have the appropriate effect on processing. We have not removed the need for lexical knowledge, but rather moved it from an enumerative and declarative database (the lexicon) into the elements of the system (the weights) that are responsible for processing. It should be noted that recurrent networks are not finite state automata, but have computational properties similar—but not identical—to stack-based automata (Boden and Wiles 2000; Boden and Blair 2003; Peters and Ritchie 1973; Rodriguez, Wiles, and Elman 1999; Rodriguez and Elman 1999; Rodriguez 2001; Siegelmann and Sontag 1995).

Elman (2009) presented a simple model that demonstrates some of the properties required to capture the kinds of lexical effects noted here. A somewhat fuller model, under development in collaboration with Ken McRae and Mary Hare, is shown in figure 5.1.

The model is inspired by and incorporates elements of a number of important prior models that have related properties. These include models of language processing (McClelland, St. John, and Taraban 1989; St. John and McClelland 1990; St. John 1992), schemas and sequential thought processes (Rumelhart et al. 1988), semantic cognition (Rogers et al. 2004), and action planning (Botvinick and Plaut 2004). One way of thinking of this model is as an attempt to take the important insights regarding schemas, scripts, frames, and stories (Abelson 1981; Minsky 1974; Norman and Rumelhart 1981; Schank and Abelson 1977) and instantiate those insights in a computational architecture that allows for richer processing than was possible using the earlier tools from the AI toolbox.

The goal of the model is to learn the contingent relationships between activities and participants that are involved in events that unfold over time. The model has a view of the world (i.e., inputs) that allows it to identify the relevant entities (including, in this simplified version, agents, patients, instruments, and locations) and actions that participate in momentary *activities* (e.g., *John enters the restaurant*). These activities are connected in sequence to form *events* (e.g., *John enters the restaurant; He sits down at a table; He orders food; He cuts the food with a knife; He eats the food;... ; He leaves*). At any given point in time, the task given to the model

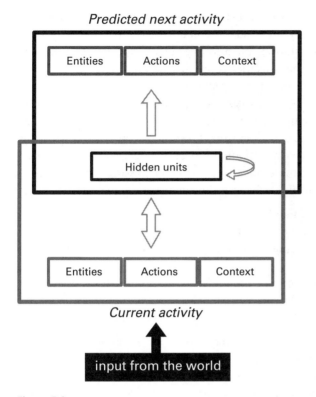

Figure 5.1

An activity consists of a collection of entities, actions, and context, all presented simultaneously as input from the world. The "current activity" portion of the network learns the patterns of co-occurrence that are typical of given activities (e.g., *Jill cuts her steak with a knife and fork in the restaurant*). The recurrent connections between the entity-action-context units and the hidden units allow the network to complete input patterns in which not all information is explicitly specified. Thus, *Jill cuts her steak* would lead to the activation of *knife* and *restaurant*. An event consists of a sequence of activities, presented in temporal order and duration appropriate to the event. The "predicted next activity" portion of the network learns the temporal relations between activities that are part of an event. This part of the network also learns to do pattern completion in the temporal domain. Thus, if the current activity is *Jill enters the restaurant*, the predicted next activity layer would anticipate subsequent activities, and would activate, in sequence, *Jill orders a steak*, then *Jill cuts her steak with a knife and fork*, then *Jill eats her steak*, and so on.

is to learn the mutual constraints between co-occurring entities and actions (this is accomplished by the model's reproducing on its output layer the details of the current activity), and also to generate expectancies regarding subsequent activities that together compose this event (accomplished by the model's predictions of future activities).

During learning, the model may see only a subset of possible entities and actions that make up an activity, or a subset of the sequence of activities that make up an event. Over time, however, accumulated experience allows the network to learn about the overarching generalizations that exist between and across events, and to fill in missing information as required. This leads to the following characteristics:

1. Pattern completion within activities and within events
2. Typicality and prototype effects
3. Soft and graded constraints on roles, participants, activities, locations, and so on
4. Ability to flexibly combine and merge novel combinations of events
5. Ability to support inferences—under the right conditions
6. Ability to capture effects of perspective on event representation

The details of the model and simulations of its behavior are described elsewhere (Elman, McRae, and Hare, in preparation), and given the scope of this chapter, I present two simulations to illustrate these properties. Central to all of these is the fact that the model implements a constraint-satisfaction network. The constraints operate both at a given point in time, reflecting patterns of co-occurrence of actions and activity participants, and across time, reflecting the succession of activities that arise over time as an event unfolds.

In figure 5.2, we see the activation of various entities and actions that result from the model's being presented with *John does something with his food*. This activity occurs in a sequence shortly after John has entered a restaurant. The activation levels are indicated by the height of curves on the ordinate, and their change over time is indicated along the abscissa. The model infers that food is involved throughout the activity, that John begins by cutting the food (so *cut* and *knife* are active), and that, as time goes along, the cutting action diminishes and the eating action increases. The model has thus not only filled in the implied elements but has captured the ordered temporal relationships between activities that make up the overall event.

These properties provide a straightforward account of the findings reported in Metusalem et al. (2012), in which words that were unexpected

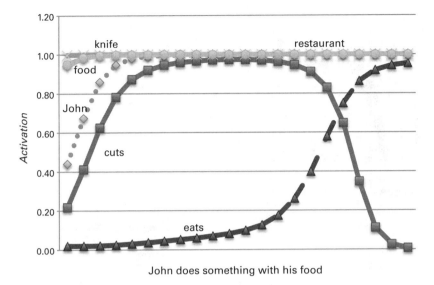

Figure 5.2
Activations of participants and actions that result when the network is presented with the ambiguous input *John does something with his food*. Drawing on prior experience, the network assumes that John first cuts his food (units for *cut* and *knife* are activated); the network has also learned that cutting is followed by eating (activation for *cut* decreases and *eat* increases).

but consistent with the larger event being described elicited smaller amplitude N400s than words that were both unexpected and unrelated to the event. In figure 5.3, we see the time-varying activation of various elements as an event is being described. At the end, the activation of *medal*, which is event relevant even though it is locally unexpected given the linguistic context, is higher than *bleach*, which is both unexpected and event unrelated.

In other simulations, we find that the model is able to use its knowledge of events, with the resulting activation of event-relevant participants that may not be named, to make inferences that affect subsequent ambiguous statements. For example, compare the following two sequences:

1. *John cut wood in the forest.*
    Suddenly, he cut himself by mistake.
    What happened to John?

2. *John cut food in the restaurant.*
    Suddenly, he cut himself by mistake.
    What happened to John?

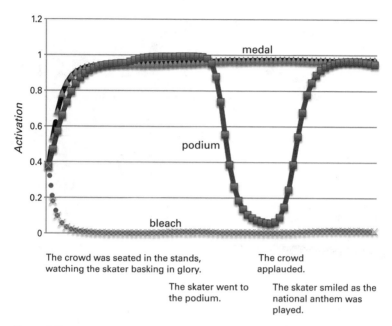

Figure 5.3
Time-varying activation of three entities (*medal*, *podium*, *bleach*) as the network processes four activities associated with the event "getting a medal at the Olympics." In the final activity, none of the three entities is mentioned. However, *podium* is reactivated (following a drop during the crowd applauding activity), because that location has previously been associated with the awarding of medals. *Medal* is also activated, even though it is contextually unexpected given the linguistic context, because it is consistent with the event. In contrast, *bleach*, which is equally unexpected in this linguistic context, is completely inactive because it is inappropriate to the event. These patterns replicate the findings of Metusalem et al. (2012).

In the context of cutting wood in the forest, the model has learned that axes are typically used. It also knows, independently (from other event knowledge), that cutting oneself with an axe can be fatal. Cutting food in a restaurant, on the other hand, involves knives; and cutting oneself with a knife is less lethal. When confronted with the final (identical) query, the model activates *John died* in the first scenario and *John bled* in the second case. The bridging inference is made possible by the model's ability not only to fill in the missing pieces (cutting wood in the forest probably involves an axe, etc.), but then to use the inferred but unmentioned information to resolve a subsequent ambiguity (*What happened?*).

5   Discussion

Although the possibility of lexical knowledge without a lexicon might seem odd, the core ideas that motivate this proposal are not new. Many elements appear elsewhere in the literature. These include the following.

(1) The meaning of a word is rooted in our knowledge of both the material and the social world. The material world includes the world around us as we experience it (i.e., it is embodied), possibly indirectly. The social world includes cultural habits and artifacts; in many cases, these habits and artifacts have significance only by agreement (i.e., they are conventionalized). Similar points have been made by many others, notably Wittgenstein (1966), Hutchins (1994) and Fauconnier (1997), and Fauconnier and Turner (2002).

(2) Context is always with us. The meaning of a word is never "out of context," although we might not always know what the context is (particularly if we fail to provide one). This point has been made by many, including Kintsch (1988), Langacker (1987), McClelland et al. (1989), and van Berkum and colleagues (van Berkum, Brown, Zwiterselood, Kooijman, and Hagoort 2005; van Berkum, Zwitserlood, Hagoort, and Brown 2003). This insight is also found in computational models of meaning that emphasize multiple co-occurrence constraints between words in order to represent them as points in a high dimensional space, such as LSA (Latent Semantic Analysis) (Landauer and Dumais 1997), HAL (Burgess and Lund 1997), or probabilistic models (Griffiths and Steyvers 2004). The dynamical approach here also emphasizes the time course of processing that results from the incremental nature of language input.

(3) Events play a major role in organizing our experience. Event knowledge is used to drive inference and access memory, and it affects the categories we construct. An event may be defined as a set of participants, activities, and outcomes that are bound together by causal interrelatedness. An extensive literature argues for this, aside from the studies described here, including work by Minsky (1974), Schank and Abelson (1977), and Zacks and Tversky (2001); see also Shipley and Zacks's (2008) book for a comprehensive collection on the role of event knowledge in perception, action, and cognition.

(4) Dynamical systems provide a powerful framework for understanding biologically based behavior. The nonlinear and continuous valued nature of dynamical systems allows them to respond in a graded manner under

some circumstances, while in other cases their responses may seem more binary. Dynamical analyses figure prominently in the recent literature in cognitive science, including work by Smith and Thelen (1993, 2003; Thelen and Smith 1994), Spencer and Schöner (2003), Spivey (2007; Spivey and Dale 2004), and Tabor (2004; Tabor and Tanenhaus 2001; Tabor et al. 1997).

The model presented here is a first step in trying to offer an alternative mechanism for representing lexical knowledge. Rather than assuming that knowledge is enumerated in a database, the model moves the knowledge into the dynamics that result from a word's effect on processing. These dynamics are lawful and predictable. They might even be called systematic, although probably not under the definition offered by F&P. Under the model's version of systematicity, such things as context effects, graded polysemy, and lexically idiosyncratic behavior are not seen as exceptions that require ad hoc explanation. Rather, these properties are predicted by the model. At the same time, the model is able to generalize and abstract across commonalities of behavior. This is "having your cake and eating it too."

The model is incomplete in important ways. It is disembodied and so lacks the conceptual knowledge about events that comes from direct experience (of course, this deficiency is equally true of F&P symbolic models). The model assumes a prior identification of some entities as agents, and others as patients, instruments, and so on. Proposals have in fact been put forward about how this might be done (e.g., Chang, Dell, and Bock 2006; Chang 2012; Gao, Newman, and Scholl 2009; Gao and Scholl 2011). An interesting future direction will be to extend the model in similar ways.

How does this view affect the way we do business (or at least, study words)? Although I have argued that many of the behavioral phenomena described above are not easily incorporated into a mental lexicon as traditionally conceived, it is possible that solutions exist. A parallel architecture of the sort described by Jackendoff (2002), for example, if it permitted direct and immediate interactions among the syntactic, semantic, and pragmatic components of the grammar, might be able to account for the data described earlier. The important question would then be how to devise tests to distinguish between these proposals. This remains an open question for the moment, at least until such counterproposals are advanced.

However, theories are also evaluated for their ability to offer new ways of thinking about old problems, and to provoke new questions that would not be otherwise asked. Let me suggest two positive consequences to the sort of words-as-cues dynamical model I am outlining (see also Elman 2009, for a fuller discussion).

The assumption that only certain information goes in the lexicon, and that the lexicon and other knowledge sources respect modular boundaries with limited and late occurring interactions, drives a research program that discourages looking for evidence of richer and more immediate interactions. For example, that selectional restrictions might be dynamic and context sensitive is fundamentally not an option within the Katz and Fodor (1963) framework. The words-as-cues approach, in contrast, suggests that such interdependencies should be expected. Indeed, there should be many such interactions among lexical knowledge, context, and nonlinguistic factors, and these might occur early in processing. Many researchers in the field have already come to this point of view. It is a conclusion that, despite considerable empirical evidence, has taken longer to arrive than it might have, given a different theoretical perspective.

A second consequence of this perspective is that it encourages a more unified view of phenomena that are often treated (de facto, if not in principle) as unrelated. Syntactic ambiguity resolution, lexical ambiguity resolution, pronoun interpretation, text inference, and semantic memory (to choose but a small subset of domains) are studied by communities that do not always communicate well, and researchers in these areas are not always aware of findings from other areas. Yet these domains have considerable potential for informing each other. That is because, although they ultimately draw on a common conceptual knowledge base, that knowledge base can be accessed in different ways, and this in turn affects what is accessed. Consider how our knowledge of events might be tapped in a priming paradigm, compared with a sentence-processing paradigm. Because prime-target pairs are typically presented with no discourse context, one might expect that a transitive verb prime might evoke a situation in which the fillers of both its agent and patient roles are equally salient. Thus, *arresting* should prime *cop* (typical arrestor) and also *crook* (typical arrestee). Indeed, this is what happens (Ferretti, McRae, and Hatherell 2001). Yet this same study also demonstrated that when verb primes were embedded in sentence fragments, the priming of good agents or patients was contingent on the syntactic frame within which the verb occurred. Primes of the form *She arrested the…* facilitated naming of *crook*, but not *cop*. Conversely, the prime *She was arrested by the…* facilitated naming of *cop* rather than *crook*.

These two results demonstrate that although words in isolation can serve as cues to event knowledge, they are only one such cue. The grammatical construction within which they occur provides independent evidence regarding the roles played by different event participants (Goldberg 2003). And, of course, the discourse context may provide further constraints on how an event is construed. Thus, as Race et al. (2008) found, although *shoppers* might typically save money and *lifeguards* save children, in the context of a disaster, both agents will be expected to save children.

Eliminating the lexicon is indeed radical surgery, and it is an operation that at this point many will not agree to. At the very least, however, I hope that by demonstrating that lexical knowledge without a lexicon is possible, others will be encouraged to seek out additional evidence for ways in which the many things that language users know are brought to bear on the way language is processed.

Acknowledgments

This work was supported by NIH grants HD053136 and MH60517 to Jeff Elman, Ken McRae, and Mary Hare; NSERC grant OGP0155704 to KM; NIH Training Grant T32-DC000041 to the Center for Research in Language (UCSD); and by funding from the Kavli Institute of Brain and Mind (UCSD). Much of the experimental work reported here and the insights regarding the importance of event knowledge in sentence processing are the fruit of a long-time productive collaboration with Mary Hare and Ken McRae. I am grateful to them for many stimulating discussions and, above all, for their friendship. I am grateful to Jay McClelland for the many conversations we have had over three decades. His ideas and perspectives on language, cognition, and computation have influenced my thinking in many ways. Finally, a tremendous debt is owed to Dave Rumelhart, whose (1979) suggestion that words do not have meaning but rather are cues to meaning inspired the proposal outlined here.

References

Abelson, R. P. 1981. Psychological status of the script concept. *American Psychologist* 36 (7):715–729.

Altmann, G. T. M., and Y. Kamide. 2007. The real-time mediation of visual attention by language and world knowledge: Linking anticipatory (and other) eye movements to linguistic processing. *Journal of Memory and Language* 57 (4):502–518.

Bicknell, K., J. L. Elman, M. Hare, K. McRae, and M. Kutas. 2010. Effects of event knowledge in processing verbal arguments. *Journal of Memory and Language* 63 (4):489–505.

Boden, M., and A. Blair. 2003. Learning the dynamics of embedded clauses. *Applied Intelligence* 19 (1–2):51–63.

Boden, M., and J. Wiles. 2000. Context-free and context-sensitive dynamics in recurrent neural networks. *Connection Science: Journal of Neural Computing, Artificial Intelligence, and Cognitive Research* 12 (3–4):197–210.

Botvinick, M., and D. C. Plaut. 2004. Doing without schema hierarchies: A recurrent connectionist approach to normal and impaired routine sequential action. *Psychological Review* 111 (2):395–429. doi:10.1037/0033-295X.111.2.395.

Bresnan, J. 2006. Is syntactic knowledge probabilistic? Experiments with the English dative alternation. In *Roots: Linguistics in Search of Its Evidential Base*, ed. S. Featherston and W. Sternefeld. Berlin: Mouton de Gruyter.

Burgess, C., and K. Lund. 1997. Modeling parsing constraints with high-dimensional context space. *Language and Cognitive Processes* 12:177–210.

Chang, F. 2012. Towards a unified approach to acquisition, production, and comprehension. In *Proceedings of the 34th Annual Meeting of the Cognitive Science Society*, ed. N. Miyake, D. Peebles, and R. P. Cooper. Austin, TX: Cognitive Science Society.

Chang, F., G. S. Dell, and K. Bock. 2006. Becoming syntactic. *Psychological Review* 113 (2):234–272. doi: 10.1037/0033-295X.113.2.234.

Chomsky, N. 1965. *Aspects of the Theory of Syntax*. Cambridge, MA: MIT Press.

Crain, S., and M. Steedman. 1985. On not being led up the garden path: The use of context by the psychological parser. In *Natural Language Processing: Psychological, Computational, and Theoretical Perspectives*, ed. D. Dowty, L. Karttunen, and A. Zwicky, 320–358. Cambridge: Cambridge University Press.

Dowty, D. 1991. Thematic proto-roles and argument selection. *Language* 67: 547–619.

Elman, J. L. 1990. Finding structure in time. *Cognitive Science: A Multidisciplinary Journal* 14 (2):179–211.

Elman, J. L. 2009. On the meaning of words and dinosaur bones: Lexical knowledge without a lexicon. *Cognitive Science* 33:1–36. doi:10.1111/j.1551-6709.2009.01023.x.

Elman, J. L., M. Hare, and K. McRae. 2005. Cues, constraints, and competition in sentence processing. In *Beyond Nature-Nurture: Essays in Honor of Elizabeth Bates*, ed. M. Tomasello and D. Slobin, 111–138. Mahwah, NJ: Erlbaum.

Elman, J. L., K. McRae, and M. Hare. In preparation. A computational model of event knowledge and sentence processing.

Fauconnier, G. 1997. *Mappings in Thought and Language.* New York: Cambridge University Press.

Fauconnier, G., and M. Turner. 2002. *The Way We Think: Conceptual Blending and the Mind's Hidden Complexities.* New York: Basic Books.

Ferreira, F., and J. M. Henderson. 1990. Use of verb information in syntactic parsing: Evidence from eye movements and word-by-word self-paced reading. *Journal of Experimental Psychology: Learning, Memory, and Cognition* 16 (4):555–568.

Ferretti, T. R., M. Kutas, and K. McRae. 2007. Verb aspect and the activation of event knowledge. *Journal of Experimental Psychology: Learning, Memory, and Cognition* 33 (1):182–196.

Ferretti, T. R., K. McRae, and A. Hatherell. 2001. Integrating verbs, situation schemas, and thematic role concepts. *Journal of Memory and Language* 44:516–547.

Fodor, J. A. 2002. The lexicon and the laundromat. In *The Lexical Basis of Sentence Processing,* ed. P. Merlo and S. Stevenson, 75–94. Amsterdam: John Benjamins.

Fodor, J. A., and Z. W. Pylyshyn. 1988. Connectionism and cognitive architecture: A critical analysis. In *Connections and Symbols: A Cognition Special Issue,* ed. S. Pinker and J. Mehler. Cambridge, MA: MIT Press.

Fodor, J. D. 1995. Thematic roles and modularity. In *Cognitive Models of Speech Processing,* ed. G. T. M. Altmann, 434–456. Cambridge, MA: MIT Press.

Frazier, L., and K. Rayner. 1982. Making and correcting errors during sentence comprehension: Eye movements in the analysis of structurally ambiguous sentences. *Cognitive Psychology* 14 (2):178–210.

Gao, T., G. E. Newman, and B. J. Scholl. 2009. The psychophysics of chasing: A case study in the perception of animacy. *Cognitive psychology* 59 (2):154–179. doi: 10.1016/j.cogpsych.2009.03.001.

Gao, T., and B. J. Scholl. 2011. Chasing vs. stalking: Interrupting the perception of animacy. *Journal of Experimental Psychology: Human Perception and Performance* 37 (3):669–684. doi: 10.1037/a0020735.

Garnsey, S. M., N. J. Pearlmutter, E. Meyers, and M. A. Lotocky. 1997. The contribution of verb-bias and plausibility to the comprehension of temporarily ambiguous sentences. *Journal of Memory and Language* 37:58–93.

Goldberg, A. E. 2003. Constructions: A new theoretical approach to language. *Trends in Cognitive Sciences* 7 (5):219–224.

Griffiths, T. L., and M. Steyvers. 2004. A probabilistic approach to semantic representation. Paper read at the 24th Annual Conference of the Cognitive Science Society, George Mason University.

Haiman, J. 1980. Dictionaries and encyclopedias. *Lingua* 50:329–357.

Hare, M., J. L. Elman, and K. McRae. 2000. Sense and structure: Meaning as a determinant of verb categorization preferences. Paper read at CUNY Sentence Processing Conference, March 30–April 1, 2000, La Jolla, California.

Hare, M., J. L. Elman, and K. McRae. 2003. Sense and structure: Meaning as a determinant of verb subcategorization preferences. *Journal of Memory and Language* 48:281–303.

Hare, M., J. L. Elman, T. Tabaczynski, and K. McRae. 2009. The wind chilled the spectators but the wine just chilled: Sense, structure, and sentence comprehension. *Cognitive Science* 33:610–628.

Hare, M., K. McRae, and J. L. Elman. 2004. Admitting that admitting verb sense into corpus analyses makes sense. *Language and Cognitive Processes* 18:181–224.

Holmes, V. M. 1987. Syntactic parsing: In search of the garden path. In *Attention and Performance 12: The Psychology of Reading*, ed. Max Coltheart, 587–599. Hove: Erlbaum.

Hutchins, E. 1994. *Cognition in the World*. Cambridge, MA: MIT Press.

Jackendoff, R. 1983. *Semantics and Cognition*. Cambridge, MA: MIT Press.

Jackendoff, R. 2002. *Foundations of Language: Brain, Meaning, Grammar, and Evolution*. Oxford: Oxford University Press.

Kamide, Y., G. T. M. Altmann, and S. L. Haywood. 2003. The time-course of prediction in incremental sentence processing: Evidence from anticipatory eye movements. *Journal of Memory and Language* 49 (1):133–156.

Kamide, Y., C. Scheepers, and G. T. M. Altmann. 2003. Integration of syntactic and semantic information in predictive processing: Cross-linguistic evidence from German and English. *Journal of Psycholinguistic Research* 32 (1):37–55.

Katz, J. J., and J. A. Fodor. 1963. The structure of a semantic theory. *Language* 39 (2):170–210.

Kehler, A. 2002. *Coherence, Reference, and the Theory of Grammar*. Palo Alto, CA: CSLI Publications.

Kehler, A., L. Kertz, H. Rohde, and J. L. Elman. 2008. Coherence and coreference revisited. *Journal of Semantics* 25:1–44.

Kennison, S. M. 1999. American English usage frequencies for noun phrase and tensed sentence complement-taking verbs. *Journal of Psycholinguistic Research* 28 (2):165–177.

Kertz, L., A. Kehler, and J. Elman. 2006. Grammatical and coherence-based factors in pronoun interpretation. Paper read at the 28th Annual Conference of the Cognitive Science Society, Vancouver, Canada.

Kintsch, W. 1988. The role of knowledge in discourse comprehension: A construction-integration model. *Psychological Review* 95 (2):163–182.

Lakoff, G. 1971. Presuppositions and relative well-formedness. In *Semantics*, ed. D. Steinberg and L. Jakobovitz, 329–340. Cambridge: Cambridge University Press.

Lakoff, G. 1987. *Women, Fire, and Dangerous Things: What Categories Reveal about the Mind*. Chicago: University of Chicago Press.

Landauer, T. K., and S. T. Dumais. 1997. A solution to Plato's problem: The Latent Semantic Analysis theory of the acquisition, induction, and representation of knowledge. *Psychological Review* 104:211–240.

Langacker, R. W. 1987. *Foundations of Cognitive Grammar*, vol. 1. Stanford: Stanford University Press.

Levin, B., and M. R. Hovav. 2005. *Argument Realization: Research Surveys in Linguistics*. Cambridge: Cambridge University Press.

Marcus, G. 2001. *The Algebraic Mind*. Cambridge, MA: MIT Press.

Matsuki, K., T. Chow, M. Hare, J. L. Elman, C. Scheepers, and K. McRae. 2011. Event-based plausibility immediately influences on-line language comprehension. *Journal of Experimental Psychology: Learning, Memory, and Cognition* 37 (4):913–934.

McClelland, J. L., and D. E. Rumelhart. 1981. An interactive activation model of context effects in letter perception. Part 1: An account of basic findings. *Psychological Review* 86:287–330.

McClelland, J. L., M. F. St. John, and R. Taraban. 1989. Sentence comprehension: A parallel distributed processing approach. *Language and Cognitive Processes* 4: 287–336.

McRae, K., T. R. Ferretti, and L. Amyote. 1997. Thematic roles as verb-specific concepts. *Language and Cognitive Processes: Special Issue on Lexical Representations in Sentence Processing* 12:137–176.

McRae, K., M. Hare, J. L. Elman, and T. R. Ferretti. 2005. A basis for generating expectancies for verbs from nouns. *Memory and Cognition* 33:1174–1184.

McRae, K., M. J. Spivey-Knowlton, and M. K. Tanenhaus. 1998. Modeling the influence of thematic fit (and other constraints) in on-line sentence comprehension. *Journal of Memory and Language* 38 (3):283–312.

Metusalem, R., M. Kutas, M. Hare, K. McRae, and J. L. Elman. 2012. Generalized event knowledge activation during online sentence comprehension. *Journal of Memory and Language* 66: 545–567.

Minsky, M. 1974. A framework for representing knowledge. MIT-AI Laboratory Memo 306. Massachusetts Institute of Technology.

Mitchell, D. C. 1987. Lexical guidance in human parsing: Locus and processing characteristics. In *Attention and Performance XII: The Psychology of Reading*, ed. M. Coltheart, 601–618. Hillsdale, NJ: Erlbaum.

Mitchell, D. C., and V. M. Holmes. 1985. The role of specific information about the verb in parsing sentences with local ambiguity. *Journal of Memory and Language* 24:542–559.

Moens, M., and M. Steedman. 1988. Temporal ontology and temporal reference. *Computational Linguistics* 14 (2):15–28.

Norman, D. A., and D. E. Rumelhart. 1981. The LNR approach to human information processing. *Cognition* 10 (1):235–240.

Onifer, W, and D. A. Swinney. 1981. Accessing lexical ambiguities during sentence comprehension: Effects of frequency of meaning and contextual bias. *Memory and Cognition* 9 (3):225–236.

Peters, S., and R. W. Ritchie. 1973. On the generative power of transformational grammars. *Information Sciences* 6:49–83.

Pickering, M. J., and M. J. Traxler. 1998. Plausibility and recovery from garden-paths: An eye-tracking study. *Journal of Experimental Psychology: Learning, Memory, and Cognition* 24:940–961.

Race, D., N. Klein, M. Hare, and M. K. Tanenhaus. 2008. What do shoppers expect to save? Agents influence patients via events, not associative priming. Poster presented at the 20th Annual CUNY Conference on Human Sentence Processing, Chapel Hill, North Carolina.

Rodriguez, P. 2001. Simple recurrent networks learn context-free and context-sensitive languages by counting. *Neural Computation* 13 (9):2093–2118.

Rodriguez, P., and J. L. Elman. 1999. Watching the transients: Viewing a simple recurrent network as a limited counter. *Behaviormetrika* 26 (1):51–74.

Rodriguez, P., J. Wiles, and J. L. Elman. 1999. A recurrent neural network that learns to count. *Connection Science* 11 (1):5–40.

Rogers, T. T., M. A. Lambon Ralph, P. Garrard, S. Bozeat, J. L. McClelland, J. R. Hodges, and K. Patterson. 2004. Structure and deterioration of semantic memory: A neuropsychological and computational investigation. *Psychological Review* 111 (1):205–235. doi: 10.1037/0033-295X.111.1.205.

Rohde, H., A. Kehler, and J. L. Elman. 2006. Event structure and discourse coherence biases in pronoun interpretation. Paper read at the 28th Annual Conference of the Cognitive Science Society, Vancouver, Canada.

Roland, D., and D. Jurafsky. 1998. How verb subcategorization frequencies are affected by corpus choice. Paper read at COLING-ACL 1998, Montreal, Canada.

Roland, D., and D. Jurafsky. 2002. Verb sense and verb subcategorization probabilities. In *The Lexical Basis of Sentence Processing: Formal, Computational, and Experimental Issues*, ed. P. Merlo and S. Stevenson, 325–347. Amsterdam: John Benjamins.

Rumelhart, D. E. 1979. Some problems with the notion that words have literal meanings. In *Metaphor and Thought*, ed. A. Ortony, 71–82. Cambridge: Cambridge University Press.

Rumelhart, D. E., P. Smolensky, J. L. McClelland, and G. E. Hinton. 1988. Schemata and sequential thought processes in PDP models. In J. L. McClelland, D. E. Rumelhart, and the PDP Research Group, *Parallel Distributed Processing: Explorations in the Microstructure of Cognition*, vol. 2: *Psychological and Biological Models*, 7–57. Cambridge, MA: MIT Press.

Schank, R. C., and R. P. Abelson. 1977. *Scripts, Plans, Goals, and Understanding: An Inquiry into Human Knowledge Structures*. Hillsdale, NJ: Erlbaum.

Schmauder, A. R., and M. C. Egan. 1998. The influence of semantic fit on on-line sentence processing. *Memory and Cognition* 26 (6):1304–1312.

Shipley, T. F., and J. M. Zacks. 2008. *Understanding Events: How Humans See, Represent, and Act on Events*. Oxford: Oxford University Press.

Siegelmann, H. T., and E. D. Sontag. 1995. On the computational power of neural nets. *Journal of Computer and System Sciences* 50 (1):132–150.

Smith, L. B., and E. Thelen. 1993. *A Dynamic Systems Approach to Development: Applications*. Cambridge, MA: MIT Press.

Smith, L.B., and E. Thelen. 2003. Development as a dynamic system. *Trends in Cognitive Sciences* 7 (8):343–348.

Spencer, J. P., and G. Schöner. 2003. Bridging the representational gap in the dynamic systems approach to development. *Developmental Science* 6 (4):392–412.

Spivey, M. J. 2007. *The Continuity of Mind*. Oxford: Oxford University Press.

Spivey, M. J., and R. Dale. 2004. *On the Continuity of Mind: Toward a Dynamical Account of Cognition*. San Diego, CA: Elsevier Academic Press.

St. John, M. 1992. The story gestalt: A model of knowledge-intensive processes in text comprehension. *Cognitive Science* 16:271–306.

St. John, M., and J. L. McClelland. 1990. Learning and applying contextual constraints in sentence comprehension. *Artificial Intelligence* 46:217–257.

Tabor, W. 2004. Effects of merely local syntactic coherence on sentence processing. *Journal of Memory and Language* 50:355–370.

Tabor, W., C. Juliano, and M. K. Tanenhaus. 1997. Parsing in a dynamical system: An attractor-based account of the interaction of lexical and structural constraints in sentence processing. *Language and Cognitive Processes* 12 (2–3):211–271.

Tabor, W., and M. K. Tanenhaus. 2001. Dynamical systems for sentence processing. In *Connectionist Psycholinguistics*, ed. Morten H. Christiansen and Nick Chater, 177–211. Westport, CT: Ablex.

Taraban, R., and J. L. McClelland. 1988. Constituent attachment and thematic role assignment in sentence processing: Influences of content-based expectations. *Journal of Memory and Language* 27 (6):597–632.

Thelen, E., and L. B. Smith. 1994. *A Dynamic Systems Approach to the Development of Cognition and Action*. Cambridge, MA: MIT Press.

Trueswell, J. C., M. K. Tanenhaus, and Christopher Kello. 1993. Verb-specific constraints in sentence processing: Separating effects of lexical preference from garden-paths. *Journal of Experimental Psychology: Learning, Memory, and Cognition* 19 (3):528–553.

van Berkum, J. J. A., C. M. Brown, P. Zwitserlood, V. Kooijman, and P. Hagoort. 2005. Anticipating upcoming words in discourse: Evidence from ERPs and reading times. *Journal of Experimental Psychology: Learning, Memory, and Cognition* 31 (3):443–467.

van Berkum, J. J. A., P. Zwitserlood, P. Hagoort, and C. M. Brown. 2003. When and how do listeners relate a sentence to the wider discourse? Evidence from the N400 effect. *Cognitive Brain Research* 17 (3):701–718.

Van Petten, C, and M. Kutas. 1987. Ambiguous words in context: An event-related potential analysis of the time course of meaning activation. *Journal of Memory and Language* 26 (2):188–208.

Weinreich, U. 1962. Lexicographic definition in descriptive semantics. In *Problems in Lexicography*, ed. F. W. Householder and W. Saporta. Bloomington, IN: Indiana University Research Center in Anthropology, Folklore, and Linguistics.

Wittgenstein, L. 1966. *Philosophical Investigations*. Oxford: Macmillan.

Zacks, J. M., and B. Tversky. 2001. Event structure in perception and conception. *Psychological Bulletin* 127 (1):3–21.

6 Getting Real about Systematicity

Stefan L. Frank

1 Introduction

1.1 Systematicity and Reality

In the twenty-five years since its inception, the systematicity debate has suffered from remarkably weak empirical grounding. For a large part, the debate has relied on purely theoretical arguments, mostly from the classicists' side (e.g., Aizawa 1997a; Fodor and Pylyshyn 1988; Phillips 2000), but occasionally from the connectionist camp as well (Bechtel 1993; Van Gelder 1990). And although there have been many attempts to empirically demonstrate (lack of) systematicity in connectionist models, it remains doubtful how these demonstrations bear upon reality, considering that they are always restricted to hand-crafted, miniature domains. This is the case irrespective of whether they are presented by supporters of connectionist systematicity[1] (Bodén 2004; Brakel and Frank 2009; Chang 2002; Christiansen and Chater 1994; Elman 1991; Farkaš and Crocker 2008; Fitz and Chang 2009; Frank 2006a,b; Frank and Čerňanský 2008; Frank, Haselager, and van Rooij 2009; Hadley, Rotaru-Varga, Arnold, and Cardei 2001; Jansen and Watter 2012; McClelland, St. John, and Taraban 1989; Miikkulainen 1996; Monner and Reggia 2011; Niklasson and Van Gelder 1994; Voegtlin and Dominey 2005; Wong and Wang, 2007) or by those who are more skeptical (Marcus 2001; Phillips 1998; Van der Velde, Van der Voort van der Kleij, and De Kamps 2004).

My goal in this chapter is to approach the systematicity problem in a fully empirical manner, by directly comparing a connectionist and a symbolic sentence-processing model in a (more or less) realistic setting.[2] As far as this chapter is concerned, getting real about systematicity means three things. First, connectionists can no longer get away with presenting models that function only within some unrealistic toy domain. To the extent that the systematicity issue is relevant to real-life cognitive systems, connectionists should be able to demonstrate that (alleged) instances of systematicity do not depend crucially on the artificial nature of the simulation.

Second, I also aim to raise the bar for classicists, who need to back up their claim empirically that symbol systems are necessarily systematic. Aizawa (1997b) argues that compositionality is not a sufficient condition for systematicity, and, indeed, to the best of my knowledge it has never been empirically demonstrated that symbol systems are any more systematic than neural networks. Nevertheless, even many connectionists accept the premise that symbol systems explain systematicity.

Third, rather than defining particular levels of systematic behavior based on the specifics of training input and novel examples (as in, e.g., Hadley 1994a,b), the question of how systematic cognition really is will be avoided altogether. People learn language from what is "out there" and, subsequently, comprehend and produce more language "out there." Hence, the generalization abilities of the models presented here are investigated by training and testing both models on a large sample of sentences from natural sources. There is no invented, miniature language, and no assumptions are made about which specific syntactic construction in the training data should result in which specific systematic generalizations.

1.2 Statistical Modeling of Language

While the systematicity debate in philosophy and cognitive science revolved around theoretical arguments and unrealistic examples, actual progress was being made in the field of computational linguistics. The development of statistical methods for learning and processing natural language resulted in many successful algorithms for tasks such as sentence parsing, translation, and information retrieval. Recently, there has been a growing interest in applying such models to explain psychological phenomena in human language comprehension (e.g., Boston, Hale, Patil, Kliegl, and Vasishth 2008; Brouwer, Fitz, and Hoeks 2010; Levy 2008), production (e.g., Levy and Jaeger 2007), and acquisition (e.g., Bod and Smets 2012; Borensztajn, Zuidema, and Bod 2009b). The systematicity controversy tends not to arise here; for one because computational linguists are often concerned more with practical than with theoretical issues. Also, and perhaps more importantly, these models are typically symbolic and thereby dodge the systematicity critique.

What recent models from computational linguistics share with connectionist ones is their statistical nature: they are concerned with the problem of extracting useful statistics from training data in order to yield optimal performance on novel input. As Hadley (1994a) pointed out, this issue of correct generalization to previously unseen examples is exactly what systematicity is all about. Hence, one might expect statistical computational linguistics to be vulnerable to the same systematicity critique as connectionism. On the other hand, these statistical models for natural language processing are symbol systems in the sense of Fodor and Pylyshyn (1988), so, according to some, their systematic abilities are beyond any doubt.

1.3 Overview

In what follows, I will directly compare the systematic abilities of two sentence-processing models. The two models are of fundamentally different types: one is a thoroughly connectionist recurrent neural network (RNN); the other is a truly symbolic probabilistic phrase-structure grammar (PSG). Both are trained on a large number of naturally occurring English sentences, after which their performance is evaluated on a (much smaller) set of novel sentences. In addition, I investigate how they handle ungrammatical word strings. The ideally systematic model would have no problem at all with correct sentences but would immediately "collapse" when it encounters ungrammaticality. As it turns out, neither model is perfectly systematic in this sense. In fact, the two models behave quite similarly, although the symbolic model displays slightly stronger systematicity. As will be discussed, these results suggest that, when dealing with real-world data, generalization or systematicity may not be relevant to assess a model's cognitive adequacy.

2 Simulations

2.1 Model Training Data

As discussed in the introduction, connectionist models are typically trained on an artificial miniature language, whereas models from computational linguistics are broad-coverage, being able to deal with sentences from natural sources. Although the latter approach was used here for both types of models, the task was made more manageable for the neural network by reducing the size of the language: the vocabulary was restricted to 7,754 word types (including the comma and the sentence-final period, which are treated as regular words) that occur with high frequency in the written-text part of the British National Corpus (BNC). The training data consisted of all 702,412 sentences from the BNC (comprising 7.6 million word tokens) that contain only words from the vocabulary. This is the same data set as used by Fernandez Monsalve, Frank, and Vigliocco (2012), Frank (2013), and Frank and Thompson (2012).

The connectionist model was trained on just these sentences. In contrast, the symbolic model, being a probabilistic grammar, needs to be induced from a so-called treebank: a collection of sentences with syntactic tree structures assigned. To obtain these, the selected BNC sentences were parsed by the Stanford parser (Klein and Manning 2003). The resulting treebank served as the training data for the grammar.

2.2. Models

2.2.1 Recurrent neural network

RNNs have formed the standard connectionist model of sentence processing ever since the seminal paper by Elman (1990). However, such models are difficult to scale up and were therefore always limited to unrealistic, miniature languages. In order to train the current model on the 7.6-million-word data set, we separated the training process into three distinct stages, as illustrated in figure 6.1.

First, each word type was represented by a high-dimensional vector, based on the frequencies with which the words occur adjacently in the training data. More specifically, word co-occurrence frequencies were collected in a matrix with 7,754 rows (one per word type) and $2 \times 7,754$ columns (corresponding to the directly preceding and following word types). These frequencies were transformed into pointwise mutual information values, after which the 400 columns with the highest variance were selected, yielding a 400-element vector per word type. These representations encode some of the paradigmatic, distributional relations between the words. For example, words from the same syntactic category tend to be represented by more similar vectors than words from different categories (cf. Frank 2013).

Second, the selected BNC sentences (in the form of sequences of word vector representations) served as training input and target output for the recurrent part of the network. As is common in simulations using Simple Recurrent Networks (e.g., Elman 1990, 1991; Frank 2006a, among many others), it was trained to predict, at each point in each sentence, what the next input will be. More specifically, for each sentence-so-far w_1, ..., w_t, the input sequence consisted of the words' vector representations, and the target output was the vector representing the sentence's next word w_{t+1}. The training data set was presented to this part of the network five times and standard backpropagation was used to update the connection weights.

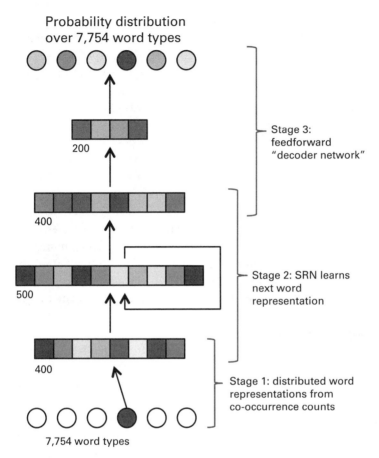

Probability distribution
over 7,754 word types

Stage 3:
feedforward
"decoder network"

200

400

Stage 2: SRN learns
next word
representation

500

400

Stage 1: distributed word
representations from
co-occurrence counts

7,754 word types

Figure 6.1
Architecture and training stages of the RNN model (reproduced from Fernandez
Monsalve et al. 2012).

After this second training stage, the outputs of the recurrent part of the
network form a mix of distributed, 400-element word vectors, somehow
representing the possible continuations w_{t+1} (and their probabilities) of the
input sequence so far. In order to interpret this mix of word vectors, a
two-layer feedforward "decoder" network was applied in the third stage of
model training. After each sequence w_1, ..., w_t, the decoder network took
the outputs of the trained recurrent part of the network as its input, and
learned to activate only one of 7,754 units, corresponding to the upcoming
word w_{t+1}. In practice, of course, the network will never actually activate
just a single output unit because many different continuations are possible
at each point in a sentence.

The training data were presented two times to this decoder network and standard backpropagation was used for training. An important difference with Stage-2 training is that the decoder network's output units have softmax activation functions, ensuring that output activations are always nonnegative and sum to one. That is, they form a probability distribution. To be precise, if $w_1, ..., w_t$ represents the first t words of a sentence, then the network's output activation pattern forms the model's estimated probability distribution $P(w_{t+1}|w_1, ..., w_t)$ over all word types. So, for each word type, the model estimates a probability that it will be the next word given the sentence so far. These estimates are accurate to the extent that the model was trained successfully.

As should be clear from the exposition above, the RNN model uses only nonsymbolic, distributed, numerical representations and processes. Crucially, the model is noncompositional in the sense that the representation of a word sequence does not contain representations of the individual words. There are no symbolic components at all. Thus, according to Fodor and Pylyshyn (1988), the model should be unable to display any systematicity.

2.2.2 Phrase-structure grammar

The PSG, needless to say, operates very differently from the RNN. To begin with, it is based on linguistic assumptions about hierarchical constituent structure: a sentence consists of phrases, which consists of smaller phrases, and so on, until we get down to individual words.

The grammar operationalizes this idea by means of context-free production rules. These rules are induced from a treebank, which in this case is formed by the sentences selected from the BNC together with their syntactic tree structures. Figure 6.2 shows one of the 702,412 training items. From this single example, we can observe that a noun phrase (NP) can consist of either a singular noun (NN) or a determiner (DT) followed by a singular noun. Hence, the two production rules "NP → NN" and "NP → DT NN" appear in the grammar.

The leaves of the tree are formed by words. Each word token belongs to one of a number of syntactic categories (also known as parts-of-speech) which forms its direct "parent" node in the tree. So, for example, the tree of figure 6.2 provides evidence that the word *at* can be a preposition (i.e., there is a production rule "IN → *at*").

In a probabilistic grammar like the one used here, each rule is assigned a probability conditioned upon the rule's left-hand side. Based on just the single example from figure 6.2, each of the two NP-rules receives a probability of 0.5, indicating that, when faced with an NP, it becomes either

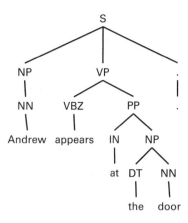

Figure 6.2
The syntactic tree structure assigned by the Stanford parser to the BNC sentence *Andrew appears at the door*. Syntactic category labels: NN (singular noun), VBZ (third-person singular present-tense verb), IN (proposition), DT (determiner). Phrasal labels: S (sentence), NP (noun phrase), VP (verb phrase), PP (propositional phrase).

NN or DT NN with equal probability. Needless to say, the 702,412 tree structures in the training data set give rise to much more fine-grained and accurate probabilities.

In practice, more advanced statistical techniques are applied to induce the PSG used here. First, the probability of a rule is conditional not only on its left-hand side but also on larger parts of the tree (see Frank and Bod 2011 for details). This is to say that the probability of producing "DT NN" depends not only on the identity of their parent node (NP) but also on their (great)grandparents (PP and VP in the example above). This makes the grammar more sensitive to structure and greatly improves its performance (Fernandez Monsalve et al. 2012). Second, probabilities are not purely based on frequencies in the treebank: to improve the probability estimates of very low-frequency events, additional smoothing of probabilities is applied.

The grammar was induced by Roark's (2001) algorithm, and implementation thereof, using his default settings. Next, Roark's incremental parser was used to estimate conditional word probabilities $P(w_{t+1}|w_1, \ldots, w_t)$, just like the RNN did. These follow from the probabilities of sentence-initial word sequences because:

$$P(w_{t+1}|w_1, \ldots, w_t) = \frac{P(w_1, \ldots, w_t)}{P(w_1, \ldots, w_{t+1})}$$

Here, $P(w_1, ..., w_t)$ is the probability of a sentence-so-far, which equals the sum of probabilities of all complete sentences that begin with $w_1, ..., w_t$. The probability of a complete sentence equals the sum of probabilities of all its possible syntactic tree structures. The probability of a tree structure, in turn, is the product of probabilities of the rules used in its construction. In this manner, a PSG can be used to estimate the probability of a word given the sentence so far. As will be discussed in the next section, these conditional probability estimates are central to model evaluation.

In contrast to the RNN, the PSG is very much a symbol system. It applies symbolic operations over discrete units (words, syntactic category labels, and phrasal labels) and has compositional representations: a sentence's tree structure consists of representations of subtrees, of syntactic and phrasal categories, and of the sentence's words. Because of the model's probabilistic nature, numerical processing is also involved to compute the probabilities, but this does not diminish the system's symbolic character. Hence, Fodor and Pylyshyn (1988) would claim that the PSG necessarily displays systematicity.

2.3. Model Evaluation

2.3.1 Evaluation sentences
To evaluate the models' ability to generalize to realistic input, 361 sentences were selected from three novels (for details, see Frank, Fernandez Monsalve, Thompson, and Vigliocco in press). These novel sentences contained a total of 5,405 word tokens (including commas and periods), all of which are present in the vocabulary of 7,754 high-frequency words.

A properly systematic model should not only be able to deal with new sentences, but also be able to reject ungrammatical input. To test how the models handle ungrammatical word strings, scrambled versions of the original 361 sentences were created. This was done by choosing, from each sentence, two words at random and swapping their positions. This is done n times, for $n = 0$ (i.e., the grammatical sentence) up to $n = 9$, creating ten different sets of test data. Table 6.1 shows an example of one sentence and its nine scrambled versions. Note that every sentence ends with a period, which remains in its sentence-final location.

The probability of the word string being grammatical decreases with larger n, and most often even $n = 1$ already yields an ungrammatical string.[3] A very rigid language model would estimate zero probability for an ungrammatical string, but such an estimate would necessarily be incorrect. After

Table 6.1
Example of an evaluation sentence and its nine increasingly scrambled versions.

Scrambling level	Sentence
0	andrew closed the office door on the way out .
1	andrew closed the office door on the out way .
2	andrew on the office door closed the out way .
3	andrew on the office door the closed out way .
4	closed on the office door the andrew out way .
5	closed on the out door the andrew office way .
6	closed the the out door on andrew office way .
7	closed the andrew out door on the office way .
8	the the andrew out door on closed office way .
9	the the andrew way door on closed office out .

all, impossible things by definition do not happen, so any input that actually occurs must have had a nonzero probability.

2.3.2 Evaluation measure

The question remains how to quantify the models' ability to generalize. Earlier evaluation measures, such as those proposed by Frank (2006a) and Christiansen and Chater (1999), require knowledge of all grammatical next-word predictions, so these measures can only be used if the true grammar is known, as is the case when using artificial languages. For natural language, the true grammar is (arguably) unknowable[4] or possibly even non-existent. Therefore, the evaluation measure applied here uses only the model-estimated next-word probabilities, $P(w_{t+1}|w_1, ..., w_t)$, for the *actual* next word w_{t+1}. The larger these probabilities, the more accurate were the model's expectations and, therefore, the better the model captured the statistics of the language.

Rather than using the conditional probability itself, it is transformed by the (natural) logarithm, that is, we take $\log(P(w_{t+1}|w_1, ..., w_t))$, which ranges from negative infinity (when $P(w_{t+1}|w_1, ..., w_t) = 0$) to zero (when $P(w_{t+1}|w_1, ..., w_t) = 1$). The negative of this value is an information-theoretic measure known as *surprisal*, which expresses the amount of information conveyed by word w_{t+1}. A word's surprisal is also of cognitive interest because it is believed to be indicative of the amount of "mental effort" required to understand the word in sentence context (Hale 2001; Levy 2008). Indeed, surprisal values have been shown to correlate positively with word-reading times (e.g., Boston et al. 2008; Demberg and Keller 2008; Fernandez Monsalve et al. 2012; Frank and Bod 2011; Frank and Thompson 2012; Smith and Levy 2013; among many others).

Note that surprisal (i.e., mental effort and amount of information) is infinitely large if a word appears that was estimated to have zero probability. A more appropriate way to put this is that a model is infinitely wrong if something happens that it considers impossible. In practice, however, both models always estimate strictly positive probability for each word type at any point in the sentence. In the PSG, this is because of the smoothing of probabilities. In the RNN, this is because the connection weights have finite values.

3 Results

The left-most panel of figure 6.3 shows how the next-word probabilities that are estimated by the models decrease as the RNN and PSG are made to process increasingly scrambled sentences.[5] For correct sentences (i.e., $n = 0$), the PSG performs slightly better than the RNN. As expected, both models make worse predictions as the input contains an increasing number of grammatical errors. This effect is stronger for the PSG than for the RNN. Although the difference between the two models is small for lower levels of scrambling, all differences were statistically significant (all $t_{5341} > 3.19$; $p < 0.002$, in paired t-tests) because of the large number of data points.

Figure 6.3 also presents the coefficient of correlation between the RNN's and PSG's estimates of $\log(P(w_{t+1}|w_1, ..., w_t))$, as a function of scrambling

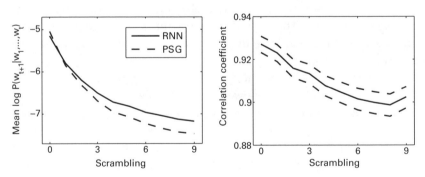

Figure 6.3
Left: average log-probability estimate as a function of scrambling level. Right: coefficient of correlation between RNN's and PSG's log-probability estimates, as a function of scrambling level (dashed lines indicate the 95 percent confidence interval).

level. This correlation is clearly very strong, and although it weakens as sentences are scrambled more, it seems to stabilize at around $r = 0.9$.

4   Discussion

Compared to the connectionist RNN model, the symbolic PSG generalizes better to grammatical sentences and (correctly) estimates lower next-word probabilities for scrambled sentences. Both these results suggest that the symbolic model is indeed more systematic than the connectionist one. However, the difference between the two models' performance is one of degree rather than kind. Both models show a fairly slow decrease (instead of a sudden drop) in prediction accuracy as scrambling level increases. Crucially, the correlation between the two models' next-word probability estimates is very strong, which means that the models behave similarly.

Why would models that are so different in their underlying assumptions nevertheless show such remarkably similar behavior? Note that they are faced with the same complex computational problem: to extract linguistic patterns from 7.6 million word tokens and to apply the discovered statistics when processing novel sentences. In terms of Marr's (1982) famous levels of description, the models differ strongly at the algorithmic level but they are similar at the computational level, which specifies the task to be performed. As Marr argued, task requirements, rather than representations and algorithms, often form the most important factor in shaping behavior. Since the models need to perform the same task, they display similar observed behavior.

Interestingly, to the (small) extent that the models do differ, the RNN seems to be closer to cognitively reality: a comparison between the two models' surprisal estimates and human word-reading times (over the same test sentences) revealed that the RNN's surprisals explained significantly more variance in reading time, whether these were collected by self-paced reading (Fernandez Monsalve et al. 2012) or by eye-tracking (Frank and Thompson 2012). This is in line with earlier findings using different models and sentences (Frank and Bod 2011). A similar result emerged from a sentence-reading experiment in which participants' brain activity was recorded by EEG. Although both RNN- and PSG-based surprisal values predicted the size of the N400 event-related potential component, the RNN model yielded more accurate predictions (Frank, Otten, Galli, and Vigliocco 2013). In fact, the PSG did not account for any unique variance in N400 size over and above the RNN's predictions. This strongly suggests that, despite being slightly less systematic, the RNN is the more accurate cognitive model.

Compared to earlier demonstrations of models' (lack of) systematicity, the simulations presented here were based on a much more complex and realistic task. Nevertheless, it was still far from what people face in the real world. For example, the amount of training data was far smaller than what children learn language from, and the input stream was noise-free and presegmented into words. Also, the models "merely" needed to learn syntax; they could ignore phonology, prosody, co-speech gesture, pragmatic constraints, the discourse setting, and any extralinguistic context. If noisy and ambiguous information from all these different sources must be integrated in real time, this puts further pressure on the system by narrowing the gap between how well it *needs to* perform and how well it *can* perform. It is only within this narrow gap that any difference in systematic abilities can occur.

So, to what extent do the models display systematicity? In absence of any gold standard of systematicity to compare with the models' performance, this question cannot be answered. All we know is that the RNN and PSG are similarly (un)systematic. If symbol systems are indeed necessarily systematic (Fodor and McLaughlin 1990), then the PSG must be systematic and, therefore, so is the RNN (or, at least, it is "almost" systematic, if some arbitrary cutoff point is introduced just below the PSG's level of performance). Conversely, if systematicity is indeed beyond the reach of connectionist models (Fodor and Pylyshyn 1988), then the RNN cannot be systematic and, therefore, neither is the PSG (or, at best, it is minimally systematic).

It may seem unlikely that a PSG could fail to be systematic. After all, provided that *Mary* and *John* belong to the same syntactic category (which is the case in the PSG used here), a grammar trained on *John loves Mary* necessarily generalizes to *Mary loves John*, thereby realizing Fodor and Pylyshyn's (1988) standard example of systematic behavior. Moreover, grammars as specified by production rules have traditionally been viewed as typical examples of symbol systems, with full-fledged systematic abilities against which generalization by connectionist models is evaluated (e.g., Christiansen and Chater 1994; Frank 2006a; Van der Velde et al. 2004). As argued by Aizawa (1997b), however, even classical symbols systems are systematic only when combined with a proper mechanism for manipulating the symbols. Perhaps our PSG lacks such a systematic algorithm, but this does raise the question of why that would be so. Roark's (2001) algorithms, which were used here for inducing the grammar and obtaining surprisal estimates, certainly do not intend to reduce the grammar's ability to generalize to new input (i.e., to display systematicity).

Possibly, then, the PSG's statistical nature somehow reduces its systematic abilities to that of a connectionist model. Although this may indeed be the case, it is likely that there is no viable alternative. Arguably, only a statistical system can learn to generalize appropriately from very complex natural data, without failing when faced with a highly unexpected event, in a noisy, ambiguous, widely varying real-world environment that provides a massive number of potentially inconsistent cues. As a case in point, it is only since the advent of statistical methods that computational linguistics has made significant progress in automatic natural language processing (see Manning and Schütze 1999). And indeed, the probabilistic nature of human cognition is increasingly recognized in cognitive science (see Oaksford and Chater 2007). So, if we do need to choose between a statistical model and a systematic model, the safest bet may well be against systematicity.

5   Conclusion

As far as systematicity is concerned, there may not be any important difference between connectionist and statistical symbolic models, as long as the models are powerful enough to perform real-world tasks. This is not to say that the difference between model types is not of interest. On the contrary, it is quite relevant to cognitive science which representations and algorithms best describe the language-processing system. There will certainly be significant differences among models regarding, for example, their learning and processing efficiency, their ability to connect to non-linguistic cognitive modalities, their performance under adverse conditions, and the ability to explain (psycho)linguistic phenomena and neuropsychological disorders. It is those kind of issues, rather than systematicity, on which the discussion about the value of connectionism should focus. When facing reality, systematicity is not something worth worrying about.

Acknowledgments

I am thankful to Gideon Borensztajn for his comments on an earlier version of this chapter. The research presented here was funded by the European Union Seventh Framework Programme (FP7/2007–2013) under grant number 253803. The UCL Legion High Performance Computing Facility and associated support services were instrumental in the completion of this work.

Notes

1. As a possible exception, the model by Borensztajn, Zuidema, and Bod (2009a) does learn from real-world data (2,000 utterances produced by a single child). However, it can be argued that the model forms a neural implementation of a symbolic parser, in which case it does not constitute a demonstration of eliminative connectionism (in the sense of Pinker and Prince 1988).

2. This chapter is only about language processing, arguably the main battleground for the systematicity debate, but I believe most of the same conclusions hold for human cognition in general.

3. Here, grammaticality is a subjective notion. There is no (known) "true" grammar to objectively decide whether a word string is grammatical.

4. Although there is a sense in which speakers of a language know its grammar, this knowledge is mostly procedural in nature (see Bybee 2003). That is, people know how to *use* their language but have little or no conscious access to its grammar.

5. For a fair comparison between models, clitics were excluded because the two models treat these differently: a word like *isn't* is considered a single word by the RNN, whereas the PSG parser splits it into *is* and *n't*. After removing such cases, there are 5,344 probability estimates for each model and level of scrambling.

References

Aizawa, K. 1997a. Exhibiting versus explaining systematicity: A reply to Hadley and Hayward. *Minds and Machines* 7:39–55.

Aizawa, K. 1997b. Explaining systematicity. *Mind and Language* 12 (2):115–136.

Bechtel, W. 1993. The case for connectionism. *Philosophical Studies: An International Journal for Philosophy in the Analytic Tradition* 71:119–154.

Bod, R., and M. Smets. 2012. Empiricist solutions to nativist puzzles by means of unsupervised TSG. In *Proceedings of the Workshop on Computational Models of Language Acquisition and Loss*, pp. 10–18. Stroudsburg, PA: Association for Computational Linguistics.

Bodén, M. 2004. Generalization by symbolic abstraction in cascaded recurrent networks. *Neurocomputing* 57:87–104. doi:10.1016/j.neucom.2004.01.006.

Borensztajn, G., W. Zuidema, and R. Bod. 2009a. The hierarchical prediction network: Towards a neural theory of grammar acquisition. In *Proceedings of the 31st Annual Conference of the Cognitive Science Society*, ed. N. A. Taatgen and H. Van Rijn, 2974–2979. Austin, TX: Cognitive Science Society.

Borensztajn, G., W. Zuidema, and R. Bod. 2009b. Children's grammars grow more abstract with age: Evidence from an automatic procedure for identifying the productive units of language. *Topics in Cognitive Science* 1 (1):175–188. doi:10.1111/j.1756 -8765.2008.01009.x.

Boston, M. F., J. Hale, U. Patil, R. Kliegl, and S. Vasishth. 2008. Parsing costs as predictors of reading difficulty: An evaluation using the Potsdam Sentence Corpus. *Journal of Eye Movement Research* 2:1–12.

Brakel, P., and S. L. Frank. 2009. Strong systematicity in sentence processing by simple recurrent networks. In *Proceedings of the 31st Annual Conference of the Cognitive Science Society*, ed. N. A. Taatgen and H. Van Rijn, 1599–1604. Austin, TX: Cognitive Science Society.

Brouwer, H., H. Fitz, and J. Hoeks. 2010. Modeling the noun phrase versus sentence coordination ambiguity in Dutch: Evidence from surprisal theory. In *Proceedings of the 2010 Workshop on Cognitive Modeling and Computational Linguistics*, pp. 72–80. Stroudsburg, PA: Association for Computational Linguistics.

Bybee, J. 2003. Cognitive processes in grammaticalization. In *The New Psychology of Language*, vol. II. ed. M. Tomasello, 145–167. Mahwah: Erlbaum.

Chang, F. 2002. Symbolically speaking: A connectionist model of sentence production. *Cognitive Science* 26:609–651.

Christiansen, M. H., and N. Chater. 1994. Generalization and connectionist language learning. *Mind and Language* 9:273–287.

Christiansen, M. H., and N. Chater. 1999. Toward a connectionist model of recursion in human linguistic performance. *Cognitive Science* 23:157–205. doi:10.1016/ S0364-0213(99)00003-8.

Demberg, V., and F. Keller. 2008. Data from eye-tracking corpora as evidence for theories of syntactic processing complexity. *Cognition* 109 (2):193–210. doi:10.1016/j .cognition.2008.07.008.

Elman, J. L. 1990. Finding structure in time. *Cognitive Science* 14 (2):179–211.

Elman, J. L. 1991. Distributed representations, simple recurrent networks, and grammatical structure. *Machine Learning* 7:195–225.

Farkaš, I., and M. W. Crocker. 2008. Syntactic systematicity in sentence processing with a recurrent self-organizing network. *Neurocomputing* 71:1172–1179.

Fernandez Monsalve, I., S. L. Frank, and G. Vigliocco. 2012. Lexical surprisal as a general predictor of reading time. In *Proceedings of the 13th conference of the European chapter of the Association for Computational Linguistics*, ed. W. Daelemans, 398–408. Stroudsburg, PA: Association for Computational Linguistics.

Fitz, H., and F. Chang. 2009. Syntactic generalization in a connectionist model of sentence production. In *Connectionist Models of Behaviour and Cognition II: Proceedings of the 11th Neural Computation and Psychology Workshop*, ed. J. Mayor, N. Ruh, and K. Plunkett, 289–300. River Edge, NJ: World Scientific Publishing.

Fodor, J. A., and B. P. McLaughlin. 1990. Connectionism and the problem of systematicity: Why Smolensky's solution doesn't work. *Cognition* 35 (2):183–204.

Fodor, J. A., and Z. W. Pylyshyn. 1988. Connectionism and cognitive architecture: A critical analysis. *Cognition* 28:3–71.

Frank, S. L. 2006a. Learn more by training less: Systematicity in sentence processing by recurrent networks. *Connection Science* 18 (3):287–302. doi:10.1080/09540090600768336.

Frank, S. L. 2006b. Strong systematicity in sentence processing by an Echo State Network. In *Proceedings of the 16th International Conference on Artificial Neural Networks, Part I*, Lecture Notes in Computer Science, ed. S. D. Kollias, A. Stafylopatis, W. Duch, and E. Oja, 505–514. Berlin: Springer.

Frank, S. L. 2013. Uncertainty reduction as a measure of cognitive effort in sentence comprehension. *Topics in Cognitive Science* 5 (3):475–494. doi: 10.1111/tops.12025.

Frank, S. L., and R. Bod. 2011. Insensitivity of the human sentence-processing system to hierarchical structure. *Psychological Science* 22 (6):829–834. doi:10.1177/0956797611409589.

Frank, S. L., and M. Čerňanský. 2008. Generalization and systematicity in echo state networks. In *Proceedings of the 30th Annual Conference of the Cognitive Science Society*, eds. B. C. Love, K. McRae, and V. M. Sloutsky, 733–738. Austin, TX: Cognitive Science Society.

Frank, S. L., I. Fernandez Monsalve, R. L. Thompson, and G. Vigliocco. In press. Reading-time data for evaluating broad-coverage models of English sentence processing. *Behavior Research Methods*.

Frank, S. L., W. F. G. Haselager, and I. van Rooij. 2009. Connectionist semantic systematicity. *Cognition* 110 (3):358–379. doi:10.1016/j.cognition.2008.11.013.

Frank, S. L., L. J. Otten, G. Galli, and G. Vigliocco. 2013. Word surprisal predicts N400 amplitude during reading. In *Proceedings of the 51st Annual Meeting of the Association for Computational Linguistics*, 878–883. Stroudsburg, PA: Association for Computational Linguistics.

Frank, S. L., and R. L. Thompson. 2012. Early effects of word surprisal on pupil size during reading. In *Proceedings of the 34th Annual Conference of the Cognitive Science Society*, ed. N. Miyake, D. Peebles, and R. P. Cooper, 1554–1559. Austin, TX: Cognitive Science Society.

Hadley, R. F. 1994a. Systematicity in connectionist language learning. *Mind and Language* 9 (3):247–272.

Hadley, R. F. 1994b. Systematicity revisited: Reply to Christiansen and Chater and Niklasson and van Gelder. *Mind and Language* 9:431–444.

Hadley, R. F., A. Rotaru-Varga, D. V. Arnold, and V. C. Cardei. 2001. Syntactic systematicity arising from semantic predictions in a Hebbian-competitive network. *Connection Science* 13 (1):73–94.

Hale, J. T. 2001. A probabilistic early parser as a psycholinguistic model. In *Proceedings of the Second Meeting of the North American Chapter of the Association for Computational Linguistics*, vol. 2, pp. 159–166. Stroudsburg, PA: Association for Computational Linguistics.

Jansen, P. A., and S. Watter. 2012. Strong systematicity through sensorimotor conceptual grounding: an unsupervised, developmental approach to connectionist sentence processing. *Connection Science* 24:25–55.

Klein, D., and C. D. Manning. 2003. Accurate unlexicalized parsing. In *Proceedings of the 41st Meeting of the Association for Computational Linguistics*, 423–430. Stroudsburg, PA: Association for Computational Linguistics.

Levy, R. 2008. Expectation-based syntactic comprehension. *Cognition* 106: 1126–1177.

Levy, R., and T. F. Jaeger. 2007. Speakers optimize information density through syntactic reduction. In *Advances in Neural Information Processing Systems 19: Proceedings of the 2006 Conference*, ed. B. Schlökopf, J. Platt, and T. Hoffman, 849–856. Cambridge, MA: MIT Press.

Manning, C. D., and H. Schütze. 1999. *Foundations of Statistical Natural Language Processing*. Cambridge, MA: MIT Press.

Marcus, G. F. 2001. *The Algebraic Mind: Integrating Connectionism and Cognitive Science*. Cambridge, MA: MIT Press.

Marr, D. 1982. *Vision*. San Francisco: W. H. Freeman.

McClelland, J. L., M. F. St. John, and R. Taraban. 1989. Sentence comprehension: A parallel distributed processing approach. *Language and Cognitive Processes* 4: 287–335.

Miikkulainen, R. 1996. Subsymbolic case-role analysis of sentences with embedded clauses. *Cognitive Science* 20:47–73.

Monner, D. D., and J. A. Reggia. 2011. Systematically grounding language through vision in a deep, recurrent neural network. In *4th International Conference on Artificial General Intelligence*, Lecture Notes in Computer Science, vol. 6830, 112–121. Berlin: Springer.

Niklasson, L. F., and T. Van Gelder. 1994. On being systematically connectionist. *Mind & Language* 9:288–302.

Oaksford, M., and N. Chater. 2007. *Bayesian Rationality: The Probabilistic Approach to Human Reasoning.* Oxford: Oxford University Press.

Phillips, S. 1998. Are feedforward and recurrent networks systematic? Analysis and implications for a connectionist cognitive architecture. *Connection Science* 10: 137–160.

Phillips, S. 2000. Constituent similarity and systematicity: The limits of first-order connectionism. *Connection Science* 12:45–63.

Pinker, S., and A. Prince. 1988. On language and connectionism: Analysis of a parallel distributed processing model of language acquisition. *Cognition* 28:73–193.

Roark, B. 2001. Probabilistic top-down parsing and language modeling. *Computational Linguistics* 27:249–276.

Smith, N. J., and R. Levy. 2013. The effect of word predictability on reading time is logarithmic. *Cognition* 128:302–319. doi:10.1016/j.cognition.2013.02.013.

Van der Velde, F., G. T. Van der Voort van der Kleij, and M. De Kamps. 2004. Lack of combinatorial productivity in language processing with simple recurrent networks. *Connection Science* 16:21–46.

Van Gelder, T. 1990. Compositionality: A connectionist variation on a classical theme. *Cognitive Science* 14:355–384.

Voegtlin, T., and P. F. Dominey. 2005. Linear recursive distributed representations. *Neural Networks* 18:878–895.

Wong, F. C. K., and W. S.-Y. Wang. 2007. Generalisation towards combinatorial productivity in language acquisition by simple recurrent networks. In *International Conference on Integration of Knowledge Intensive Multi-Agent Systems*, 139–144. Piscataway, NJ: IEEE.

7   Systematicity and the Need for Encapsulated Representations

Gideon Borensztajn, Willem Zuidema, and William Bechtel

Debates about systematicity originated with Fodor and Pylyshyn's (1988) challenge to connectionist models of cognition in the 1980s (Hinton and Anderson 1981; Rumelhart and McClelland 1986; McClelland and Rumelhart 1986). They identified systematicity as a feature not just of language but also of thought in general, and argued that connectionist networks lacked the resources to account for it unless they implemented more traditional symbolic architectures. In the ensuing twenty-five years, connectionist or neural network models (as they are more commonly referred to today) have developed in a host of ways. Many of the most interesting models of cognitive capacities are ones that abandon the supervised framework of the backpropagation learning algorithm that figured centrally in the 1980s and employ units that more closely resemble actual neurons (Maass, Natschläger, and Markram 2003; Eliasmith et al. 2012). The debates between connectionists and symbolic theorists have largely disappeared from cognitive science itself. The main contemporary debate is between Bayesian optimality models (Griffiths et al. 2010) and connectionists who adopt an emergentist approach (McClelland et al. 2010).

Systematicity has thus been left as an issue for philosophers. In part, this is because much of contemporary cognitive science has focused on phenomena (sensory processing and categorization, working memory, motor control, decision making) for which systematicity has appeared to be less central. It is further the result of an implicit assumption among neural network modelers that where systematicity is an issue, they can invoke the same emergentist perspective they adopt toward Bayesian models: systematicity too can be accounted for as an emergent property of underlying simpler neural network processes. We contend that this is a mistake on both counts. Systematicity applies to cognition beyond language processing; we will discuss it in the context of sensory processing and related processes of categorization. We will begin, however, with language processing, arguing that the emergentist perspective seriously impoverishes the neural network approach. By neglecting the role of representations that encapsulate information and can be operated on while remaining invariant to contextual factors, the emergentist thread in neural network approaches denies itself the resources needed to capture systematic components of cognition.

Although there have been important neural network models that do not eschew encapsulated representations (Grossberg 1982; Miikkulainen 1993), for many proponents and opponents of neural networks the fundamental commitment seems to be to associationist processing in which associations between all units are learned through the gradual strengthening or weakening of their weights. We argue, however, that this should not be the core commitment of the neural network approach and, in many contexts, should be abandoned. Rather, the fundamental commitment should be a constraint imposed on modeling cognitive behavior, restricting what interpretations are permitted of the primitive units of the system. Such interpretations must depend exclusively on internal, autonomously executed processes, and not imposed externally or globally. In other words, all interactions must be local, and all meanings must eventually be grounded in the external input to the network. This constraint rules out symbolic rules with variables, but does not prohibit operations over *encapsulated* representations (representations not directly affected by processing elsewhere in the system).

Encapsulated units share with variables that they can act as placeholders and bind to appropriate units on their inputs. They are different in that the extensional scope of a variable is globally specified in advance, whereas the extensional scope and meaning of an encapsulated representation is learnable from experience through interaction with the external environment. Neural network models that allow learning encapsulated representations can thus acknowledge that a linguistic category like "noun phrase" is cognitively real and that representations of categories can play causal roles in linguistic behavior. In symbolic models, in contrast, the category must be stipulated to exist prior to experience (and, indeed, linguistic categories are often claimed to be innate; see Thornton and Wexler 1999), whereas in emergentist connectionist models category membership is recognized only by an external observer and plays no causal role. We will offer an example of a neural network that implements encapsulated representations toward the end of the chapter.

Since debates about systematicity have been hampered by the lack of a clear statement of the phenomenon, we begin by offering a distinctive account of how it should be understood. We then demonstrate the limitations of one of the most influential attempts to account for systematicity with a relatively unstructured connectionist network—Elman's recurrent neural networks. Our goal, however, is not to reject neural networks but to argue for an approach that takes insights from neurobiology seriously and at the same time accounts for real empirical phenomena that have been widely (though wrongly in our view) seen as necessitating symbolic models. To do this we will introduce Hawkins's Memory Prediction Framework, at the core of which is a hierarchical system of encapsulated representations.

Of course, encapsulated representations are only part of the story; equally important is accounting for how they are temporarily combined into structurally complex representations. This is especially important when dealing with the productivity and systematicity of language. We argue that this requires a more flexible kind of connectivity than is offered by pair-wise associative connections. In the literature this is known as dynamic binding (Hummel and Biederman 1992). In the last section of the chapter, we sketch how this can be implemented between encapsulated representations.

1   The Appropriate Concept of Systematicity: A Causal Role for Categories

Fodor and Pylyshyn introduce systematicity largely by means of an example: any cognitive system capable of representing *John loves Mary* must also be able to represent *Mary loves John* and other related thoughts. The question is: what makes thoughts related? The intuitive idea seems to be that related thoughts are those that organize different representations into the same syntactic structure. Treating systematicity as if it provided a clear standard, Fodor and Pylyshyn quickly move on to the shortcomings of neural networks—arguing that by relying merely on associations between representations, they are incapable of respecting systematicity.[1] Their diagnosis of this failure is that neural network models fail to employ a compositional syntax that combines simpler representations into more complex, structured representations. This is what traditional symbolic models in cognitive science achieved by treating representations as symbols that are operated on by rules such as those of logic or formal grammars. What is crucial about rules is that they contain variables that can be instantiated by different symbols. Thus Fodor and Pylyshyn contend that, by eschewing rules and relying only on associations, neural network modelers have ignored the lessons of the cognitive revolution and have returned to behaviorism.

As we will discuss in the next section, neural network theorists have tried to show that their models can exhibit systematicity. Part of the problem in evaluating these proposals is that the criterion of systematicity has not been articulated in a sufficiently precise way that it can be invoked operationally to evaluate candidate models. Hadley (1994) offered one of the most influential attempts to operationalize systematicity, capturing the important insight that the ability to generalize is fundamental to systematic behavior. Accordingly, he focused on the degree of generalization required to handle novel test items after a network has been trained, while distinguished *weak* from *strong* systematicity. A system is weakly systematic if it can generalize to novel sentences in which a word appears in a grammatical position it did not occupy in a training sentence but which it occupied in other training sentences. A system exhibits strong systematicity if, in addition, "it can correctly process a variety of novel simple and novel embedded sentences containing previously learned words in positions where they *do not appear* in the training corpus (i.e., the word within the novel sentence does *not appear in that same syntactic position* within any *simple* or *embedded* sentence in the training corpus)" (250–251).

Hadley's criteria have been invoked in evaluations of the type of network that we will discuss in the next section (Hadley 1994; Christiansen and Chater 1994). However, they fall short of what should be required for systematicity in two respects. First, they overlook the fact that systematicity presupposes the existence of classes of linguistic expressions that can be substituted for each other. The requirement for strong systematicity is not that a system should handle any expression in which a word is placed in a new syntactic position (e.g., *John Mary Mary*) but only ones in which it is placed in an appropriate syntactic position. When working with artificial languages, as do many neural network studies of systematicity, these "appropriate" positions are defined by the formal grammar used. However, if the account of systematicity is to be applied to natural language (and natural thinking), then the substitution classes must be identified explicitly in characterizing systematicity. Second, by focusing on lexical categories exclusively and not larger units, Hadley's account is too restricted. Systematicity in natural language also involves the class membership of phrases of multiple words (e.g., anyone who understands *The brother of John loves Mary* can also understand *John loves the brother of Mary*), and accounting for this fact requires postulating constituents of a particular category (e.g., the noun phrase *The brother of John*) to contain constituents of potentially the same category (e.g., *John*). We suggest that this ability to generalize over constituents (rather than over words) is evidence for internal hierarchical representations of language in humans.

What these considerations reveal is that systematicity needs to be characterized as the property of cognition that allows categories (treated as substitution classes), category membership, and hierarchical category structure to play causal roles. With respect to syntax in language, then, we propose to define systematicity as the property that, given the constituents of part of a sentence, their substitution class membership alone predicts the class membership of possible subsequent constituents. To the degree that natural language is systematic, it possesses sets of constituents (words or phrases) that behave as substitution classes. (The qualification "to the degree that" recognizes that categories are often not sharply defined but graded.) This characterization allows for compound constituents to be composed from simpler ones (e.g., by recursively nesting relative clauses) while keeping membership in the same substitution class.

The crucial feature of our characterization of systematicity is its appeal to substitution classes. These are classes whose members can all be treated alike. When it comes to substituting one for another, context is not taken into account—membership in the class is context invariant. This is the key component that symbolic accounts provide through rules with variables and which, we will argue, can be achieved in neural networks that allow for encapsulated representations. But before doing so, we will show that neural networks that maintain a mistaken commitment to purely associationist principles lack encapsulation, and therefore, we suggest, fail to account for systematicity.

2   The Limitations of Recurrent Neural Networks as Models of Systematicity

Bechtel and Abrahamsen (2002) differentiated three responses neural network theorists have offered to the systematicity challenge: (1) implementing in networks rules operating on variables (Shastri and Ajjanagadde 1993), (2) constructing representations that compress compositional structure (Pollack 1990) that can be operated on by other networks (Chalmers 1990), and (3) employing procedural knowledge to process external symbols (Elman 1990, 1991, 1993). We focus on the last as it represents the most radical attempt to apply associationist neural network approaches to the systematicity challenge. (For a discussion of the limitations of the compressed representation approach, see Borensztajn 2011.)

Output

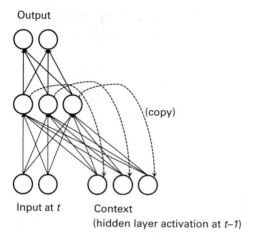

Input at *t*   Context
(hidden layer activation at *t–1*)

Figure 7.1
Simple recurrent network developed by Elman (1990) in which the activations on the hidden units during the previous cycle of processing are copied onto a set of input units to be sent forward on the next cycle of processing.

To accommodate the fact that processing language requires a memory of the words that came before, Elman developed the simple recurrent network (SRN). Each successive word is presented on one set of input units, and the previous activation on the hidden units is copied onto other input units (figure 7.1). This design provided sufficient memory of the network's response to previous words that it was able to predict with high accuracy the next word when trained on a corpus of sentences constructed from a phrase-structure grammar. Since the grammar allowed for both multiple words within each grammatical category and for multiple sentence types, the network could not be expected to predict correctly the actual next word in a sentence. Rather, it was evaluated by whether the output units activated corresponded to the range of possible continuations given the grammar. For example, the network correctly predicted that after being presented sequentially with *boy* and *lives*, the next input would be a period, whereas after *boy* and *sees* it could be a period, a singular or plural noun, or a proper name.

To explain how the network is able to achieve the level of performance it does, Elman argues that it *implicitly* represents the syntactic information needed to perform the task and offers strategies to reveal how this information is encoded in the activation pattern distributed over the hidden units. Elman (1990) employed hierarchical cluster analysis (HCA) to show that words originating from the same grammatical category, as specified by the grammar (e.g., *human* or *transitive verb*), produced similar activation patterns on hidden units.

We focus on Elman's early studies since they engendered the prevalent view among connectionists that *explicit* representations of categories are not necessary; rather, *implicit* knowledge of grammatical categories (Elman 1990) and constituent structure (Elman 1991) suffice to explain systematic behavior. This view explicitly rejects our contention that systematic behavior (if properly defined) necessitates a causal, hence *explicit*, role for categories. More recent studies with the SRN go beyond Elman's initial studies and emphasize correlations between certain aspects of behavior of SRN models and human linguistic behavior, for instance in the processing of different types of relative clauses (e.g., Christiansen and Chater 1999; MacDonald and Christiansen 2002). We do not discuss these here because they do not offer fundamentally different insights on the question of how grammatical knowledge is encoded in the SRN, which is central to our claims about systematicity (but see Borensztajn 2011 for a critical discussion of the "leaky recursion" argument).

Our concern about Elman's (1990) strategy is that he extracts the implicit knowledge of the SRN by averaging over all sentence contexts in which a given word occurs and performing the HCA analysis on the resulting average. This means that the regions in the state space corresponding to particular grammatical (word) categories are categories only in the eyes of the observer: they are never actually "consumed" by the SRN. Network operations are performed on the actual activations produced by individual category members, corresponding to different contexts of the same word, and not on some computed average activation; activation of any individual item may actually lie outside the computed region for the "implicit category." Implicit representations, thus constructed, can play no causal role in the dynamics of the network; hence they lack explanatory power. In contrast, a representation functions causally only if the representational vehicle figures in a mechanism that uses it (Bechtel 2009). In a similar vein, Kirsh (1991) argues that what counts as a representation is determined by the processing system that is accessing its content, and not by any external agency. Since the SRN has no way of using the representations that are constructed in the HCA, it would be a mistake to attribute the grammatical knowledge that can be inferred from the HCA (by an external observer) to the SRN.[2]

Similar considerations apply to Elman's (1991) claims with regard to the SRN's implicit knowledge of constituent structure. To analyze how an SRN

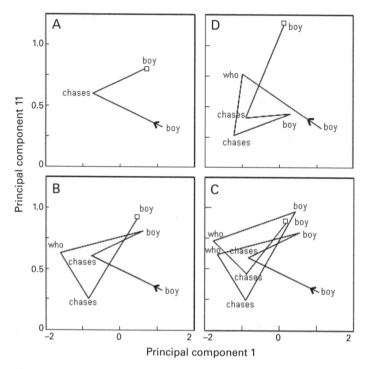

Figure 7.2

In terms of principal components developed from an analysis of activations in hidden units, Elman shows how the SRN processes related sentences. He argues that the displacement of the activation value generated by a given word in different contexts accounts for the network's ability to predict subsequent words correctly.

deals with the hierarchical structure of a sentence, he examined the trajectory through the space of possible activation patterns as each successive word was provided. Since in his network architecture this space has seventy dimensions, Elman turned to principal component analysis to provide a reduced dimensional analysis of the network's performance. Figure 7.2 presents an example of such an analysis, showing in two dimensions the changing activation over hidden units as the network processes four sentences using the same vocabulary with differently constructed relative clauses. Elman focused on how the representation of a given word was displaced in state space when it appears in a different grammatical construction (compare the different locations of activation for *boy* when it is the object of different levels of embedding in figures 7.2b and 7.2c). He concluded: "displacement provides a systematic way for the network to encode the depth of embedding" (1991, 111).

Although this analysis suggests that the SRN represents hierarchical structure, there is no account of how the network could use this representation. To make use of displacement information, the network would have to be able to perform a meta-analysis by comparing the state of the hidden units at different time steps. Perhaps, it would be possible to build into the network a mechanism designed to extract differential information, for instance by adding units that act as "difference detectors" between hidden states at different times. But for the network to use such information in order to determine long-term dependencies, it would need to be provided with information about the level of embedding (i.e., a *stack*), and this would require implementing a grammar that uses category representations. Of course, this would violate the spirit of Elman's project, which was to show that a network could exhibit appropriate performance without these representations. A similar objection applies to more recent attempts to model human language processing without relying on hierarchical structure, such as Frank, Bod, and Christiansen's (2012) proposal of a switching mechanism between parallel sequential streams. For a network to be able to use such a mechanism, it requires a means to keep track of which stream the processing is in and how to switch between them.

Our objection to this point is not that the network has failed to perform appropriately but that claims that connectionist networks implicitly represent linguistic categories or constituency do not suffice for demonstrating systematicity as we characterized it in the previous section. That requires representations of substitution classes all of whose members are treated alike. The hidden unit representations formed in Elman's SRNs are different for each instance, and there is no representation of the class that the system is able to use to make the same inference about each.

One could possibly object that SRNs are able to achieve human level performance without requiring systematicity as we have characterized it, and without explicit representations of substitution classes. There is evidence, however, that in some contexts humans do rely on representations of substitution classes. We briefly note two examples. First, humans acquire new words or phrases and new grammatical constructions from just one encounter and generalize them to a wide range of additional uses. SRNs, on the other hand, generalize over words that occur in similar sentence contexts, but not to different usages of a new word. Elman (1990) offers an example in which he shows that if a new word, "ZOG," replaces the word "man" in every position it occurs and the new set of sentences is presented to the network, the HCA analysis of the resulting activations assigns "ZOG" a similar location as "man." Elman's gloss on this is illuminating: "The network expects man, or something very much like it," and relates the result to context effects in word recognition. This suggests that, rather than learning the new word (learning was disabled), the network ignored the input and determined its response from context.[3] Generalization from a single encounter requires a procedure to assign the new word or phrase to a substitution class or to apply a new construction to all instances of the substitution class. We know of no demonstration that SRN networks can do this.

Second, one can show that humans are sensitive to constituent classes in the processing of garden path sentences such as *The horse raced past the barn fell*. The initial phrase *The horse raced* can be parsed in two ways, treating *raced* as the main verb or as part of a relative clause. The temptation to treat *raced* as the main verb is so powerful that most hearers do not even consider the alternative possibility until they hit *fell*. Then they must backtrack and construct the alternative. This indicates that humans are sensitive to the two possible syntactic categories and treat them as separate. Although to our knowledge the SRN has not been tested with a grammar that allows such constructions, given its mode of operation we expect that it would generate a single activation pattern for the first four words and predict both continuations, each in accordance with its frequency in the training corpus. It has no procedure to back up, reconsider the category assignment of "raced," and develop a new parse.

Elman's strategy was to show that procedural knowledge would enable neural networks to exhibit the behavior associated with systematicity. Our suggestion in the last two paragraphs is that this may not be possible in some cases without employing explicitly represented substitution classes. We might be wrong about this—clever network design might achieve sufficient levels of performance that it is indistinguishable from human performance. But, as we will argue in subsequent sections, to pursue this strategy is to unnecessarily limit the neural network approach, since systematicity as we have characterized it can be realized in the types of neural networks actually found in our brain. Our goal in highlighting the shortcomings of the SRN approach is not, as was Fodor and Pylyshyn's, to advance an objection to neural networks tout court. We are not arguing for rules operating over variables. Rather, we are arguing for employing networks capable of generating encapsulated representations. To illustrate such networks, we introduce Hawkins's Memory Prediction Framework (Hawkins and Blakeslee 2004).

3   Encapsulated Representations in Hawkins's Memory Prediction Framework (MPF)

In developing his theoretical framework, Hawkins drew on what we learned about the architecture of the mammalian neocortex during the second half of the twentieth century, largely from studies recording the electrical activity of neurons in different cortical regions in various mammals while responding to stimuli. Starting from Hubel and Wiesel's (1962, 1968) studies of cats and monkeys that identified neurons in the rear of the brain (primary visual cortex or V1) that responded to bars of light or edges with particular orientations at specific locations, researchers proceeded forward in the brain to identify higher regions that responded to more complex features such as illusory contours (V2), shapes (V4), and ultimately the identity of objects (anterior inferotemporal cortex or AIT) (for a historical review of this research, see Bechtel 2008, ch. 3). Drawing together information about the specific pattern of neural projections between thirty-three areas known to be involved in visual processing, Felleman and van Essen (1991) developed a hierarchical account in which lower areas such as V1 send projections forward to areas at the next level (V2) while receiving much more numerous backward projections from those areas (figure 7.3). We will refer to the resulting hierarchy as the hierarchy of brain regions.

As shown on the right in figure 7.3, individual neurons in successively higher levels have larger receptive fields (regions in the sensory field to which they respond when appropriate stimuli are presented). Whereas a neuron in V1 responds to an edge only in a specific small region of the visual field, a neuron in the anterior inferotemporal cortex (AIT) responds to a given object irrespective of where it appears. Moreover, as one moves up in the hierarchy, neurons respond to increasingly complex stimuli and respond invariantly regardless of the particulars of the current presentation of the stimulus. Cells in AIT have been reported that respond to any image of Bill Clinton, whether it is a line drawing or a photograph (Kreiman, Fried, and Koch 2002).

Beyond the hierarchy of brain regions, each region of the neocortex is organized into six anatomically distinctive layers parallel to its surface. Crosscutting these are columns of neurons like those shown in figure 7.4. Many neural network accounts of brain function jump directly from

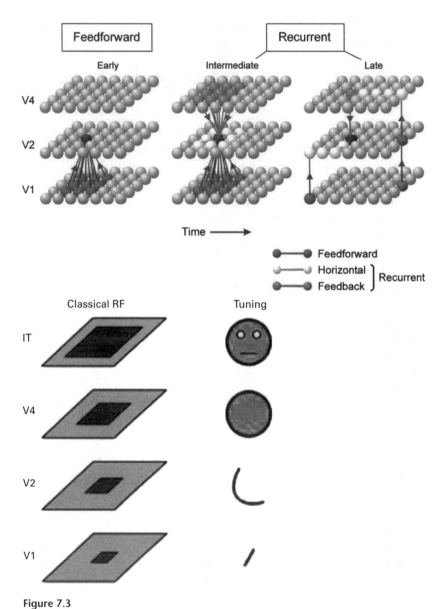

Figure 7.3
The feedforward and recurrent processing through three visual areas is shown on the left. The receptive field of individual neurons and the types of stimuli to which they are responsive is shown on the right. (From Roelfsema 2006.)

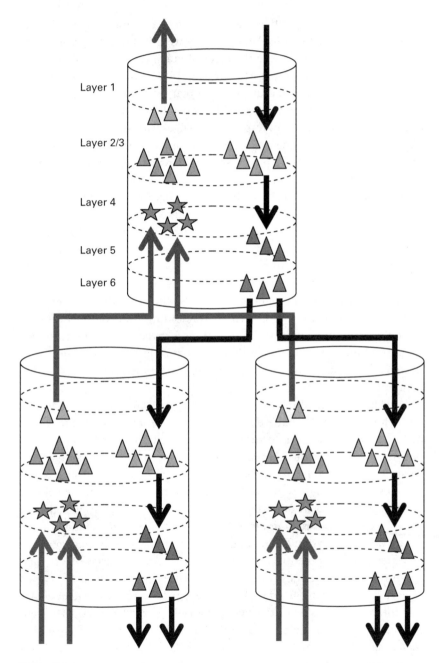

Figure 7.4

Three minicolumns in two connected brain regions. Upward-pointing arrows represent the forward flow from earlier cortical areas, whereas downward-pointing arrows represent recurrent projections back from higher areas.

individual neurons to the brain areas in which they are located, skipping the level of cortical columns. Yet since they were first noted by Lorente de Nó (1938), columns have been the focus of much research. In his study of the cat's somatosensory cortex, Mountcastle (1957) postulated that columns are the basic computational units of the brain. Hubel and Wiesel (1962) found that if they inserted an electrode perpendicular into the cell surface, all neurons within a column responded to edges with the same orientation and processed by the same eye, whereas neurons in neighboring columns either responded to edges at progressively different angles or switched from inputs to one eye to inputs to the other.

Hubel and Wiesel (1962) furthermore recognized that columns consist of a cluster of "minicolumns." They interpreted the cells that all responded to a stimulus at a given orientation as constituting a minicolumn. Minicolumns contain on the order of 100 neurons, whereas approximately 70 minicolumns that respond to a given part of the visual field irrespective of orientation constitute a macro- or hypercolumn. Later work further clarified that throughout the cortex the minicolumns are organized into topological maps, such that adjoining minicolumns respond to similar stimuli. Inhibitory interneurons between these minicolumns support winner-take-all competitions, so that only the minicolumn with the greatest input remains active and sends action potentials to other brain areas.

Hawkins's principal focus is on how columns connect to one another, but before turning to that we look briefly at the information processing made possible by the architecture within minicolumns and their organization into macrocolumns. Minicolumns are best understood as organized collections of neurons that collaborate in processing information taken in from a variety of sources. About 90 percent of the axons within a column connect to neurons in other columns, either at the same or at higher or lower levels. Within each minicolumn, layers 2, 3, 5, and 6 contain many pyramidal cells. Pyramidal neurons are the primary cells that generate excitatory action potentials when sufficient input is received on their dendrites. These layers also contain many inhibitory interneurons, which typically receive input from other minicolumns in a given macrocolumn and inhibit the activity of the pyramidal cells. Layer 1 contains few cells but instead contains many axons running parallel to the cortical surface, including those originating in cells in layers 2 and 3 of the column. Layer 4 contains a distinctive cell type, stellate cells. As shown in figure 7.4, this layer is where inputs arrive from lower regions of the hierarchy of brain regions. Axons from cells in layer 4 project to neurons in layer 3, whose axons then project to those in layer 2, and from there to a higher region. While the upward flow of information stays mostly within a single column, with multiple columns converging on one at a higher level, the downward flow diverges more. When a given column receives projections back from higher regions, these arrive at layer 1 and project downward in many columns to layers 2 and 3, eventually reaching layers 5 and 6 (Mountcastle 1997; Buxhoeveden and Casanova 2002). Layers 5 and 6 send projections to the thalamus and lower brain regions as well as cortical regions lower in the hierarchy. Layer 1 also receives delayed feedback connections from the thalamus, originating from columns that were previously active, resulting in a complex pattern of activation in lower areas.

Hawkins's account is distinctive in that it emphasizes how the columnar organization of the cortex serves to accomplish two major feats: (1) learning and storing temporal sequences of patterns that can then be used for prediction and (2) encapsulating the output of these sequences, so that they can be used as an 'alphabet' in higher regions. Hawkins proposes that a sequence, "a set of patterns that generally accompany each other but not always in a fixed order" (Hawkins and Blakeslee 2004, 86), can be learned by a Hebbian process of strengthening synaptic weights between currently active and recently active minicolumns (via the thalamic feedback loop). This results in stored associations between sets of features from different macrocolumns that frequently go together over time (e.g., edges in one part of the visual field that extend those in another). With these connections in place, activity in one of these minicolumns can elicit or inhibit activity in others, regenerating the sequence of coherent patterns at one level. These patterns provide input to minicolumns at the next level, with Hebbian learning leading to specific minicolumns responding to a given sequence. Hawkins hypothesizes that the fact that both the spatial and temporal extents of inputs that drive responses increase as one moves up the hierarchy of brain regions allows higher brain areas to respond to longer enduring stimuli or to stimuli composed of multiple temporal components that might be processed sequentially in lower areas in the hierarchy. Recently, imaging studies have provided some empirical support for this hypothesis (Hasson et al. 2008).

The second feat concerns the encapsulation of the details of a sequence that is formed in a lower region from processes at hierarchically higher levels. Hawkins hypothesizes that lower regions pass on category labels or "names" for the sequences they have recognized. Such a name—"a group of cells whose collective firing represents the set of objects in the sequence" (Hawkins and Blakeslee 2004, 129)—represents a constant pattern for the inputs of minicolumns in higher regions. He suggests a plausible mechanism for how the cortex generates constant names that stay "on" for the entire duration of a sequence (152). This involves simultaneously activating layer 2 cells of multiple columns that are members of a sequence, driven by connections from a hierarchically higher region of cortex that provides it a name. At the same time, an inhibitory mechanism, involving certain layer 3 cells, suppresses the throughput of the time-varying details of the stimulus to the higher region when the sequence is predicted successfully, so that subsequent brain areas have access only to the names activated at the previous level in the hierarchy. The columns at each level in the hierarchy of brain regions generate names for the patterns received on their input units, which are then encapsulated: further processing draws on them, not on the patterns that generated them.

We can now show how the organization we have described realizes what Hawkins characterizes as a Memory Prediction Framework. He construes projections from sequences stored in columns at higher levels back to lower levels as serving to predict subsequent input, as the sequence unfolds in time to the lower level. Hawkins is not alone in emphasizing the role of recurrent projections in prediction (Dayan et al. 1995; Rao and Ballard 1999; Llinás 2001; Hohwy, Roepstorff, and Friston 2008; Huang and Rao 2011; Clark in press). What is distinctive is his characterization of the organization of cortex that facilitates encapsulated representations which then figure in this process. As we have noted, he construes the output of a lower region as a name for the sequence of patterns received on the input level. Because of the fan-in relation between levels, the name is necessarily more abstract than the sequence it names—it names a number of specific sequences. For example, a square as recognized in V4 can have its edges at different points in the visual field; accordingly, the name *square* generated in V4 will project back to columns in V1 other than those from which it received input on a given occasion. Since the V4 name is encapsulated, it has no sensitivity to these differences; when it is activated, it sends projections back to all columns that feed into it. This feedback enters the lower-level column in layers 3 and 2, and serves to increase the activity of the pyramidal cells therein. Since the minicolumns at a given level often compete with each other, this can serve to alter the competition at the lower level. The result of this top-down activity can bias processing in favor of interpretation of an ambiguous input or even generate activity seriously at odds with the input being received from still lower areas and ultimately from the senses.

4   Structured Cognition with Encapsulated Representations

In the previous section, we showed how the hierarchical, columnar organization of the cortex provides a basis for encapsulated representations that are invariant. Our contention is that this provides a crucial resource for explaining the systematicity of thought. Encapsulated representations are structures that can be further composed but have a distinct identity from the varying instances that elicit them. In more traditional computer science vocabulary, they are pointers to their various instantiations. As processing moves down the hierarchy, encapsulated representations are unpacked into sequences of patterns. But they can also be targets of operations specific to them.

To illustrate the advantages of encapsulated representations, consider the representations developed in response to a linguistic corpus. The same hierarchical columnar organization can be deployed both for visual representations and the construction of encapsulated representations of temporal patterns that correspond to the syntactic units of a language. Categories such as *noun phrase* and *relative clause* become available to the processing system as encapsulated units while preserving the ability to generate instances of linguistic sequences that unfold in time (e.g., *det adj noun*) when processing downward through the columnar structure of cortex. Our contention is that encapsulated representations provide the means for addressing systematicity since the relations between encapsulated representations are distinct from the relations between lower-level inputs and are available to be used in information-processing operations.

What we haven't shown yet is how primitive encapsulated representations can give rise to the complex and structured representations that are used in parsing sentences. This seems to require a combinatorial mechanism that can flexibly group representations into complex structures. "Structured cognition" is perceived by some as orthogonal to the spirit of connectionism (e.g., McClelland et al. 2010; Marcus et al. 1999). Indeed, activation spreading through the pair-wise associative connections in traditional neural networks is inherently context sensitive and so cannot provide the aforementioned mechanism. Encapsulated representations, however, allow for operations that are directed specifically toward them, although the challenge in deploying the right operations is not simple. Given that constituents of a particular category can contain constituents of the same category (as occur, for example, in embedded clauses), the mechanism must allow recursive groupings between representations that are possibly stored at the same level in the hierarchy of brain regions. Moreover, it must do so in the face of changing representations—a consequence of the fact that these representations are not innate and symbolic, but are learned from experience. Thus, operations on encapsulated representations must involve a flexible and indirect means of communication that allows for transmitting variable representations (the "names") through fixed lines of communication.

To motivate the type of operations required, consider the phenomenon of productivity—the ability of language users to produce arbitrarily many novel sentences based on only a limited number of stored words and procedures for combining them. Several theorists have proposed that the productivity of language is supported by some form of *dynamic binding* of network units (e.g., Hummel and Holyoak 1997; van der Velde, van der Voort van der Kleij, and de Kamps 2004). The general idea behind dynamic binding is that units can be flexibly grouped in novel configurations without requiring preexisting, dedicated binding neurons. To this end, most models of dynamic binding employ "tags," with different proposals reserving this role for the "oscillation phase" (i.e., in synchronous binding; von der Malsburg 1982), or "enhanced activity," corresponding to attention (in serial binding; Roelfsema 2006). We cannot explore in any detail how dynamic binding might be realized for the language domain, but the Hierarchical Prediction Network proposed by Borensztajn (2011) offers a suggestion. The network employs complex primitive units (shown as "treelets" in figure 7.5) that should be seen as encapsulated sequences (representing, for example, graded syntactic categories), which can dynamically bind to other sequences or to word units. At the core of the network, as shown in figure 7.5, is a "switchboard," a neural hub that routes the encapsulated representations to specific subunits, using an address that is stored in the root of the unit. In this model, the names of the sequences are interpreted as addresses in a high-dimensional vector space and forwarded through the switchboard as a spike pattern to a subunit of another sequence with the appropriate address, allowing for pointers to the sequences to be used as building blocks in higher-order sequences.

The high-dimensional vector space constitutes a "name space," and can be viewed as the equivalent of the topological maps observed in the visual and auditory cortex. However, it is proposed that in the syntactic domain the topology is realized in a reverse hierarchy (Hochstein and Ahissar 2002). In this view, relations between syntactic categories (the "names") are expressed as relative distances in a topological space, and grammar acquisition amounts to self-organization of the topology: when the network

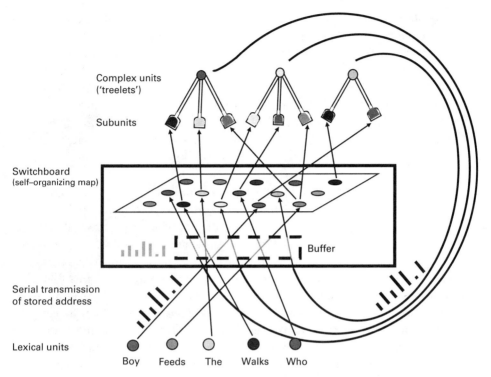

Figure 7.5

The switchboard, as used in the Hierarchical Prediction Network (HPN) (Borensztajn 2011). Circles indicate lexical units; the "treelets" indicate complex primitive units, representing sequences of patterns that can learn to assume the role of context-free rules.

is trained with sentences from an artificial language, regions gradually emerge in the topology that correspond to conventional syntactic categories, such as nouns and verbs (Borensztajn, Zuidema, and Bod 2009).

A version of Borensztajn's Hierarchical Prediction Network has been implemented in a computational model, but it is important to consider how it might be implemented in the brain. One possibility draws on a feature that is widely recognized in neural systems—that brains exhibit oscillations at a wide range of frequencies, from <0.1 Hz, observed in fMRI, to >100 Hz, observed in EEG. These are the product not of action potentials but of electrical potentials across neural membranes that oscillate between being hypo- and hyperpolarized states. The oscillations in different brain regions synchronize, typically in a metastable manner in which patterns of synchronized activity emerge, but soon disintegrate as a result of endogenous processes. The function of these bouts of synchrony is not entirely clear, but it has been proposed that they represent channels of communication between brain regions—when two neurons are simultaneously hypopolarized, an action potential generated by one is more likely to elicit an action potential in the other (Buzsáki 2006; Abrahamsen and Bechtel 2011; Bechtel in press). Saalmann, Pinsk, Wang, Li, and Kastner (2012) have recently provided evidence that the pulvinar (a part of the thalamus) regulates such synchrony and thereby helps coordinate the opening and closing of transient communication channels required for dynamic binding.

Although our discussion of HPN has been necessarily sketchy, it serves to indicate that encapsulated representations, once generated, could be operated on in ways that produce systematic behavior. As required by our definition of systematicity, prediction of subsequent words (and constituents) in the sentence depends on the categorical representation at the root of a treelet alone, and not on preceding context coming from units that bind to the treelet's subunits. As the primitive representations range over sequences, they allow for generalization over constituents (which is one aspect of systematicity that has often been ignored) and thereby automatically generate hierarchical representations. Note, however, that the process subserving hierarchical representations is sequential (i.e., it is dynamic binding), in line with evidence from many imaging and psycholinguistic studies (as reviewed, e.g., by Frank, Bod, and Christiansen, 2012).

5   Conclusions

One of the shortcomings in the debates about systematicity initiated by Fodor and Pylyshyn has been the failure to articulate adequate accounts of systematicity. The most developed operational criterion, by Hadley, appeals to the unarticulated idea of substitution classes of constituent representations. Symbolic accounts provide these through the use of variables within rules, but these could only be accommodated in neural network accounts that attempt to explicitly implement global symbols and rules. We argue for a strategy that involves encapsulated representations consistent with the local control that is fundamental to neural network models. Our characterization of systematicity, like that of Hadley, still defines the explanandum—systematicity in human cognition and behavior—in terms of components of the explanans—the formal models that explicitly involve categories. We do not claim, therefore, to be able to prove that only models with encapsulated categories can account for that body of empirical data; we do claim, however, that none of the existing emergentist models that we are aware of accounts for the data satisfactorily. The associationist perspective realized in these networks results in all representations being context sensitive, and the degree of success these networks achieve in domains that seem to require systematicity is in fact due to this context sensitivity. Accordingly, they do not generate encapsulated categories and cannot generalize over such categories.

The architecture of the mammalian cortex, however, provides a way to generate encapsulated representations. The trick, as employed in Hawkins's Memory Prediction Framework, is to take advantage of the different types of processing available within and between cortical minicolumns. Minicolumns occur at each level of the hierarchy of brain regions and produce outputs that are encapsulated from the variation at lower levels that provides their input. Once generated, these representations can be operated on in ways appropriate to them that still involve only local processes. Once these operations are applied, the process of unpacking representations at higher levels in the hierarchy onto sequences of patterns at lower levels assures their application to any members of the category. This, we claim, is what Fodor and Pylyshyn should have had in mind in arguing for the systematicity of thought, and this is what can be provided by neural networks appropriately inspired by what is known about the mammalian neocortex.

Notes

1. As Fodor and McLaughlin (1990) make clear in a subsequent paper, the issue is not whether a neural network might happen to exhibit systematicity because of the happenstance of its training, but whether it is required to do so as a result of its fundamental design.

2. To demonstrate this point, Borensztajn (2011) applied the same cluster analysis to the data on which Elman's (1990) model was trained, and obtained a nearly identical hierarchy of categories. This reveals that the hidden unit activations merely provide a redescription of the temporal sequence of inputs, but do not, contrary to what Elman suggested, add a category structure to it.

3. More recently, Christiansen and MacDonald (2009) have compared the insertion of a novel word and a learned word from a different category into a test corpus for an SRN and claimed that the network correctly generalized to the novel word but not the word from another category. But in this case as well, there is no evidence that the network really learned the properties of the novel word rather than simply ignoring it and thus treating it as essentially a random input.

References

Abrahamsen, A., and W. Bechtel. 2011. From reactive to endogenously active dynamical conceptions of the brain. In *Philosophy of Behavioral Biology*, ed. T. Reydon and K. Plaisance, 329–366. New York: Springer.

Bechtel, W. 2008. *Mental Mechanisms*. London: Routledge.

Bechtel, W. 2009. Constructing a philosophy of science of cognitive science. *Topics in Cognitive Science* 1 (3):548–569.

Bechtel, W. In press. The endogenously active brain: The need for an alternative cognitive architecture. *Philosophia Scientia*.

Bechtel, W, and A. Abrahamsen. 2002. *Connectionism and the Mind: Parallel Processing, Dynamics, and Evolution in Networks*, 2nd ed. Oxford: Blackwell.

Borensztajn, G. 2011. The neural basis of structure in language: Bridging the gap between symbolic and connectionist models of language processing. Amsterdam: Institute for Language, Logic, and Computation, Universiteit van Amsterdam.

Borensztajn, G., W. Zuidema, and R. Bod. 2009. The hierarchical prediction network: Towards a neural theory of grammar acquisition. In *Proceedings of the 31st Annual Conference of the Cognitive Science Society*, 2974–2979. Austin, TX: Cognitive Science Society.

Buxhoeveden, D. P., and M. F. Casanova. 2002. The minicolumn hypothesis in neuroscience. *Brain* 125 (5):935–951.

Buzsáki, G. 2006. *Rhythms of the Brain*. Oxford: Oxford University Press.

Chalmers, D. J. 1990. Syntactic transformations on distributed representations. *Connection Science* 2:53–62.

Christiansen, M. H., and N. Chater. 1994. Generalization and connectionist language learning. *Mind and Language* 9 (3):273–287.

Christiansen, M. H., and N. Chater. 1999. Toward a connectionist model of recursion in human linguistic performance. *Cognitive Science* 23 (2):157–205.

Christiansen, M. H., and M. C. MacDonald. 2009. A usage-based approach to recursion in sentence processing. *Language Learning* 59:126–161.

Clark, A. In press. Whatever next? Predictive brains, situated agents, and the future of cognitive science. *Behavioral and Brain Sciences*.

Dayan, P., G. E. Hinton, R. M. Neal, and R. S. Zemel. 1995. The Helmholtz machine. *Neural Computation* 7:889–904.

Eliasmith, C., T. C. Stewart, X. Choo, T. Bekolay, T. DeWolf, Y. Tang, and D. Rasmussen. 2012. A large-scale model of the functioning brain. *Science* 338 (6111): 1202–1205.

Elman, J. L. 1990. Finding structure in time. *Cognitive Science* 14:179–211.

Elman, J. L. 1991. Distributed representations, simple recurrent networks, and grammatical structure. *Machine Learning* 7:195–224.

Elman, J. L. 1993. Learning and development in neural networks: The importance of starting small. *Cognition* 48:71–99.

Felleman, D. J., and D. C. van Essen. 1991. Distributed hierarchical processing in the primate cerebral cortex. *Cerebral Cortex* 1:1–47.

Fodor, J. A., and B. McLaughlin. 1990. Connectionism and the problem of systematicity: Why Smolensky's solution doesn't work. *Cognition* 35:183–204.

Fodor, J. A., and Z. W. Pylyshyn. 1988. Connectionism and cognitive architecture: A critical analysis. *Cognition* 28:3–71.

Frank, S. L., R. Bod, and M. H. Christiansen. 2012. How hierarchical is language use? *Proceedings of the Royal Society of London, Series B: Biological Sciences*.

Griffiths, T. L., N. Chater, C. Kemp, A. Perfors, and J. B. Tenenbaum. 2010. Probabilistic models of cognition: Exploring representations and inductive biases. *Trends in Cognitive Sciences* 14 (8):357–364.

Grossberg, S. 1982. *Studies of Mind and Brain: Neural Principles of Learning, Perception, Development, Cognition, and Motor Control*. Boston Studies in the Philosophy of Science, vol. 70. Dordrecht: D. Reidel.

Hadley, R. F. 1994. Systematicity in connectionist language learning. *Mind and Language* 9 (3):247–272.

Hasson, U., E. Yang, I. Vallines, D. J. Heeger, and N. Rubin. 2008. A hierarchy of temporal receptive windows in human cortex. *Journal of Neuroscience* 28 (10): 2539–2550.

Hawkins, J., and S. Blakeslee. 2004. *On Intelligence*. New York: Times Books.

Hinton, G. E., and J. A. Anderson, eds. 1981. *Parallel Models of Associative Memory*. Hillsdale, NJ: Erlbaum.

Hochstein, S., and M. Ahissar. 2002. View from the top: Hierarchies and reverse hierarchies in the visual system. *Neuron* 36 (5):791–804.

Hohwy, J., A. Roepstorff, and K. Friston. 2008. Predictive coding explains binocular rivalry: An epistemological review. *Cognition* 108 (3):687–701.

Huang, Y., and R. P. N. Rao. 2011. Predictive coding. *Wiley Interdisciplinary Reviews: Cognitive Science* 2 (5):580–593.

Hubel, D. H., and T. N. Wiesel. 1962. Receptive fields, binocular interaction, and functional architecture in the cat's visual cortex. *Journal of Physiology* 160: 106–154.

Hubel, D. H., and T. N. Wiesel. 1968. Receptive fields and functional architecture of monkey striate cortex. *Journal of Physiology* 195:215–243.

Hummel, J. E., and I. Biederman. 1992. Dynamic binding in a neural network for shape recognition. *Psychological Review* 99 (3):480–517.

Hummel, J. E., and K. J. Holyoak. 1997. Distributed representations of structure: A theory of analogical access and mapping. *Psychological Review* 104 (3):427–466.

Kirsh, D. 1991. Today the earwig, tomorrow man? *Artificial Intelligence* 47: 161–184.

Kreiman, G., I. Fried, and C. Koch. 2002. Single-neuron correlates of subjective vision in the human medial temporal lobe. *Proceedings of the National Academy of Sciences of the United States of America* 99 (12):8378–8383.

Llinás, R. R. 2001. *I of the Vortex: From Neurons to Self.* Cambridge, MA: MIT Press.

Maass, W., T. Natschläger, and H. Markram. 2003. A model for real-time computation in generic neural microcircuits. *Advances in Neural Information Processing Systems* 15:213–220.

MacDonald, M. C., and M. H. Christiansen. 2002. Reassessing working memory: Comment on Just and Carpenter (1992) and Waters and Caplan (1996). *Psychological Review* 109 (1):35–54, discussion 55–74.

Marcus, G. F., S. Vijayan, S. B. Rao, and P. M. Vishton. 1999. Rule learning by seven-month-old infants. *Science* 283 (5398):77–80.

McClelland, J. L., M. M. Botvinick, D. C. Noelle, D. C. Plaut, T. T. Rogers, M. S. Seidenberg, and L. B. Smith. 2010. Letting structure emerge: Connectionist and dynamical systems approaches to cognition. *Trends in Cognitive Sciences* 14 (8):348–356.

McClelland, J. L., D. E. Rumelhart, and the PDP Research Group. 1986. *Parallel Distributed Processing: Explorations in the Microstructure of Cognition*, vol. 2: *Psychological and Biological Models*. Cambridge, MA: MIT Press.

Miikkulainen, R. 1993. *Subsymbolic Natural Language Processing: An Integrated Model of Scripts, Lexicon, and Memory.* Cambridge, MA: MIT Press.

Mountcastle, V. B. 1957. Modality and topographic properties of single neurons of cat's somatic sensory cortex. *Journal of Neurophysiology* 20:408–434.

Mountcastle, V. B. 1997. The columnar organization of the neocortex. *Brain* 120 (4):701–722.

Nó, L. R. de. 1938. The cerebral cortex: Architecture, intracortical connections, and motor projections. In *Physiology of the Nervous System*, ed. J. Fulton, 291–301. Oxford: Oxford University Press.

Pollack, J. B. 1990. Recursive distributed representation. *Artificial Intelligence* 46: 77–105.

Rao, R. P. N., and D. H. Ballard. 1999. Predictive coding in the visual cortex: A functional interpretation of some extra-classical receptive-field effects. *Nature Neuroscience* 2 (1):79–87.

Roelfsema, P. R. 2006. Cortical algorithms for perceptual grouping. *Annual Review of Neuroscience* 29 (1):203–227.

Rumelhart, D. E., J. L. McClelland, and the PDP Research Group. 1986. *Parallel Distributed Processing: Explorations in the Microstructure of Cognition*, vol. 1: *Foundations*. Cambridge, MA: MIT Press.

Saalmann, Y. B., M. A. Pinsk, L. Wang, X. Li, and S. Kastner. 2012. The pulvinar regulates information transmission between cortical areas based on attention demands. *Science* 337 (6095):753–756.

Shastri, L., and V. Ajjanagadde. 1993. From simple associations to systematic reasoning: A connectionist representation of rules, variables, and dynamic bindings using temporal synchrony. *Behavioral and Brain Sciences* 16:417–494.

Thornton, R., and K. Wexler. 1999. *Principle B, VP Ellipsis, and Interpretation in Child Grammar*. Current Studies in Linguistics 31. Cambridge, MA: MIT Press.

van der Velde, F., G. T. van der Voort van der Kleij, and M. de Kamps. 2004. Lack of combinatorial productivity in language processing with simple recurrent networks. *Connection Science* 16 (1):21–46.

von der Malsburg, C. 1982. The correlation theory of brain function. Internal Report 81-2, Department of Neurobiology, Max-Planck Institute for Biophysical Chemistry, Göttengen, Germany.

8 How Limited Systematicity Emerges: A Computational Cognitive Neuroscience Approach

Randall C. O'Reilly, Alex A. Petrov, Jonathan D. Cohen, Christian J. Lebiere, Seth A. Herd, and Trent Kriete

1 Introduction

In this chapter, we address the claims made by Fodor and Pylyshyn (1988) (FP88 hereafter). We strike a middle ground between classic symbolic and connectionist perspectives, arguing that cognition is less systematic than classicists claim, but that connectionist, neural-processing-based theories have yet to explain the extent to which cognition is systematic. We offer a sketch of an emerging understanding of the basis of human systematicity in terms of interactions between specialized brain systems, leveraging the computational principles identified and empirical work done in the quarter-century since the target work was published. We identify a full spectrum of processing mechanisms, arrayed along the continuum between context-sensitivity and combinatorial, systematic processing, each associated with different parts of the human brain. We find that attempting to understand the role of these different brain areas through the lens of systematicity results in a rich picture of human cognitive abilities.

FP88 make two central claims about what a classical symbol processing system must be capable of, which define a classical model:

1. *Mental representations have combinatorial syntax and semantics.* Complex representations ("molecules") can be composed of other complex representations (compositionality) or simpler "atomic" ones, and these combinations behave sensibly in terms of the constituents.

2. *Structure sensitivity of processes.* There is a separation between form and content, exemplified in the distinction between syntax and semantics, and processes can operate on the form (syntax) while ignoring the semantic content.

Taken together, these abilities enable a system to be fully *systematic* and *compositional.* Systematicity comes directly from the ability to process the form or structure of something, independent of its specific contents: if you can process sentences with a given syntax (e.g., Noun Verb Object) then you can process any constituent words in such sentences—you do not have to relearn the syntax all over again for each new word. In Chomsky's famous example, you can tell that "Colorless green ideas sleep furiously" is grammatically correct because you can encode its structural form, independent of the (lack of) meaning, while "Furiously sleep ideas green colorless" is not grammatically correct. FP88 made the point that connectionist models of that time failed to exhibit these features, and thus were insufficient models of the full power of human cognition (Fodor and Pylyshyn 1988; Fodor and McLaughlin 1990; McLaughlin 1993). This debate remains active to this day, with various critical commentaries (Aizawa 1997; Cummins 1996; Hadley 1994; Horgan and Tienson 1996; Matthews 1997; van Gelder 1990), anthologies (Macdonald and Macdonald 1995), and a book-length treatment (Aizawa 2003). Recently, Bayesian symbolic modelers have raised similar critiques of neural network models (Kemp and Tenenbaum 2008; Griffiths, Chater, Kemp, Perfors, and Tenenbaum 2010), which are defended in return (McClelland, Botvinick, Noelle, Plaut, Rogers, Seidenberg, and Smith 2010).

Qualitatively, there are two opposing poles in the space of approaches one can take in attempting to reconcile FP88 and subsequent critiques with the fact that the human brain is, in fact, made of networks of neurons. One could argue that this systematic, compositional behavior is a defining feature of human cognition, and figure out some way that networks of neurons can implement it (the "mere implementation" approach). Alternatively, one could argue that the kind of systematicity championed by FP88 is actually not an accurate characterization of human cognition, and that a closer examination of actual human behavior shows that people behave more as would be expected from networks of neurons, and not as would be expected from a classical symbol processing system (the "dismissive" approach). Few connectionist researchers have shown much enthusiasm for the project of merely implementing a symbolic system, although proof-of-concept demonstrations do exist (Touretzky 1990). Instead, there have been numerous attempts to demonstrate systematic generalization with neural networks (Bodén and Niklasson 2000; Chalmers 1990; Christiansen and Chater 1994; Hadley 1997; Hadley and Hayward 1997; Niklasson and van Gelder 1994; Smolensky 1988, 1990b; Smolensky and Legendre 2006). Also, careful examinations of language (Johnson 2004) and various aspects of human behavior have questioned whether human language, thought, and behavior really are as systematic as it is commonly assumed (van Gelder and Niklasson 1994).

An intermediate approach is to attempt to implement a symbolic system using neural networks with the intent of finding out which symbolic aspects of systematicity are plausible from a neural perspective and which are not (Lebiere and Anderson 1993). This attempt to implement the Adaptive Control of Thought—Rational (ACT-R) cognitive architecture using standard neural network constructs such as Hopfield networks and feedforward networks resulted in a considerable simplification of the architecture. This included both the outright removal of some of its most luxuriant symbolic features as neurally implausible, such as chunks of information in declarative memory that could contain lists of items and production rules that could perform arbitrarily complex pattern-matching over those chunks. More fundamentally, neural constraints on the architecture led to a modular organization that combines massive parallelism within each component (procedural control, declarative memory, visual processing, etc.) with serial synchronization of information transfers between components. That organization in turn has been validated by localization of architectural modules using neural imaging techniques (Anderson 2007). In general, this hybrid approach has resulted in an architecture that largely preserves the systematicity of the original one while greatly improving its neural plausibility. It should be pointed out, though, that systematicity in ACT-R is limited by both the skills and knowledge needed to perform any of the tasks in which it is demonstrated, and more fundamentally by the combination of the symbolic level with a subsymbolic level that controls every aspect of its operations (procedural action selection, information retrieval from memory, etc.).

The reason the systematicity debate has persisted for so long is that both positions have merit. In this chapter, we take a "middle way" approach, arguing that purely systematic symbol-processing systems do not provide a good description of much of human cognition, but that nevertheless there are some clear examples of people approximating the systematicity of symbol-processing systems, and we need to understand how the human brain can achieve this feat. Going further, we argue that a careful consideration of all the ways in which the human brain can support systematicity actually deals with important limitations of the pure symbol-processing approach, while providing a useful window into the nature of human cognition. From a neural mechanisms perspective, we emphasize the role that interactions between brain systems—including the more "advanced" brain areas, and specifically the prefrontal cortex/basal ganglia (PFC/BG) system—play in enabling the systematic aspects of human cognition. In so doing, we move beyond the limitations of traditional "connectionist" neural network models, while remaining committed to only considering neural mechanisms that have strong biological support.

Although the overall space of issues relevant to this systematicity debate is quite high-dimensional and complex, one very important principal component can be boiled down to a trade-off between *context-sensitivity and combinatoriality*. At the extreme context-sensitivity end of the spectrum, the system maintains a lookup table that simply memorizes each instance or exemplar, and the appropriate interpretation or response to it. Such a system is highly context sensitive, and thus can deal with each situation on a case-by-case basis, but is unable to generalize to novel situations. At the other end, the system is purely combinatorial and processes each separable feature in the input independently, without regard for the content in other feature channels. Such a purely combinatorial system will readily generalize to novel inputs (as new combinations of existing features), but is unable to deal with special cases, exceptions, or any kind of nonlinear interactions between features. It seems clear that either extreme is problematic and that we need a more balanced approach. This balance can be accomplished in two ways. First, one could envisage representations and information-processing mechanisms with intermediate degrees of context-sensitivity. Second, one could envisage a combination of processing systems that specialize on each of these distinct ends of the spectrum. These two strategies are not incompatible and can be combined. In this chapter, we argue that the brain incorporates functional subsystems that fall along various points of the spectrum, with evolutionarily older areas being strongly context sensitive and newer areas, notably the prefrontal cortex, being more combinatorial (though still not completely combinatorial). This limited combinatoriality is expected to produce limited systematicity in behavior. We argue that human cognition exhibits precisely this kind of limited systematicity.

The limits of human systematicity have been pointed out before (Johnson 2004; van Gelder and Niklasson 1994). Here we limit ourselves to three well-known examples from vision, language, and reasoning. Our first example is shown in figure 8.1. The context surrounding the middle letter of each word is critical for disambiguating this otherwise completely ambiguous input. A purely combinatorial system would be unable to achieve this level of context-sensitivity. Our second example is from the domain of language and illustrates the interplay between syntax and semantics. Consider the sentences:

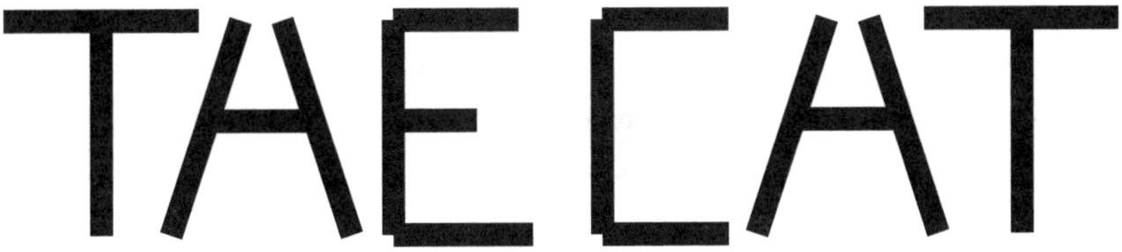

TAE CAT

Figure 8.1
Example of the need for at least some level of context-sensitivity, to disambiguate ambiguous input in middle of each word. This disambiguation happens automatically and effortlessly in people.

(1a)   Time flies like an arrow.
(1b)   Fruit flies like a banana.

Again, people automatically take the context into account and interpret ambiguous words such as "like" and "flies" appropriately based on this context. Our final example is from the domain of logical reasoning. Formal logic is designed to be completely context invariant and content free. Yet, psychological studies with the so-called Wason card selection task have shown that human reasoning is strongly sensitive to concrete experience. People can easily decide who to card at a bar given a rule such as "You can only drink if you are over 21," but when given the same logical task in abstract terms, their performance drops dramatically (Griggs and Cox 1982; Wason and Johnson-Laird 1972). Even trained scientists exhibit strong content effects on simple conditional inferences (Kern, Mirels, and Hinshaw 1983). More examples from other domains (e.g., the underwater memory experiments of Godden and Baddeley 1975) can easily be added to the list, but the above three suffice to illustrate the point. Human cognition is strongly context sensitive.

The standard classicist response to such empirical challenges is to refer to the competence–performance distinction (Aizawa 2003)—the idea that people are clearly capable of systematicity even if they sometimes fail to demonstrate it in particular circumstances. However, commercial symbolic AI systems are explicitly designed to have as few performance-related limitations as possible, and yet they face well-known difficulties in dealing with commonsense knowledge and practical reasoning tasks that people perform effortlessly. Arguably, these difficulties stem from the fact that a purely syntactic, formal representational system bottoms out in a sea of meaningless "atoms" and is undermined by the symbol grounding problem (Harnad 1990).

On the other hand, the classicist position also has merit. In some circumstances, it is desirable to be as context *insensitive* as possible. Perhaps the strongest examples come from the domain of deductive inference. Changing the meaning of a term halfway through a logical proof leads to the fallacy of equivocation. Consider the following fallacious argument:

(2a)   A feather is light.
(2b)   What is light cannot be dark.
(2c)   *Therefore, a feather cannot be dark.

Here the word "light" appears in two different (context-dependent) senses in the two premises, which breaks the inferential chain. All tokens of a symbol in logic must have identical meaning throughout the proof or else the proof is not valid. Despite their natural tendency for context specificity, we can appreciate Aristotle's basic insight that the validity of deductive inference depends solely on its form and not on its content. We can learn to do logic, algebra, theoretical linguistics, and other highly abstract and formal disciplines. This fact requires explanation, just as the pervasive tendency for context-sensitivity requires explanation. Classical connectionist theories explain context-sensitivity well, but have yet to provide a fully satisfying explanation of the limited systematicity that people demonstrate.

We see the trade-off between context-sensitivity and combinatoriality as emblematic of the systematicity debate more generally. The literature is dominated by attempts to defend positions close to the extremes of the continuum. Our position, by contrast, recognizes that human cognition seems better characterized as a combination of systems operating at different points along this continuum, and for good reason: it works better that way. Thus, FP88 are extreme in advocating that human cognition should be characterized as purely combinatorial. Taken literally, the pure symbol-processing approach fails to take into account the considerable context-sensitivity that people leverage all the time that makes us truly smart, giving us that elusive common sense that such models have failed to capture all these years (and indeed Fodor himself has more recently noted that context-sensitivity of most aspects of human cognition is among the clearest and most notable findings of cognitive psychology; Fodor 2001). In other words, FP88 focus on the sharp, pristine "competence" tip of the cognitive iceberg, ignoring all the rich contextual complexity and knowledge embedded below the surface, which can be revealed in examining people's actual real-world performance. On the other side, basic 1980s-style connectionist networks are strongly weighted toward the context-sensitivity side of the spectrum, and fail to capture the considerable systematicity that people can actually exhibit, for example, when confronting novel situations or systematic domains such as syntactic processing or mathematics. For example, while McClelland and colleagues have shown that such networks can capture many aspects of the regularities and context-sensitivities of English word pronunciation (Plaut, McClelland, Seidenberg, and Patterson 1996), they also had to build into their network a precisely hand-tuned set of input features that balanced context-sensitivity and combinatoriality—in other words, the modelers, not the network, solved important aspects of this trade-off. Furthermore, such models are nowhere near capable of exhibiting the systematicity demonstrated in many other aspects of human cognition (e.g., in making grammaticality judgments on nonsense sentences, as in Chomsky's example).

As an example of the need to integrate multiple aspects of human cognition, Anderson and Lebiere (2003) proposed a test for theories of cognition called the Newell test. It consisted of a dozen criteria spanning the full range from pure combinatoriality (e.g., "behave as an almost arbitrary function of the environment") to high context-sensitivity (e.g., "behave robustly in the face of error, the unexpected, and the unknown"). They evaluated two candidate theories, ACT-R and classical connectionism, and found them both scoring well against some criteria and poorly against others. Strengths and weaknesses of the two theories were mostly complementary, indicating that human cognition falls at some intermediate point on the combinatorial–context-sensitive spectrum.

Just as we find extremism on the context-sensitivity versus combinatoriality dimension to be misguided, we similarly reject extremist arguments narrowly focused on one level of Marr's famous three-level hierarchy of computation, algorithm, and implementation. Advocates of symbol-processing models like to argue that they capture the computational level behavior of the cognitive architecture and that everything else is "mere implementation." From the other side, many neuroscientists and detailed neural modelers ignore the strong constraints that can be obtained by considering the computational and algorithmic competencies that people exhibit, which can guide top-down searches for relevant neural-processing mechanisms. We argue for a balanced view that does not single out any privileged level of analysis. Instead, we strive to integrate multiple constraints across levels to obtain a convergent understanding of human cognitive function (Jilk, Lebiere, O'Reilly, and Anderson 2008).

This convergent, multilevel approach is particularly important given our central claim that different brain areas lie at different points on the context-sensitivity versus combinatoriality continuum (and differ in other important ways as well)—the biological data (at the implementational level) provide strong constraints on the nature of the computations in these different brain areas. In contrast, a purely computational-level account of this nature would likely be underconstrained in selecting the specific properties of a larger set of specialized processing systems. Thus, most purely computational-level accounts, such as that of FP88, tend to argue strongly for a single monolithic computational-level system as capturing the essence of human cognition, whereas we argue above that such an approach necessarily fails to capture the full spectrum of human cognitive functionality.

In the following, we present a comprehensive overview of a variety of ways in which neural networks in different parts of the brain can overcome a strong bias toward context-sensitive, embedded processing that comes from the basic nature of neural processing. From both an evolutionary and online processing perspective (processing recapitulates phylogeny?), we argue that more strongly context-sensitive processing systems tend to be engaged first, and if they fail to provide a match, then progressively more combinatorial systems are engaged, with complex sequential information processing supported by the PFC/BG system providing a "controlled processing" system of last resort.

This is similar to the roles of the symbolic and subsymbolic levels in hybrid architectures such as ACT-R. The subsymbolic level is meant to replicate many of the adaptive characteristics of neural frameworks. For instance, the activation calculus governing declarative memory includes mechanisms supporting associative retrieval such as spreading activation, as well as context-sensitive pattern matching such as partial matching based on semantic similarities corresponding directly to distributed representations in neural networks. A mechanism called blending (Lebiere 1999) aggregates together individual chunks of information in a way similar to how neural networks blend together the individual training instances that they were given during learning. Together with others that similarly control procedural flow, these mechanisms constitute the highly context-sensitive, massively parallel substrate that controls every step of cognition. If they are successful in retrieving the right information and selecting the correct action, processing just flows with little awareness or difficulty (for instance, when the right answer to a problem just pops into one's head). But if they fail, then the mostly symbolic, sequential level takes over, deploying painstaking backup procedures at considerable effort to maintain the proper context information and select the right processing step at each moment.

Our most systematic, combinatorial computational model of this PFC/ BG system demonstrates how an approximate, limited form of indirect variable binding can be supported through observed patterns of interconnectivity among two different PFC/BG areas (Kriete, Noelle, Cohen, and O'Reilly submitted). We have shown that this model can process items in roles they have never been seen in before, a capability that most other neural architectures entirely fail to exhibit. We then argue how this basic indirection dynamic can be extended to handle limited levels of embedding and recursion, capabilities that appear to depend strongly on the most anterior part of the PFC (APFC or frontopolar PFC, BA10; Christoff, Prabhakaran, Dorfman, Zhao, Kroger, Holyoak, and Gabrieli 2001; Bunge, Helskog, and Wendelken 2009; Koechlin, Ody, and Kouneiher 2003; Stocco, Lebiere, O'Reilly, and Anderson 2012). Thus, overall, we identify a full spectrum of processing mechanisms, arrayed along the continuum between context-sensitivity and combinatorial, systematic processing, and associated with different parts of the human brain. We find that attempting to understand the role of these different brain areas through the lens of systematicity results in a rich picture of human cognitive abilities.

2   Biological Neural Network Processing Constraints

Neuroscience has come a very long way in the intervening years since Fodor and Pylyshyn's (1988) seminal article. Yet, fundamentally, it has not moved an inch from the core processing constraints that were understood at that time and captured in the first generation of neural network models. What has changed is the level of detail and certainty with which we can assert that these constraints hold. Fundamentally, information processing in the neocortex takes place through weighted synaptic connections among neurons that adapt through local activity-dependent plasticity mechanisms. Individual pyramidal neurons in the neocortex integrate roughly 10,000 different synaptic inputs, generate discrete action potential spikes, and send these along to a similar number of downstream recipients, to whom these hard-won spikes are just a tiny drop in a large bucket of other incoming spikes. And the process continues, with information flowing bidirectionally and being regulated through local inhibitory interneurons, helping to ensure things do not light up in an epileptic fit.

Somehow, human information processing emerges from this very basic form of neural computation. Through amazing interventions like the ZIP molecule (Shema, Haramati, Ron, Hazvi, Chen, Sacktor, and Dudai 2011), which resets the learned arrangement of excitatory synaptic channels (and many other convergent experiments), we know with high confidence that learning and memory really do boil down to these simple local synaptic changes. Just as the early neural network models captured, processing and memory are truly integrated into the same neural substrate. Indeed, everything is distributed across billions of neurons and trillions of such synapses, all operating in parallel. These basic constraints are not in dispute by any serious neuroscientist working today.

The implications of this computational substrate favor context-sensitive, embedded processing, in contrast to the pure combinatoriality of the symbol processing paradigm. First, neurons do not communicate using symbols, despite the inevitable urge to think of them in this way (O'Reilly 2010). Spikes are completely anonymous, unlabeled, and nearly insignificant at an individual level. Thus, the meaning of any given spike is purely a function of its relationship to other spikes from other neurons, in the moment and over the long course of learning that has established the pattern of synaptic weights. In effect, neurons live in a big social network, learning slowly who they can trust to give them reliable patterns of activation. They are completely blind to the outside world, living inside a dark sea, relying completely on hearsay and murmurs to try to piece together some tiny fragment of "meaning" from a barrage of seemingly random spikes. That this network can do anything at all is miraculous, and the prime mover in this miracle is the learning mechanism, which slowly organizes all these neurons into an effective team of information-processing drones. Armed with many successful learning models and a clear connection between known detailed features of synaptic plasticity mechanisms and effective computational learning algorithms (O'Reilly, Munakata, Frank, Hazy, et al. 2012), we can accept that all this somehow manages to work.

The primary constraints on neural information processing are that each neuron is effectively dedicated to a finite pattern-detection role, where it sifts through the set of spikes it receives, looking for specific patterns and firing off spikes when it finds them. Because neurons do not communicate in symbols, they cannot simply pass a symbol across long distances among many other neurons, telling everyone what they have found. Instead, each step of processing has to rediscover meaning, slavishly, from the ground up, over time, through learning. Thus, information processing in the brain is fully embedded in dedicated systems. There is no such thing as "transparency"; it is the worst kind of cronyism and payola network, an immense bureaucracy. Everything is who you know—who you are connected to. We (at least those of us who love freedom and independence) would absolutely hate living inside our own brains.

This kind of network is fantastic for rapidly processing specific information, dealing with known situations and quickly channeling things down well-greased pathways—in other words, context-sensitive processing. However, as has been demonstrated by many neural network models (Plaut et al. 1996), exceptions, regularities, interactions, main effects—all manner of patterns can be recognized and processed in such a system, with sufficient learning.

From an evolutionary perspective, it is not hard to see why this is a favored form of information processing for simpler animals. We argue that the three more evolutionarily ancient brain structures—the basal ganglia, cerebellum, and hippocampus—all employ a "separator" processing dynamic, which serves to maximize context-sensitivity and minimize possible interference from other possibly unrelated learning experiences. In each of these areas, the primary neurons are very sparsely active, and thus tend to fire only in particular contexts. However, the most evolutionarily recent brain area, the neocortex, has relatively higher levels of neural activity, and serves to integrate across experiences and extract statistical regularities that can be combinatorially recombined to process novel situations. In prior work, the extreme context-sensitivity of the sparse representations in the hippocampus has been contrasted with the overlapping, more systematic combinatorial representations in the neocortex (McClelland, McNaughton, and O'Reilly 1995), yielding the conclusion that both of these systems are necessary and work together to support the full range of human cognition and memory functionality.

Next, we show how, against this overall backdrop of context-sensitive, embedded neural processing, information can be systematically transformed through cascades of pattern detectors, which can extract and emphasize some features, while collapsing across others. This constitutes the first of several steps toward recovering approximate symbol-processing systematicity out of the neural substrate.

3   The Systematicity Toolkit Afforded by Different Neural Systems

Here we enumerate the various cognitive-level capabilities that contribute to human systematicity and discuss how we think they are deployed to enable people to sometimes approximate combinatorial symbol processing. The crux of FP88's argument rests on the observation that people

exhibit a level of systematicity that is compatible with the symbol processing model, and not with traditional connectionist models. Technically,
systematicity is a relation among entities that are internal to the cognitive
system. The *systematicity of representation* is a relation among certain representations, the *systematicity of inference* is a relation among the capacities
to perform certain inferences, and so forth (Aizawa 2003; Johnson 2004).
As these internal relations cannot be observed directly, the systematicity
hypothesis can be tested only indirectly. Researchers have reached a broad
consensus that *generalization*—the ability to apply existing knowledge to
some kind of novel case—is the primary evidence for systematicity. As the
structural overlap between the existing knowledge and the novel case can
vary along a continuum, generalization comes in degrees. By implication,
systematicity also comes in degrees (Hadley 1994; Niklasson and van
Gelder 1994). Thus, it is counterproductive to view the systematicity debate
as a dichotomous choice between two irreconcilable opposites. A more
balanced view seems much more appropriate. In support of this view, the
remainder of this chapter enumerates the sources of graded generalization
that exist in neural networks and articulates how they contribute to the
increasingly systematic patterns of generalization demonstrated by people.

3.1   Categorical Abstraction (Neocortex)

Networks of neurons, typically in the context of a hierarchical organization
of representations, can learn to be sensitive to some distinctions in their
inputs while ignoring others. The result is the formation of a categorical
representation that abstracts over some irrelevant information while focusing on other relevant dimensions of variation. When processing operates
on top of such categorical abstractions, it can be highly systematic, in that
novel inputs with appropriate features that drive these categorical representations can be processed appropriately. Examples include commonsense
categories ("dog," "cat," "chair," etc.), and also less obvious but important
categories such as "up," "down," and so on. We know, for example, that
the ventral visual stream, likely common to most mammals, systematically
throws away spatial information and focuses contrasts on semantically
relevant visual categorization (Ungerleider and Mishkin 1982; Goodale and
Milner 1992). The abstract "symbolic" categories of small integer numbers
have been demonstrated to exist in at least some form in monkeys and
other animals, including in PFC recordings (Nieder, Freedman, and Miller
2002). In all of these cases, abstraction only works if an input has certain
features that drive learned synaptic pathways that lead to the activation
of a given abstract category representation. Thus, this form of generaliza

tion or systematicity implies a certain scope or basin of feature space over which it operates. But this can nevertheless be rather broad; "thing" and "one" are both rather severe abstractions that encompass a very broad scope of inputs. Categorical abstraction thus yields representations that can be used more systematically, since they are effectively stripped of context. Furthermore, it is possible to use top-down attentional processes to emphasize (or even create) certain feature activations in order to influence the categorization process and make it considerably more general— this is an important "hook" that the PFC can access, as we describe later.

One key limitation of abstraction is that, by definition, it requires throwing away specific information. This can then lead to confusion and "binding errors" when multiple entities are being processed, because it can be difficult to keep track of which abstraction goes with which concrete entity. For example, perhaps you know someone who tends to use very general terms like "thing" and "this" and "that" in conversations—it is easy to lose track of what such people are actually saying.

3.2   Relational Abstraction (Neocortex)
This is really a subtype of categorical abstraction, but one which abstracts out the relationship between two or more items. For example, "left of" or "above," or "heavier" are all relational abstractions that can be easily learned in neural networks, through the same process of enhancing some distinctions while collapsing across others (O'Reilly and Busby 2002; Hinton 1986). Interestingly, there is often an ambiguity between which way the relationship works (e.g., for "left of," which object is to the left and which is to the right?), which must be resolved in some way. One simple way is to have a dynamic focus of attention, which defines the "subject" or "agent" of the relationship. In any case, this relational ability is likely present in parietal spatial representations, and rats routinely learn "rules" such as "turn right" in mazes of various complexity. Indeed, it may be that motor actions, which often need to be sensitive to this kind of relational information and relatively insensitive to semantic "what" pathway information, provide an important driver for learning these relational abstractions (Regier and Carlson 2001). Once learned, these relational representations provide crucial generalizable ingredients for structure-sensitive processing: they are abstract representations of structure that can drive further abstract inferences about the structural implications of some situation, irrespective of the specific "contents." For example, a relational representation of physical support, such as "the glass is on the table" can lead to appropriate inferences for what might happen if the glass gets pushed off the table.

These inferences will automatically apply to any entity on a tablelike surface (even though it may seem that babies learn this fact purely through exhaustive, redundant enumeration at their high chairs).

We think these relational and inferential reasoning processes are present in a wide range of animals and can readily be inferred from their behavior. However, there are strong limits to how many steps of such reasoning can be chained together, without the benefits of an advanced PFC. Furthermore, the binding errors and tracking problems associated with abstract representations, described above, apply here as well. Thus, these relational abstractions support making abstract inferences about the implications of structural relationships, all at an abstract level, but it requires quite a bit of extra machinery to keep track of all the specific items entering into these relationships, and requires dereferencing the abstract inference back out to the concrete level again. Again, we see the PFC and its capacity for maintaining and updating temporary variable bindings as key for this latter ability.

3.3 Combinatorial Generalization (Neocortex)

Despite a bias toward context-sensitivity, it is possible for simple neural networks to learn a basic form of combinatoriality—to simply learn to process a composite input pattern in terms of separable, independent parts (Brousse 1993; O'Reilly 2001). These models develop "slot-based" processing pathways that learn to treat each separable element separately and can thus generalize directly to novel combinations of elements. However, they are strongly constrained in that each processing slot must learn independently to process each of the separable elements, because as described above, neurons cannot communicate symbolically, and each set of synapses must learn everything on its own from the ground up. Thus, such systems must have experienced each item in each "slot" at least a few times to be able to process a novel combination of items. Furthermore, these dedicated processing slots become fixed architectural features of the network and cannot be replicated ad hoc—they are only applicable to well-learned forms of combinatorial processing with finite numbers of independent slots. In short, there are strong constraints on this form of combinatorial systematicity, which we can partially overcome through the PFC-based indirection mechanism described below. Nevertheless, even within these constraints, combinatorial generalization captures a core aspect of the kind of systematicity envisioned by FP88, which manifests in many aspects of human behavior. For example, when we prepare our participants for a novel experimental task, we tell them what to do using words that describe core cognitive processing operations with which they are already familiar (e.g., push the right button when you see an A followed by an X, left otherwise); it is only the particular combination of the operations and stimuli that is novel. In many cases, a simple slot-based combinatorial network can capture this level of generalization (Huang, Hazy, Herd, and O'Reilly, in press).

3.4 Dynamic Gating (Basal Ganglia and PFC)

The basal ganglia (BG) are known to act as a dynamic gate on activations in frontal cortex, for example in the case of action selection, where the BG can "open up the gate" for a selected action among several that are being considered (Mink 1996). Anatomically, this gating takes place through a seemingly over-complex chain of inhibitory connections, leading to a modulatory or multiplicative disinhibitory relationship with the frontal cortex. In the PFC, this dynamic operates in the context of updating working memory representations, where the BG gating signal determines when and where a given piece of information is updated and maintained (Frank, Loughry, and O'Reilly 2001; O'Reilly and Frank 2006). In many ways, this is equivalent to a logic gate in a computer circuit, where a control channel gates the flow of information through another channel (O'Reilly 2006). It enables an important step of *content-independent* processing, as in structure-sensitive processing. Specifically, the BG gate can decide where to route a given element of content information, based strictly on independent control signals, and not on the nature of that content information. In the example shown in figure 8.2, "syntactic" form information

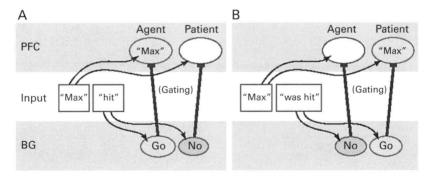

Figure 8.2
Illustration of how the basal ganglia gating dynamic with PFC can separately control the functional role assignment of other information in a content-independent fashion.

(passive vs. active verb, cued by presence or absence of keyword "was") can determine whether the preceding word is routed into an "agent" slot versus a "patient" slot in working memory. As this example makes clear, dynamic gating also helps to resolve the problem of dedicated slots for combinatorial generalization: by being able to dynamically route information into different functional slots, these slots can become more generalized, reducing the slot-explosion problem. However, it is essential to appreciate that all of this machinery must be trained up over time: the BG gating system learns through trial-and-error experience the gating strategies that lead to reward (O'Reilly and Frank 2006; Hazy, Frank, and O'Reilly 2006, 2007), and the PFC "slots" (anatomically referred to as "stripes") must learn to encode any information that they might maintain, while any other brain area that uses this maintained information must also learn to decode it (such are the basic constraints of the neural substrate, as articulated above). Thus, whatever systematicity this gating system affords must develop slowly over extensive learning experience, consistent with what we know about human symbol-processing abilities.

3.5   Active Memory Juggling and Top-Down Control (PFC/BG)

The ability to "juggle" activation states in the PFC, through the dynamic BG-mediated gating mechanism, can lead to a form of computation that escapes some of the limitations of synaptic weights (while still operating within the general confines of learning). Specifically, active maintenance plays a role like random access memory (RAM) or registers in a traditional computer architecture: whatever is being actively maintained can be rapidly updated (in a matter of a few hundreds of milliseconds), instead of requiring slow repeated learning over time. Thus, I can tell you to "pay attention to the ink color" in the ubiquitous Stroop task, and you can dynamically gate in an active representation in PFC that will drive activation of color-processing areas in posterior cortex (Herd, Banich, and O'Reilly 2006; Cohen, Dunbar, and McClelland 1990). Then, on the very next trial, you can immediately alter your behavior by gating in a "word reading" PFC representation and paying attention to the letters in the word instead of the ink color. As noted above, these PFC representations themselves have to be learned slowly over time in order to have the appropriate impact on processing elsewhere in the brain, but dynamically they can be rapidly updated and deactivated, leading to a flexibility that is absent without this PFC/BG mechanism. In principle, this kind of activation-based juggling can implement an abstract "state machine" where the active state at one point in time conditions what gets updated at the next, and relatively arbitrary sequences of such state transitions can be flexibly triggered. In the ACT-R architecture, production firing serves to update the active state of buffers, which we associate with the PFC active maintenance state (Jilk et al. 2008), demonstrating the power of this activation-based state machine for arbitrary symbolic-like processing. However, relative to ACT-R, the biology of the BG and PFC place stronger constraints on the "matching conditions" and "right-hand side" buffer update operations that result from production firing, as we discuss in greater detail below. Exactly how strong these constraints are and their implications for overall processing abilities in practice largely remains to be seen, pending development of increasingly sophisticated cognitive processing models based on this PFC/BG architecture and relevant learning mechanisms.

We have started making some progress in bridging that gap by implementing a detailed neural model of how the basal ganglia can implement the ACT-R procedural module in routing information between cortical areas associated with other ACT-R modules (Stocco, Lebiere, and Anderson 2010). Because of prior factoring of neural constraints in the evolution of the ACT-R architecture, production conditions and actions had already become naturally parallelizable, leading to a straightforward neural implementation. However, the detailed neural model reflecting the specific topology and capacity of the basal ganglia has suggested new restrictions, such as on the amount of information transfer that can occur within a single production. At the symbolic level, this is accomplished by a process of variable binding that transfers information from the condition side of the production to its action side. In terms of the neural model, that variable binding is simply realized by gating neural channels between cortical areas.

3.6 Episodic Variable Binding (Hippocampus)

The hippocampus is well known to be specialized for rapidly binding arbitrary information together in the form of a *conjunctive representation*, which can later be recalled from a partial cue (Marr 1971; McClelland et al. 1995; O'Reilly 1995; O'Reilly and Rudy 2001). This is very handy for remembering where specific objects are located (e.g., where you parked your car), the names of new people you meet, and a whole host of other random associations that need to be rapidly learned. For symbol processing, this rapid arbitrary binding and recall ability can obviously come in handy. If I tell you "John loves Mary," you can rapidly bind the relational and abstract categorical representations that are activated, and then retrieve them later through various cues ("who loves Mary?" "John loves who?"). If I go on and tell you some other interesting information about Mary ("Mary was out last night with Richard") then you can potentially start encoding and recalling these different pieces of information and drawing some inferences, while not losing track of the original facts of the situation. However, hippocampal episodic memory also has limitations—it operates one memory at a time for both encoding and retrieval (which is a consequence of its voracious binding of all things at once), and it can take some work to avoid interference during encoding, and generate sufficiently distinct retrieval cues to get the information back out. But there is considerable evidence that people make extensive use of the hippocampus in complex symbolic reasoning tasks—undoubtedly an important learned skill that people develop is this ability to strategically control the use of episodic memory. Specific areas of PFC are implicated as these episodic control structures, including medial areas of the most anterior portion of PFC (Burgess, Dumontheil, and Gilbert 2007).

3.7 Indirection-Based Variable Binding (PFC/BG)

The final, somewhat more speculative specialization we describe has the greatest power for advancing the kind of systematicity envisioned by FP88. By extending the basic BG dynamic gating of PFC in a set of two interconnected PFC areas, it is possible to achieve a form of *indirection* or representation by (neural) address, instead of representing content directly (Kriete et al. submitted) (figure 8.3). Specifically, one set of PFC stripes (region A) can encode a pattern of activity that drives gating in the BG for a different set of PFC stripes (region B); region A can then act as a "puppet master," pulling the strings for when the information contained in region B is accessed and updated. This then allows region A to encode the structural form of some complex representation (e.g., Noun, Verb, and Object roles of a sentence), completely independent of the actual content information that fills these structural roles (which is encoded in the stripes in region B). Critically, Kriete et al. showed that such a system can generalize in a much more systematic fashion than even networks using PFC/BG gating dynamics (which in turn generalized better than those without gating) (figure 8.4). Specifically, it was able to process a novel role filler item that had never been processed in that role before, because it had previously learned to encode the *BG address* where that content was stored. Thus, assuming that the PFC content stripes can encode a reasonable variety of information, learning only the addresses and not the contents can lead to a significant increase in the scope of generalization. Nevertheless, as in all the examples above, all of these representations must be learned slowly in

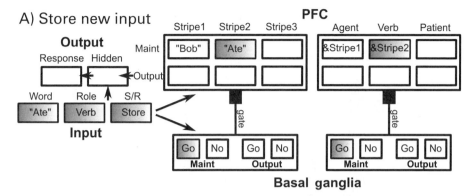

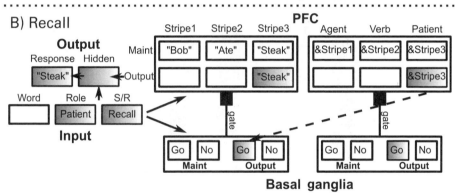

Figure 8.3

The Kriete et al. (submitted) indirection model, performing the simple sentence encoding task, demonstrating indirection in the PFC/BG working memory system. Three-word sentences are encoded one word at time, with each word associated with a role ("Agent," "Verb," or "Patient"). After encoding the sentence, the network is probed for each word using the associated roles (e.g., "What was the 'Agent' of the sentence?"). The shaded layers indicate currently active inputs. (A) One step of the encoding process for the sentence "Bob ate steak" in the PFC/BG working memory (PBWM) indirection model. The word "Ate" is presented to the network along with its current role ("Verb") and the instruction "Store" to encode this information for later retrieval. In this example, the word "Ate" is stored in Stripe2 of PFC filler stripes (left side of figure). The identity/location of Stripe2 is subsequently stored in the Verb stripe of PFC role stripes (right side of figure). The same set of events occurs for each of the other two words in the sentence (filling the agent and patient roles). (B) One step of the recall process. A role ("Patient" in the example) and the instruction "Recall" are presented as input. This drives output gating of the address information stored by that role stripe (highlighted by the dashed arrow), which in turn causes the BG units corresponding to that address to drive output gating of the corresponding filler stripe, thus outputting the contents of that stripe ("Steak").

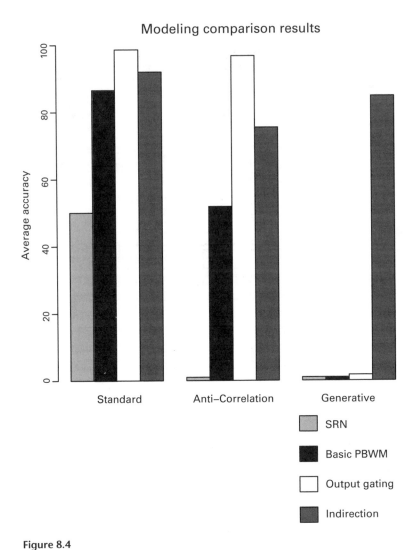

Figure 8.4
Accuracy performance of the indirection-based network juxtaposed against comparison networks, for three increasingly challenging generalization tasks. The results are grouped by task: standard, anti-correlation, and generative. Bars correspond to the four networks (from left to right): SRN, basic PBWM network with maintenance only, PBWM output gating network, and PBWMindirection network. The indirection network is the only one capable of achieving high levels of performance across all the tasks.

the first place. Our model demonstrates that, with appropriate connectivity and the same learning mechanisms used for prior PFC/BG models (O'Reilly and Frank 2006), this learning can happen naturally.

4 Putting It All Together

Having enumerated a range of different mechanisms, each of which promotes systematicity in a specific way, we now attempt to spell out some particular examples for how a complex sequence of cognitive operations, which achieves a limited approximation of classical symbol processing, could unfold through the interactions of these systems. Note that these examples are based on informed speculation, not hard data, and we do not currently have well-validated biologically based models that capture the behavior we describe. Nevertheless, we consider it plausible that this is how it is actually solved in the human brain, based on a variety of sources too numerous to explicate here. Moreover, this speculation is informed by models of similar tasks (e.g., (Lebiere 1999) in higher-level frameworks for which a correspondence to the neural architecture exists, such as ACT-R (see the section on the SAL framework below). Recently, this methodology of developing higher-level symbolic models to guide the structure and content of neural models has been applied to the complex task of sensemaking (Lebiere, Pirolli, Thomson, Paik, Rutledge-Taylor, Staszewski, and Anderson submitted).

First, consider the case of multidigit mental arithmetic, for example, multiplying 42×17. This requires a systematic sequence of cognitive operations and keeping track of partial products, which most adults can apply to arbitrary numbers (i.e., in a fully systematic, content-independent manner). Before we consider how this happens in the general case, it is important to appreciate that if the problem was 10×42, for example, one would use a much faster context sensitive special-case process to arrive at the answer—people automatically and effortlessly recognize and apply these special case solutions, demonstrating the primacy of context-sensitivity as we argued above. Furthermore, in the well-studied domain of chess experts, much of the expertise is associated with this special-case pattern recognition ability and not with optimization of a fully general-purpose algorithm, whereas symbolic computer models of chess have the exact opposite profile, optimizing a general-purpose search algorithm instead of memorizing a bunch of special cases (Chase and Simon 1973).

This fundamental distinction between cognitive and algorithmic solutions arises from the hardware available to those classes of solutions.

Traditional CPUs are able to flawlessly execute billions of operations per second, but the access to the largest memory store is considerably slower and sequential. Thus, algorithmic solutions evolved to emphasize computation over memory. Neural hardware, on the other hand, is the mirror image: an excruciatingly slow and error-prone central loop (on the order of 20Hz, about eight times slower than off-the-shelf CPUs), but an extremely powerful, context-sensitive, massively parallel access to long-term memory. Cognitive solutions, therefore, evolved to emphasize memory over computation and, when computation is necessary, attempt to cache its results as efficiently and automatically as possible.

To begin on the general-case multidigit multiplication problem, people will almost certainly start by encoding the problem into hippocampal episodic memory, so they can retrieve it when interference overtakes the system and they lose track of the original problem. The next step is to recall an overall strategy for such problems, and the BG gates an abstract encoding of this strategy into an anterior portion of dorsal-lateral PFC (DLPFC). This "strategy plan" representation then activates the first step of the strategy, in a more posterior portion of DLPFC, which then drives top-down perceptual biasing in the parietal cortex to focus attention on the ones decimal place numbers (i.e., the right-most digits). Considerable categorical abstraction is required to even extract a numerical value from a particular pattern of light and dark on the retina, and abstract relational representations are required to focus on the appropriate portions of the digits, including things like top, bottom, right, and so on.

In any case, you end up activating the sub-problem of multiplying 7×2, which should activate the answer of 14 through well-learned parietal or perhaps temporal verbally mediated representations, perhaps even with support from the hippocampus depending on your educational status and level of recent practice. Having produced this answer, you cache away this partial product either by gating it into another available stripe in PFC (perhaps in verbal and/or numeric coding areas), or by encoding it episodically in the hippocampus (or likely both, as the hippocampus is automatically encoding everything). Next, guided by the strategic plan, you move on to the tens position in the first number, multiplying 7×4, encoding the 28, and so on. After each step, the partial products must be tagged and encoded in such a way that they can later be accessed for the final addition step, which in itself may require multiple substeps, with carry-overs and so on. An indirection-based variable-binding solution may be employed here, where each partial product is encoded in a different stripe, and "tagged" with the functional role of an ordinal list of items to add. Of course, items may be added incrementally in an opportunistic, context-sensitive manner, and various permutations on an overall strategy may be employed. But clearly, considerable "activation-based juggling" of information is required, along with likely several strategic hippocampal episodic encoding and retrieval steps to maintain the partial products for subsequent processing.

At some level of description, this could be considered to be a kind of classical symbol-processing system, with the hippocampus playing the role of a "tapelike" memory system in the classical Turing model and DLPFC coordinating the execution of a mental program that sequences cognitive operations over time. We do not disagree that, at that level of description, the brain is approximating a symbol-processing system. However, it is essential to appreciate that each element in this processing system has strong neurally based constraints, such that the capacity to perform this task degrades significantly with increasing number size, in a way that is completely unlike a true symbol-processing system, which can churn along on its algorithm indefinitely, putting items on the stack and popping them off at will. In contrast, the human equivalent of the "stack" is severely limited in capacity, subject to all manner of interference, and likely distributed across multiple actual brain systems. Furthermore, as noted above, the human brain will very quickly recognize shortcuts and special cases (e.g., starting with 10×42 as an easier problem and adjusting from there), in ways that no Turing machine would be able to. Thus, the bias toward context-sensitive processing results in very rapid and efficient processing of familiar cases—a perfectly sensible strategy for a world where environments and situations are likely to contain many of the same elements and patterns over time.

Indeed, a symbolic architecture such as ACT-R operates exactly in the way described above, with the hippocampus corresponding to declarative memory and the DLPFC corresponding to the retrieval buffer through which cognitive operations would flow for execution by the procedural system. Limitations arise through the subsymbolic level controlling the operations of the symbolic level. Chunks may exist perfectly crisp and precise at the symbolic level, but their activation ebbs and flows with the pattern of occurrence in the environment, and their retrieval is approximate, stochastic, and error-prone. Similarly, productions may implement a clocklike finite state machine, but the chaining of their individual steps into a complex processing stream is dependent on the stochastic, adaptive calculus of utilities that makes flawless execution of long procedures increasingly difficult and unlikely. Other system bottlenecks at both the architectural and subsymbolic level include limited working memory, attentional bottlenecks, and limits on execution speed for every module. Thus, hybrid symbolic-subsymbolic architectures such as ACT-R provide us with an abstraction of the capacities and limitations of neural architectures that can guide their development.

5 Discussion

We conclude with a brief discussion of some additional points of relevance to our main arguments, including the importance of data on the timecourse of learning and development on understanding the nature of human systematicity, the importance of multilevel modeling and the specific case of relating the ACT-R and Leabra modeling frameworks, and how our models compare with other related models in the literature.

5.1 The Importance of Learning and Development of Systematicity

We put a lot of emphasis on the role of "learning from the ground up" as a strong constraint on the plausibility of a given cognitive framework. Empirically, one of the strongest arguments in favor of our overall approach comes from the developmental timecourse of symbolic processing abilities in people—only after years and years of learning do we develop symbolic processing abilities, and the more advanced examples of these abilities depend critically on explicit instruction (e.g., math, abstract logic). Only in the domain of language, which nevertheless certainly is dependent on a long timecourse of exposure to and learning from a rich social world of language producers, does systematicity happen in a relatively natural, automatic fashion. And as we discuss in greater detail in a moment, language development provides many possible windows into how systematicity develops over time; it is certainly not a hallmark of language behavior right from the start.

In short, we argue that learning processes, operating over years and often with the benefit of explicit instruction, enable the development of neural dynamics involving widely distributed interacting brain systems, which support these approximate symbol-processing abilities. It is not just a matter of "resource limitations" slapped on top of a core cognitive architecture that does fully general symbol processing, as argued by FP88; rather, the very abilities themselves emerge slowly and in a very graded way, with limitations at every turn. We think this perspective on the nature of human symbolic processing argues strongly against systems that build in core symbol-processing abilities as an intrinsic part of the architecture. But unlike some of our colleagues (McClelland et al.), we nevertheless agree that these approximate symbol-processing abilities *do* develop, and that they represent an important feature that any neural network framework must account for.

One of the most famous debates between connectionists and symbol-processing advocates took place in the context of the developmental data on the U-shaped curve of overregularization of past tense morphology in English. After correctly producing irregular verbs such as "went," kids start saying things like "goed," seemingly reflecting the discovery and application of the regular "rule" ("add -ed"). First, this doesn't happen until age three or four (after considerable exposure and productive success with the language), and it is a very stochastic, variable process across kids and across time. Rates of overregularization rarely exceed a few percent. Thus, it certainly is not the kind of data that one would uphold as a clear signature of systematicity. Instead, it seems to reflect some kind of wavering balance between different forces at work in the ever-adapting brain, which we argue is a clear reflection of the different balances between context-sensitivity and combinatoriality in different brain areas. Interestingly, single-process generic neural network models do not conclusively demonstrate this U-shaped curve dynamic, without various forms of potentially questionable manipulations. Some of these manipulations were strong fodder for early critiques (Rumelhart and McClelland 1986; Pinker and Prince 1988), but even later models failed to produce this curve in a purely automatic fashion without strong external manipulations. For example, the Plunkett and Marchman (1993) model is widely regarded as a fully satisfactory account, but it depends critically on a manipulation of the training environment that is similar to the one heavily criticized by Rumelhart and McClelland (1986).

5.2 Convergent Multilevel Modeling: The SAL Framework

A valuable perspective on the nature of symbolic processing can be obtained by comparing different levels of description of the cognitive architecture. The ongoing SAL (Synthesis of ACT-R and Leabra) project provides important insight here (Jilk et al. 2008). ACT-R is a higher-level cognitive architecture that straddles the symbolic-subsymbolic divide (Anderson and Lebiere 1998; Anderson, Bothell, Byrne, Douglass, Lebiere, and Qin 2004), while Leabra is a fully neural architecture that embodies the various mechanisms described above (O'Reilly et al. 2012). Remarkably, we have found that, through different sources of constraint and inspiration, these two architectures have converged on largely the same overall picture of the cognitive architecture (figure 8.5). Specifically, both rely on the PFC/ BG mechanism as the fundamental engine of cognitive sequencing from one step to the next, and this system interacts extensively with semantic and episodic declarative memory to inform and constrain the next actions selected. In ACT-R, the PFC/BG system is modeled as a production system, where production-matching criteria interrogate the contents of active memory buffers (which we associate with the PFC in Leabra). When a production fires, it results in the updating of these buffers, just as the BG updates PFC working memory in Leabra. Productions are learned through a reinforcement-based learning mechanism, which is similar across both systems.

A detailed neural model of how the topology and physiology of the basal ganglia can enable computations analog to the ACT-R production system has been developed (Stocco et al. 2010). As previously discussed, that model explains how the abstract symbolic concept of variable binding has a straightforward correspondence in terms of gating information flows between neural areas. Another major outstanding issue regarding symbolic representations is the ability to arbitrarily compose any values or structures, which in turn translates into the capacity to implement distal access to symbols (Newell 1990). The original implementation of ACT-R into neural networks (Lebiere and Anderson 1993) assumed a system of movable codes for complex chunks of information that could be decoded and their constituent parts extracted by returning to the original memory area where the composition was performed. Recent architectural developments (Anderson 2007) include the separation of the goal-related information into a goal buffer containing goal state information and an imaginal buffer containing the actual problem content. The former is associated with the working memory functionality of the prefrontal cortex whereas the latter is associated with the spatial representation and manipulation functions of the parietal cortex. This suggests that rather than using movable codes, distal access is implemented using a system of control connections that can remotely activate constructs in their original context.

5.3 Other Neural Network Approaches to Systematicity

A number of different approaches to introducing systematicity into neural network models have emerged over the years (Bodén and Niklasson 2000; Chalmers 1990; Christiansen and Chater 1994; Hadley 1997; Hadley and

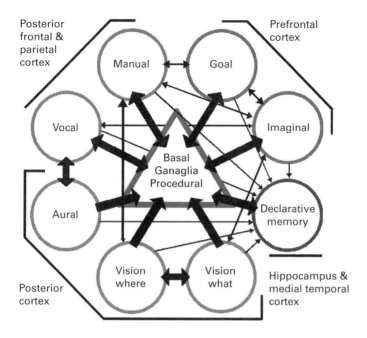

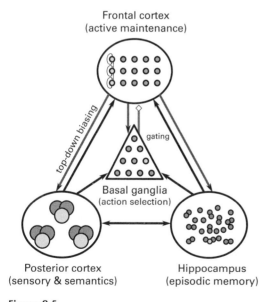

Figure 8.5
Convergent architecture between ACT-R (a) and Leabra (b) developed independently based on very different considerations.

Hayward 1997; Niklasson and van Gelder 1994; Smolensky 1988, 1990b; Smolensky and Legendre 2006). Broadly speaking, our approach is distinct from these others in focusing on a systems neuroscience perspective to the problem, both in terms of differential specializations of different brain areas and in terms of how overall symbol processing functionality can emerge through the complex interactions, over distinct time steps, between these specialized areas, as sketched above in our multidigit arithmetic example.

In terms of specific points of comparison, one of the most important mechanisms for achieving any kind of symbol processing is arbitrary variable binding, which we have argued above depends on episodic memory in the hippocampus and on the indirection-based dynamics in the PFC/BG system (Kriete et al. submitted). A number of models adopt a tensor product approach to variable binding (Plate 2008; Smolensky 1990a; Pollack 1990), which is similar in some respects to the kind of conjunctive binding achieved by the hippocampal episodic memory system. Another solution is to assume a synchrony-based binding mechanism, but we are skeptical that such a mechanism would be able to interleave multiple bindings across a phase cycle (O'Reilly and Busby 2002; O'Reilly, Busby, and Soto 2003). Furthermore, if such a mechanism were in place, it would predict a much more pervasive ability to perform arbitrary variable binding than people actually exhibit. In this respect, we think that the evidence for a long period of learning and development being required before people can even begin to demonstrate symbol-processing-like abilities is consistent with our focus on variable binding being a learned skill that involves the coordinated contributions of multiple brain areas.

As was evident in our multidigit arithmetic example, just forming a binding is only part of the problem: you also need to be able to manipulate the bound information in systematic ways. Here, we are less clear about the strong claims made by these other models: it seems that they mostly engineer various mechanisms to achieve what look to us like implementations of symbol-processing mechanisms, without a strong consideration for how such mechanisms would operate plausibly in the brain. What is conspicuously lacking is an account of how all of the complex neural processing required for these systems can be learned through experience-driven plasticity mechanisms. Our own work on this challenging problem is still in its infancy, so we certainly cannot claim to have shown how it can be learned from the ground up. Nevertheless, we remain optimistic that a learning-based approach fits best with the available human data.

6 Conclusion

After twenty-five years of earnest debate, considerable progress has been made in advancing our understanding about the nature of human systematicity. We hope that our biologically based systems neuroscience approach to these issues may provide some further insight into the nature of the human cognitive architecture and how a limited form of symbol processing can emerge through interactions between different specialized brain areas. We are excited about continuing to advance this program of research, to the point of one day showing convincingly how neural tissue can achieve such lofty cognitive functions as abstract mathematics and abstract logical reasoning.

Acknowledgments

This work was supported by NIMH grant MH079485, ONR grant N00014-13-1-0067, and the Intelligence Advanced Research Projects Activity (IARPA) via Department of the Interior (DOI) contract number D10PC20021. The US Government is authorized to reproduce and distribute reprints for governmental purposes notwithstanding any copyright annotation thereon. The views and conclusions contained hereon are those of the authors and should not be interpreted as necessarily representing the official policies or endorsements, either expressed or implied, of IARPA, DOI, or the US Government.

References

Aizawa, K. 1997. Explaining systematicity. *Mind and Language* 12:115–136.

Aizawa, K. 2003. *The Systematicity Arguments*. New York: Kluwer Academic.

Anderson, J. R. 2007. *How Can the Human Mind Occur in the Physical Universe?* New York: Oxford University Press.

Anderson, J. R., D. Bothell, M. D. Byrne, S. Douglass, C. Lebiere, and Y. Qin. 2004. An integrated theory of the mind. *Psychological Review* 111 (4):1036–1060.

Anderson, J. R., and C. Lebiere. 1998. *The Atomic Components of Thought*. Mahwah, NJ: Erlbaum.

Anderson, J. R., and C. Lebiere. 2003. Optimism for the future of unified theories. *Behavioral and Brain Sciences* 26 (5):628–640.

Bodén, M., and L. F. Niklasson. 2000. Semantic systematicity and context in connectionist networks. *Connection Science* 12:111–142.

Brousse, O. 1993. *Generativity and systematicity in neural network combinatorial learning* (Technical Report CU-CS-676-93). Boulder, CO: University of Colorado at Boulder, Department of Computer Science.

Bunge, S. A., E. H. Helskog, and C. Wendelken. 2009. Left, but not right, rostrolateral prefrontal cortex meets a stringent test of the relational integration hypothesis. *NeuroImage* 46 (1):338–342.

Burgess, P. W., I. Dumontheil, and S. J. Gilbert. 2007. The gateway hypothesis of rostral prefrontal cortex (area 10) function. *Trends in Cognitive Sciences* 11 (7):290–298.

Chalmers, D. J. 1990. Syntactic transformations on distributed representations. *Connection Science* 2:53–62.

Chase, W. G., and H. A. Simon. 1973. The mind's eye in chess. In *Visual Information Processing*, ed. W. G. Chase. New York: Academic.

Christiansen, M. H., and N. Chater. 1994. Generalization and connectionist language learning. *Mind and Language* 9 (3):273–287.

Christoff, K., V. Prabhakaran, J. Dorfman, Z. Zhao, J. K. Kroger, K. J. Holyoak, and J. D. Gabrieli. 2001. Rostrolateral prefrontal cortex involvement in relational integration during reasoning. *NeuroImage* 14 (5):1136–1149.

Cohen, J. D., K. Dunbar, and J. L. McClelland. 1990. On the control of automatic processes: A parallel distributed processing model of the Stroop effect. *Psychological Review* 97 (3):332–361.

Cummins, R. 1996. Systematicity. *Journal of Philosophy* 93 (12):591–614.

Fodor, J. 2001. *The Mind Doesn't Work That Way: The Scope and Limits of Computational Psychology*. Cambridge, MA: MIT Press.

Fodor, J. A., and B. P. McLaughlin. 1990. Connectionism and the problem of systematicity: Why Smolensky's solution doesn't work. *Cognition* 35 (2):183–204.

Fodor, J. A., and Z. W. Pylyshyn. 1988. Connectionism and cognitive architecture: A critical analysis. *Cognition* 28:3–71.

Frank, M. J., B. Loughry, and R. C. O'Reilly. 2001. Interactions between the frontal cortex and basal ganglia in working memory: A computational model. *Cognitive, Affective & Behavioral Neuroscience* 1:137–160.

Godden, D. R., and A. D. Baddeley. 1975. Context-dependent memory in two natural environments: On land and under water. *British Journal of Psychology* 66: 325–331.

Goodale, M. A. and D. Milner. 2010. Separate visual pathways for perception and action. *Trends in Neurosciences* 15 (1):20-25.

Griffiths, T. L., N. Chater, C. Kemp, A. Perfors, and J. B. Tenenbaum. 2010. Probabilistic models of cognition: Exploring representations and inductive biases. *Trends in Cognitive Sciences* 14 (8):357–384.

Griggs, R. A., and J. R. Cox. 1982. The elusive thematic materials effect in the Wason selection task. *British Journal of Psychology* 73:407–420.

Hadley, R. F. 1994. Systematicity in connectionist language learning. *Mind and Language* 9 (3):247–272.

Hadley, R. F. 1997. Cognition, systematicity, and nomic necessity. *Mind and Language* 12:137–153.

Hadley, R. F., and M. Hayward. 1997. Strong semantic systematicity from Hebbian connectionist learning. *Minds and Machines* 7:1–37.

Harnad, S. 1990. The symbol grounding problem. *Physica D: Nonlinear Phenomena* 42:335–346.

Hazy, T. E., M. J. Frank, and R. C. O'Reilly. 2006. Banishing the homunculus: Making working memory work. *Neuroscience* 139:105–118.

Hazy, T. E., M. J. Frank, and R. C. O'Reilly. 2007. Towards an executive without a homunculus: Computational models of the prefrontal cortex/basal ganglia system. *Philosophical Transactions of the Royal Society of London, Series B: Biological Sciences* 362 (1):105–118.

Herd, S. A., M. T. Banich, and R. C. O'Reilly. 2006. Neural mechanisms of cognitive control: An integrative model of Stroop task performance and fMRI data. *Journal of Cognitive Neuroscience* 18:22–32.

Hinton, G. E. 1986. Learning distributed representations of concepts. In *Proceedings of the 8th Conference of the Cognitive Science Society*, 1–12. Hillsdale, NJ: Erlbaum.

Horgan, T., and J. Tienson. 1996. *Connectionism and the Philosophy of Psychology*. Cambridge, MA: MIT Press.

Huang, T.-R., T. E. Hazy, S. A. Herd, and R. C. O'Reilly. In press. Assembling old tricks for new tasks: A neural model of instructional learning and control. *Journal of Cognitive Neuroscience*.

Jilk, D., C. Lebiere, R. C. O'Reilly, and J. R. Anderson. 2008. SAL: An explicitly pluralistic cognitive architecture. *Journal of Experimental and Theoretical Artificial Intelligence* 20 (3):197–218.

Johnson, K. 2004. On the systematicity of language and thought. *Journal of Philosophy* 101:111–139.

Kemp, C., and J. B. Tenenbaum. 2008. Structured models of semantic cognition. *Behavioral and Brain Sciences* 31:717–718.

Kern, L. H., H. L. Mirels, and V. G. Hinshaw. 1983. Scientists' understanding of propositional logic: An experimental investigation. *Social Studies of Science* 13: 136–146.

Koechlin, E., C. Ody, and F. Kouneiher. 2003. Neuroscience: The architecture of cognitive control in the human prefrontal cortex. *Science* 302:1181–1184.

Kriete, T., D. C. Noelle, J. D. Cohen, and R. C. O'Reilly. Submitted. Indirection and symbol-like processing in the prefrontal cortex and basal ganglia. *Proceedings of the National Academy of Sciences.*

Lebiere, C. 1999. The dynamics of cognition: An ACT-R model of cognitive arithmetic. *Kognitionswissenschaft* 8:5–19.

Lebiere, C., and J. R. Anderson. 1993. A connectionist implementation of the ACT-R production system. In *Proceedings of the 15th Annual Conference of the Cognitive Science Society.* Hillsdale, NJ: Erlbaum.

Lebiere, C., P. Pirolli, R. Thomson, J. Paik, M. Rutledge-Taylor, J. Staszewski, and J. Anderson. Submitted. A functional model of sensemaking in a neurocognitive architecture. *Computational Intelligence and Neuroscience.*

Macdonald, C., and G. Macdonald, eds. 1995. *Connectionism: Debates on Psychological Explanation.* Malden, MA: Blackwell.

Marr, D. 1971. Simple memory: A theory for archicortex. *Philosophical Transactions of the Royal Society of London, Series B: Biological Sciences* 262 (841):23–81.

Matthews, R. 1997. Can connectionists explain systematicity? *Mind and Language* 12:154–177.

McClelland, J. L., M. M. Botvinick, D. C. Noelle, D. C. Plaut, T. T. Rogers, M. S. Seidenberg, and L. B. Smith. 2010. Letting structure emerge: Connectionist and dynamical systems approaches to cognition. *Trends in Cognitive Sciences* 14 (8): 348–356.

McClelland, J. L., B. L. McNaughton, and R. C. O'Reilly. 1995. Why there are complementary learning systems in the hippocampus and neocortex: Insights from the successes and failures of connectionist models of learning and memory. *Psychological Review* 102 (3):419–457.

McLaughlin, B. P. 1993. The classicism/connectionism battle to win souls. *Philosophical Studies* 70:45–72.

Mink, J. W. 1996. The basal ganglia: Focused selection and inhibition of competing motor programs. *Progress in Neurobiology* 50:381–425.

Newell, A. 1990. *Unified Theories of Cognition*. Cambridge, MA: Harvard University Press.

Nieder, A., D. J. Freedman, and E. K. Miller. 2002. Representation of the quantity of visual items in the primate prefrontal cortex. *Science* 297:1708–1711.

Niklasson, L. F., and T. van Gelder. 1994. On being systematically connectionist. *Mind and Language* 9 (3):288–302.

O'Reilly, R. 2006. Biologically based computational models of high-level cognition. *Science* 314:91–94.

O'Reilly, R. C. 1995. Biological mechanisms of controlled processing: Interactions between the prefrontal cortex and the hippocampus. Carnegie-Mellon.

O'Reilly, R. C. 2001. Generalization in interactive networks: The benefits of inhibitory competition and Hebbian learning. *Neural Computation* 13: 1199–1242.

O'Reilly, R. C. 2010. The *what* and *how* of prefrontal cortical organization. *Trends in Neurosciences* 33 (8):355–361.

O'Reilly, R. C., and R. S. Busby. 2002. Generalizable relational binding from coarse-coded distributed representations. In *Advances in Neural Information Processing Systems (NIPS) 14*, ed. T. G. Dietterich, S. Becker, and Z. Ghahramani. Cambridge, MA: MIT Press.

O'Reilly, R. C., R. S. Busby, and R. Soto. 2003. Three forms of binding and their neural substrates: Alternatives to temporal synchrony. In *The Unity of Consciousness: Binding, Integration, and Dissociation*, ed. A. Cleeremans, 168–192. Oxford: Oxford University Press.

O'Reilly, R. C., and M. J. Frank. 2006. Making working memory work: A computational model of learning in the prefrontal cortex and basal ganglia. *Neural Computation* 18:283–328.

O'Reilly, R. C., Munakata, Y., Frank, M. J., Hazy, T. E., et al. 2012. *Computational Cognitive Neuroscience*. Wiki Book, http://ccnbook.colorado.edu.

O'Reilly, R. C., and J. W. Rudy. 2001. Conjunctive representations in learning and memory: Principles of cortical and hippocampal function. *Psychological Review* 108:311–345.

Pinker, S., and A. Prince. 1988. On language and connectionism: Analysis of a parallel distributed processing model of language acquisition. *Cognition* 28: 73–193.

Plate, T. A. 2008. Holographic reduced representations. *IEEE Transactions on Neural Networks* 6: 623–641.

Plaut, D. C., J. L. McClelland, M. S. Seidenberg, and K. Patterson. 1996. Understanding normal and impaired word reading: Computational principles in quasi-regular domains. *Psychological Review* 103:56–115.

Plunkett, K., and V. Marchman. 1993. From rote learning to system building: Acquiring verb morphology in children and connectionist nets. *Cognition* 48 (1):21–69.

Pollack, J. B. 1990. Recursive distributed representations. *Artificial Intelligence* 46: 77–105.

Regier, T., and L. A. Carlson. 2001. Grounding spatial language in perception: An empirical and computational investigation. *Journal of Experimental Psychology. General* 130:273–298.

Rumelhart, D. E., and J. L. McClelland. 1986. On learning the past tenses of English verbs. In *Parallel Distributed Processing*, vol. 2: *Explorations in the Microstructure of Cognition: Psychological and biological models*, J. L. McClelland, D. E. Rumelhart, and the PDP Research Group, 216–271. Cambridge, MA: MIT Press.

Shema, R., S. Haramati, S. Ron, S. Hazvi, A. Chen, T. C. Sacktor, and Y. Dudai. 2011. Enhancement of consolidated long-term memory by overexpression of protein kinase Mzeta in the neocortex. *Science* 331:1207–1210.

Smolensky, P. 1988. On the proper treatment of connectionism. *Behavioral and Brain Sciences* 11:1–74.

Smolensky, P. 1990a. Tensor product variable binding and the representation of symbolic structures in connectionist networks. *Artificial Intelligence* 46:159–216.

Smolensky, P. 1990b. Tensor product variable binding and the representation of symbolic structures in connectionist systems. *Artificial Intelligence* 46 (1–2): 159–216.

Smolensky, P., and G. Legendre. 2006. *The Harmonic Mind: From Neural Computation to Optimality-Theoretic Grammar*. Cambridge, MA: MIT Press.

Stocco, A., C. Lebiere, and J. Anderson. 2010. Conditional routing of information to the cortex: A model of the basal ganglia's role in cognitive coordination. *Psychological Review* 117:541–574.

Stocco, A., C. Lebiere, R. C. O'Reilly, and J. R. Anderson. 2012. Distinct contributions of the caudate nucleus, rostral prefrontal cortex, and parietal cortex to the execution of instructed tasks. *Cognitive, Affective, and Behavioral Neuroscience* 12 (4):611–628.

Touretzky, D. 1990. BoltzCONS: Dynamic symbol structures in a connectionist network. *Artificial Intelligence* 46 (1–2):5–46.

Ungerleider, L. G., and Mishkin, M. 1982. Two cortical visual systems. In *The Analysis of Visual Behavior*, ed. D. J. Ingle, M. A. Goodale, and R. J. W. Mansfield, 549–586. Cambridge, MA: MIT Press.

van Gelder, T. 1990. Compositionality: A connectionist variation on a classical theme. *Cognitive Science* 14 (3):355–384.

van Gelder, T., and L. F. Niklasson. 1994. Classicism and cognitive architecture. In *Proceedings of the Sixteenth Annual Conference of the Cognitive Science Society*, 905–909. Hillsdale, NJ: Erlbaum.

Wason, P. C., and P. N. Johnson-Laird. 1972. *Psychology of Reasoning: Structure and Content*. Cambridge, MA: Harvard University Press.

9 A Category Theory Explanation for Systematicity: Universal Constructions

Steven Phillips and William H. Wilson

1 Introduction

When, in 1909, physicists Hans Geiger and Ernest Marsden fired charged particles into gold foil, they observed that the distribution of deflections followed an unexpected pattern. This pattern afforded an important insight into the nature of atomic structure. Analogously, when cognitive scientists probe mental ability, they note that the distribution of cognitive capacities is not arbitrary. Rather, the capacity for certain cognitive abilities correlates with the capacity for certain other abilities. This property of human cognition is called systematicity, and systematicity provides an important clue regarding the nature of cognitive architecture: the basic mental processes and modes of composition that underlie cognition—the structure of mind.

Systematicity is a property of cognition whereby the capacity for some cognitive abilities implies the capacity for certain others (Fodor and Pylyshyn 1988). In schematic terms, systematicity is something's having cognitive capacity c_1 if and only if it has cognitive capacity c_2 (McLaughlin 2009). An often-used example is one's having the capacity to infer that John is the lover from *John loves Mary* if and only if one has the capacity to infer that Mary is the lover from *Mary loves John.*

What makes systematicity interesting is that not all models of cognition possess it, and so not all theories (particularly, those theories deriving such models) explain it. An elementary theory of mind, atomism, is a case in point: on this theory, the possession of each cognitive capacity (e.g., the inferring of John as the lover from *John loves Mary*) is independent of the possession of every other cognitive capacity (e.g., the inferring of Mary as the lover from *Mary loves John*), which admits instances of having one capacity without the other. Contrary to the atomistic theory, you don't find (English-speaking) people who can infer John as the lover (regarding the above example) without being able to infer Mary as the lover

(Fodor and Pylyshyn 1988). Thus, an atomistic theory does not explain systematicity.

An atomistic theory can be augmented with additional assumptions so that the possession of one capacity is linked to the possession of another. However, the problem with invoking such assumptions is that any pair of capacities can be associated in this way, including clearly unrelated capacities such as being able to infer John as the lover and being able to compute 27 as the cube of 3. Contrary to the augmented atomistic theory, there are language-capable people who do not understand such aspects of number. In the absence of principles that determine which atomic capacities are connected, such assumptions are ad hoc—"free parameters," whose sole justification is to take up the explanatory slack (Aizawa 2003).

Compare this theory of cognitive capacity with a theory of molecules consisting of atoms (core assumptions) and free parameters (auxiliary assumptions) for arbitrarily combining atoms into molecules. Such auxiliary assumptions are ad hoc, because they are sufficiently flexible to account for any possible combination of atoms (as a data-fitting exercise) without explaining why some combinations of atoms are never observed (see Aizawa 2003 for a detailed analysis).

To explain systematicity, a theory of cognitive architecture requires a (small) coherent collection of assumptions and principles that determine only those capacities that are systematically related and no others. The absence of such a collection, as an alternative to the classical theory (described below), has been the primary reason for rejecting connectionism as a theory of cognitive architecture (Fodor and Pylyshyn 1988; Fodor and McLaughlin 1990).

The classical explanation for systematicity posits a cognitive architecture founded upon a combinatorial syntax and semantics. Informally, the common structure underlying a collection of systematically related cognitive capacities is mirrored by the common syntactic structure underlying the corresponding collection of cognitive processes. The common semantic structure between the John and Mary examples (above) is the *loves* relation. Correspondingly, the common syntactic structure involves a process for tokening symbols for the constituents whenever the complex host is tokened. For example, in the *John loves Mary* collection of systematically related capacities, a common syntactic process may be $P \rightarrow Agent\ loves\ Patient$, where *Agent* and *Patient* subsequently expand to *John* and *Mary*. Here, tokening refers to instantiating both terminal (no further processing) and nonterminal (further processing) symbols. The tokening principle seems to support a much needed account of systematicity, because all capacities involve one and the same process; thus, having one capacity implies having the other, assuming basic capacities for representing constituents *John* and *Mary*.

Connectionists, too, can avail themselves of an analogous principle. In neural network terms, computational resources can be distributed between task-specific and task-general network components (e.g., weighted connections and activation units) by a priori specification and/or learning as a form of parameter optimization. For instance, an intermediate layer of weighted connections can be used to represent common components of a series of structurally related tasks instances, and the outer connections (the input–output interface) provide the task-specific components, so that the capacity for some cognitive function transfers to some other related function, even across completely different stimuli (see, e.g., Hinton 1990). Feedforward (Rumelhart, Hinton, and Williams 1986), simple recurrent (Elman 1990), and many other types of neural network models embody a generalization principle (see, e.g., Wilson, Marcus, and Halford 2001). In connectionist terms, acquiring a capacity (from training examples) transfers to other capacities (for testing examples).

Beyond the question of whether such demonstrations of systematicity, recast as generalization (Hadley 1994), correspond to the systematicity of humans (Marcus 1998; Phillips 1998), there remains the question of articulating the principle from which systematicity (as a kind of generalization) is a necessary, not just possible consequence. To paraphrase Fodor and Pylyshyn (1988), it is not sufficient to simply show existence—that there exists a suitably configured model realizing the requisite capacities; one also requires uniqueness—that there are no other models not realizing the systematicity property. For if there are other such configurations, then further (ad hoc) assumptions are required to exclude them. Existence/ uniqueness is a recurring theme in our explanation of systematicity.

Note that learning, generally, is not a principle that one can appeal to as an explanation of systematicity. Learning can afford the acquisition of many sorts of input–output relationships, but only some of these correspond to the required systematic capacity relationships. For sure, one can construct a suitable set of training examples from which a network acquires one capacity if and only if it acquires another. But, in general, this principle begs the question of the necessity of that particular set of training examples. Connectionists have attempted to ameliorate this problem by showing how a network attains some level of generalization for a variety of training sets. However, such attempts are far from characteristic of what humans actually get exposed to.

Some authors have claimed to offer "alternative" nonclassical compositionality methods to meet this challenge whereby complex entities are tokened without tokening their constituents. The tensor product network formalism (Smolensky 1990) is one (connectionist) example. Another (nonconnectionist) example is Gödel numbering (van Gelder 1990). However, it's unclear what is gained by this notion of nontokened compositionality (see Fodor and McLaughlin 1990). For example, any set of localized orthonormal vectors (a prima facie example of classical tokening) can be made nonlocal (i.e., nonclassical, in the above sense) with a change of basis vectors. Distributed representations are seen as more robust against degradation than local representations—the loss of a single unit does not result in the loss of an entire capacity. In any case, the local–distributive dimension is orthogonal to the classical–nonclassical tokening dimension—a classical system can also be implemented in a distributed manner simply by replicating representational resources.

Classicists explicitly distinguish between their symbols and their implementation via a *physical instantiation function* (Fodor and Pylyshyn 1988, n. 9). Though much has been made of the implementation issue, this distinction does not make a difference in providing a complete explanation for systematicity, as we shall see. Nonetheless, the implementation issue is important, because any explanation that reduces to a classical one suffers the same limitations. We mention it because we also need to show that our explanation (presented shortly) is not classical—nor connectionist (nor Bayesian, nor dynamicist), for that matter.

The twist in this tale of two theories is that classicism does not provide a complete explanation for systematicity, either, though classicism arguably fares better than connectionism (Aizawa 2003). That the classical explanation also falls short seems paradoxical. After all, the strength of symbol systems is that a small set of basic syntax-sensitive processes can be recombined in a semantically consistent manner to afford all sorts of systematically related computational capacities. Combinatorial efficacy notwithstanding, what the classical theory fails to address is the many-to-many relationship between syntax and computational capacity: more than one syntactic structure gives rise to closely related though not necessarily identical groups of capacities. In these situations, the principle of syntactic compositionality is not sufficient to explain systematicity, because the theory leaves open the (common) possibility of constructing classical cognitive models possessing some but not all members of a collection of systematically related cognitive capacities (for examples, see Aizawa 2003; Phillips and Wilson 2010, 2011, 2012). For instance, if we replace the production $P \rightarrow$ *Agent loves Patient* with productions $P_1 \rightarrow$ *Agent loves John* and $P_2 \rightarrow$ *John loves Patient*, then this alternative classical system no longer generates the instance *Mary loves Mary*. The essential challenge for classicism echoes that for connectionism: explain systematicity without excluding models admitted by the theory just because they don't support all systematically related capacities—why are those models not realized (cf.: why don't some combinations of atoms form molecules)?

So far, none of the major theoretical frameworks in cognitive science—classicist, connectionist, Bayesian, nor dynamicist—has provided a theory that fully explains systematicity. This state of affairs places cognitive science in a precarious position—akin to physics without a theory of atomic structure: without a stable foundation on which to build a theory of cognitive representation and process, how can one hope to scale the heights of mathematical reasoning? In retrospect, the lack of progress on the systematicity problem has been because cognitive scientists were working with "models" of structure (i.e., particular concrete implementations), where systematicity is a *possible* consequence, rather than "theories" of structure from which systematicity *necessarily* follows. This diagnosis led us (Phillips and Wilson 2010, 2011, 2012) to *category theory* (Eilenberg and Mac Lane 1945; Mac Lane 2000), a theory of structure par excellence, as an alternative approach to explaining systematicity.

The rest of this chapter aims to be, as much as possible, an informal, intuitive discussion of our category theory explanation as a complement to the formal, technical details already provided (Phillips and Wilson 2010, 2011, 2012). To help ground the informal discussion, though, we also include some standard formal definitions (see Mac Lane 2000 for more details) as stand-alone text, and associated *commutative diagrams*, in which entities (often functions) indicated by paths with the same start point and the same end point are equal. An oft-cited characteristic feature of category theory is the focus on the directed relationships between entities (called *arrows*, *morphisms*, or *maps*) instead of the entities themselves—in fact, categories can be defined in arrow-only terms (Mac Lane 2000). This change in perspective is what gives category theory its great generality. Yet, category theory is not arbitrary—category theory constructs come with formally precise conditions (axioms) that must be satisfied for one to avail oneself of their computational properties. This unique combination of abstraction and precision is what gives category theory its great power. However, the cost of taking a category theory perspective is that it may not be obvious how category theory should be applied to the problem at hand, nor what benefits are afforded when doing so. Hence, our purpose in this chapter is threefold: (1) to provide an intuitive understanding of our category theory explanation for systematicity; (2) to show how it differs from other approaches; and (3) to discuss the implications of this explanation for the broader interests of cognitive science.

2   What Is Category Theory?

Category theory was invented in the mid-1940s (Eilenberg and Mac Lane 1945) as a formal means for comparing mathematical structures. Originally it was regarded as a formal language for making precise the way in which two types of structures are to be compared. Subsequent technical development throughout the twentieth century has seen it become a branch of mathematics in its own right, as well as placing it on a par with set theory as a foundation for mathematics (see Marquis 2009 for a history and philosophy of category theory). Major areas of application, outside of mathematics, have been computer science (see, e.g., Arbib and Manes 1975; Barr and Wells 1990) and theoretical physics (see, e.g., Baez and Stay 2011; Coecke 2006). Category theory has also been used as a general conceptual tool for describing biological (Rosen 1958) and neural/cognitive systems (Ehresmann and Vanbremeersch 2007), yet applications in these fields are relatively less extensive.

Category theory can be different things in different contexts. In the abstract, a category is just a collection of *objects* (often labeled A, B, ...), a collection of *arrows* (often labeled f, g, ...) between pairs of objects (e.g., $f : A \to B$, where A is called the *domain* and B the *codomain* of arrow f), and a *composition operator* (denoted $\circ$) for composing pairs of arrows into new arrows (e.g., $f \circ g = h$), all in a way that satisfies certain basic rules (axioms). When the arrows are functions between sets, the composition is ordinary composition of functions, so that $(f \circ g)(x) = f(g(x))$. To be a category, every object in the collection must have an *identity arrow* (often denoted as $1_A : A \to A$); every arrow must have a domain and a codomain in the collection of objects; for every pair of arrows with matching codomain and domain objects there must be a third arrow that is their composition (i.e., if $f : A \to B$ and $g : B \to C$, then $g \circ f : A \to C$ must also be in the collection of arrows); and composition must satisfy *associativity*, that is, $(h \circ g) \circ f = h \circ (g \circ f)$, and *identity*, that is, $1_B \circ f = f = f \circ 1_A$, laws for all arrows in the collection. Sets as objects and functions as arrows satisfy all of this: the resulting category of all ("small") sets is usually called **Set**. In this regard, category theory could be seen as an algebra of arrows (Awodey 2006).

For a formal (abstract) category, the objects, arrows, and composition operator need no further specification. A simple example is a category whose collection of objects is the set $\{A,B\}$ and collection of arrows is the set $\{1_A : A \to A, 1_B : B \to B, f : A \to B, g : B \to A\}$. Since there are no other arrows in this category, compositions $f \circ g = 1_A$ and $g \circ f = 1_B$ necessarily hold. Perhaps surprisingly, many important results pertain to this level and hence apply to anything that satisfies the axioms of a category.

For particular examples of categories, some additional information is provided regarding the specific nature of the objects, arrows, and composition. Many familiar structures in mathematics are instances of categories. For example, a partially ordered set, also called a *poset*, $(P,\leq)$ is a category whose objects are the elements of the set P, and arrows are the order relationships $a \leq b$, where $a, b \in P$. A poset is straightforwardly a category, since a partial order $\leq$ is *reflexive* (i.e., $a \leq a$, hence identities) and *transitive* (i.e., $a \leq b$ and $b \leq c$ implies $a \leq c$, hence composition is defined). Checking that identity and associativity laws hold is also straightforward. The objects in a poset considered as a category have no internal parts. In other categories, the objects may also have internal structure, in which case the arrows typically preserve that structure. For instance, the category **Pos** has posets now considered as objects and order-preserving functions for arrows, i.e., $a \leq b$ implies $f(a) \leq f(b)$. For historical reasons, the arrows in a category may also be called *morphisms, homomorphisms,* or *maps* or functions when specifically involving sets.

Definition (Category). A *category* **C** consists of a class of objects $|\mathbf{C}|=\{A, B, ...\}$; and for each pair of object A, B in **C**, a set $\mathbf{C}(A,B)$ of morphisms (also called arrows, or maps) from A to B where each morphism $f : A \to B$ has A as its *domain* and B as its *codomain*, including the *identity* morphism $1_A : A \to A$ for each object A; and a composition operation, denoted $\circ$, of morphisms $f : A \to B$ and $g : B \to C$, written $g \circ f : A \to C$ that satisfies the laws of:

• *identity*, where $f \circ 1_A = f = 1_B \circ f$, for all $f : A \to B$; and
• *associativity*, where $(h \circ g) \circ f = h \circ (g \circ f)$, for all $f : A \to B$, $g : B \to C$ and $h : C \to D$.

In the context of computation, a category may be a collection of types for objects and functions (sending values of one type to values of another, or possibly the same, type) for arrows, where composition is just composition of functions. In general, however, objects need not be sets, and arrows need not be functions, as shown by the first poset-as-a-category example. For our purposes, though, it will often be helpful to think of objects and arrows as sets and functions between sets. Hence, for cognitive applications, one can think of a category as modeling some cognitive (sub)system, where an object is a set of cognitive states and an arrow is a cognitive process mapping cognitive states.

2.1 "Natural" Transformations, Universal Constructions

To a significant extent, the motivations of category theorists and cognitive scientists overlap: both groups aim to establish the principles underlying particular structural relations, be they mathematical structures or cognitive structures. In this regard, one of the central concepts is a natural transformation between structures. Category theorists have provided a formal definition of "natural," which we use here. This definition builds on the concepts of *functor* and, in turn, category. We have already introduced the concept of a category. Next we introduce functors before introducing natural transformations and universal constructions.

Functors are to categories as arrows (morphisms) are to objects. Arrows often preserve internal object structure; functors preserve category structure (i.e., identities and compositions). Functors have an object-mapping component and an arrow-mapping component.

Definition (Functor). A *functor* $F : \mathbf{C} \to \mathbf{D}$ is a map from a category $\mathbf{C}$ to a category $\mathbf{D}$ that sends each object A in $\mathbf{C}$ to an object $F(A)$ in $\mathbf{D}$; and each morphism $f : A \to B$ in $\mathbf{C}$ to a morphism $F(f) : F(A) \to F(B)$ in $\mathbf{D}$, and is structure-preserving in that $F(1_A) = 1_{F(A)}$ for each object A in $\mathbf{C}$, and $F(g \circ_{\mathbf{C}} f) = F(g) \circ_{\mathbf{D}} F(f)$ for all morphisms $f : A \to B$ and $g : B \to C$, where $\circ_{\mathbf{C}}$ and $\circ_{\mathbf{D}}$ are the composition operators in $\mathbf{C}$ and $\mathbf{D}$.

There is an intuitive sense in which some constructions are more natural than others. This distinction is also important for an explanation of systematicity, as we shall see. The concept of a *natural transformation* makes this intuition formally precise.

Natural transformations are to functors as functors are to categories. We have already seen that functors relate categories, and similarly, natural transformations relate functors. Informally, what distinguishes a natural transformation from some arbitrary transformation is that a natural transformation does not depend on the nature of each object A. This independence is also important for systematicity: basically, the cognitive system does not need to know ahead of time all possible instances of a particular transformation.

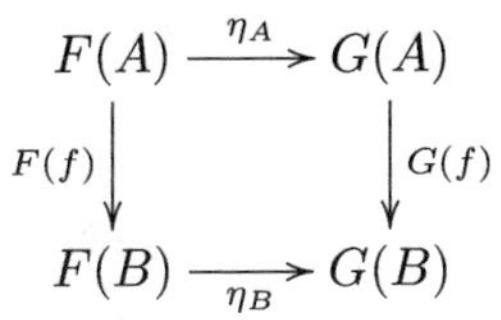

$$F(A) \xrightarrow{\eta_A} G(A)$$

$$F(f) \downarrow \qquad \qquad \downarrow G(f)$$

$$F(B) \xrightarrow{\eta_B} G(B)$$

Figure 9.1
Commutative diagram for natural transformation.

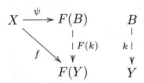

Figure 9.2
Commutative diagram for universal construction.

Definition (Natural transformation). A *natural transformation* $\eta : F \to G$ between a domain functor $F : \mathbf{C} \to \mathbf{D}$ and a codomain functor $G : \mathbf{C} \to \mathbf{D}$ consists of $\mathbf{D}$-maps $\eta_A : F(A) \to G(A)$ for each object A in $\mathbf{C}$ such that $G(f) \circ \eta_A = \eta_B \circ F(f)$. (See figure 9.1.)

For the purpose of explaining systematicity, we need something more than just that constructions are natural in this sense; constructions are also required to be *universal*, in a technical sense to be introduced next. A universal construction is basically an arrow that is "part" of every arrow in the category that models the (cognitive) domain of interest.

Definition (Universal construction). Given an object $X \in |\mathbf{C}|$ and a functor $F : \mathbf{B} \to \mathbf{C}$, a *couniversal morphism* from X to F is a pair (B, Ψ) where B is an object of $\mathbf{B}$, and Ψ is a morphism in $\mathbf{C}$, such that for every object $Y \in |\mathbf{B}|$ and every morphism $f : X \to F(Y)$, there exists a unique morphism $k : B \to Y$ such that $F(k) \circ \psi = f$. (See figure 9.2.)

A universal construction is either a couniversal morphism, or (its dual) a universal morphism (whose definition is obtained by reversing all the arrows in the definition of "couniversal").

At first it may not seem obvious how universal constructions are related to natural transformations. Note that, given a category of interest $\mathbf{D}$ and an object X in $\mathbf{D}$, X corresponds to a constant functor $X : \mathbf{C} \to \mathbf{D}$ from an arbitrary category $\mathbf{C}$ to the category of interest $\mathbf{D}$, where functor X sends every object and arrow in **C** to the object X and identity 1_X in **D**, thus yielding a natural transformation $\eta : X \to F$.

3 Systematicity: A Category Theory Explanation

All major frameworks assume some form of compositionality as the basis of their explanation for systematicity. In the classical case, it's syntactic; for the connectionist, it's functional (as we have already noted in section 1). In both cases, systematic capacity is achieved by combining basic processes. However, the essential problem is that there is no additional constraint to circumscribe only the relevant combinations. Some combinations are possible that do not support all members of a specific collection of systematically related capacities. So, beyond simply stipulating the acceptable models (i.e., those consistent with the systematicity property), additional principles are needed to further constrain the admissible models.

Our category theory explanation also relies on a form of compositionality, but not just any form. The additional ingredient in our explanation is the formal category theory notion of a universal construction. The essential point of a universal construction is that each and every member of a collection of systematically related cognitive capacities is modeled as a morphism in a category that incorporates a common universal morphism in a unique way. From figure 9.2, having one capacity $f_1 : A \to F(Y_1)$ implies having the common universal morphism $\psi : X \to F(B)$, since $f_1 = F(k_1) \circ \psi$. And, since the capacity-specific components $F(k_i)$ are uniquely given (constructed) by functor $F : \mathbf{B} \to \mathbf{C}$, and the arrows k_i in **B**, one also has capacity $f_2 : A \to F(Y_2)$, since $f_2 = F(k_2) \circ \psi$. That is, all capacities (in the domain of interest) are systematically related via the couniversal morphism ψ. Our general claim is that each collection of systematically related capacities is an instance of a universal construction. The precise nature of this universal morphism will depend (of course) on the nature of the collection of systematically related capacities in question. In this section, we illustrate several important examples.

3.1 Relations: (Fibered) Products

We return to the *John loves Mary* example to illustrate our category theory explanation. This and other instances of relational systematicity are captured by a categorical product (Phillips and Wilson 2010). A categorical product provides a universal means for composing two objects (A and B) as a third object (P) together with the two arrows (p_1 and p_2) for retrieving information pertaining to A and B from their composition P. The require-

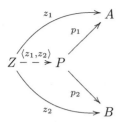

Figure 9.3
Commutative diagram for product.

ment that such a construction be universal is critical for explaining the systematicity property.

Definition (Product). A *product* of objects A and B in category **C** is an object P (also denoted $A{\times}B$) together with two morphisms (sometimes called *projections*) $p_1 : P \rightarrow A$ and $p_2 : P \rightarrow B$, jointly expressed as $(P,\ p_1,\ p_2)$ such that for every object $Z \in |C|$ and pair of morphisms $z_1 : Z \rightarrow A$ and $z_2 : Z \rightarrow B$ there exists a unique morphism $u : Z \rightarrow P$, also denoted $\langle z_1, z_2 \rangle$, such that $z_1 = p_1 \circ u$ and $z_2 = p_2 \circ u$. (See figure 9.3).

Suppose A and B correspond to the set of representations for the possible agents and patients that can partake in the *loves* relation, which includes instances such as *John loves Mary* and *Mary loves John*, and the product object P, which is the Cartesian product $A{\times}B$ (in the case of products of sets, as in this example), corresponds to the representations of those relational instances. Then any requirement to extract components A (B) from some input Z necessarily factors through $p_1(p_2)$ uniquely.

As a universal construction, products are constructed from the *product functor*. Informally, the product functor sends pairs of objects (A, B) to the product object $A{\times}B$ and pairs of arrows (f, g) to the product arrow $f{\times}g$. In this way, all possible combinations must be realized. The universality requirement rules out partial constructions, such as a triple (Q, q_1, q_2) where the object Q contains just three of the four possible pair combinations of *John* and *Mary* (and q_1 and q_2 return the first and second item of each pair), because this triple does not make the associated diagram (see figure 9.3) commute. Thus, no further assumptions are needed to exclude such cases, in contrast with the classical (or connectionist) explanation, which admits such possibilities.

One can think of the product functor as a way of constructing new wholes (i.e., $A{\times}B$) from parts (A, B). The product functor is seen as the "conceptual inverse" to the *diagonal functor* which makes wholes into parts by making copies of each object and morphism: that is, object A and morphism f are sent to pairs (A, A) and (f, f). Together, diagonal and product functors constitute an *adjunction*, a family of universal constructions (Mac Lane 2000). Adjunctions are important to our explanation of systematicity because they link representation and access (inference) without relying on the assumption that such processes be compatible. A classical explanation simply assumes compatibility between syntactically compositional representations and the processes that operate on them. Yet, as we've seen, there is more than one way of syntactic composition, and not all of them support systematicity. By contrast, in our category theory explanation, the commutativity property associated with an adjunction enforces compatibility (Phillips and Wilson 2010, 2011).

We further contrast our explanation with a proposed alternative illustrated by Gödel numbering (van Gelder 1990). This scheme depends on careful construction of a suitable transformation function that depends on the values of all possible elements (past, present, and future) that can partake in the relation. However, in general, a cognitive system cannot have knowledge of such things. At best, a cognitive system can update a set of representations to accommodate new instances to maintain correct transformation. But such allowances admit nonsystematicity: at the point prior to update, the cognizer is in a state of having some but not all systematically related capacities. Thus, such schemes do not account for systematicity.

Products address systematic capacity where there is no interaction between constituents A and B. Another case addresses quasi systematicity, where capacity extends to some but not all possible combinations of constituents. In this situation, the interaction between A and B is given by two arrows $f : A \to C$ and $g : B \to C$ to a common (constraint) object C. The universal construction for this situation is called a *pullback* (or *fibered product*, or *constrained product*). The explanation for systematicity in this case essentially parallels the one given for products: replace product with pullback (see Phillips and Wilson 2011).

3.2 Recursion: F-(co)algebras

Our explanation for systematicity with regard to recursive domains also employs a universal construction, albeit with a different kind of functor, called an *endofunctor*, where the domain and codomain are the same category (Phillips and Wilson 2012), hence its importance for recursion. A motivating example is that the capacity to find the smallest item in a list implies the capacity to find the largest item, assuming a basic capacity for distinguishing the relative sizes of items. For example, you don't find people who can select the lowest card from a deck without being able to select the highest card, assuming they understand the relative values of cards. Yet, classical theory admits recursive and nonrecursive compositional methods, for realizing these two capacities, without there being any common component processes. If one capacity is realized by a recursive method and the other by a nonrecursive method, then the two capacities are not intrinsically connected in any way—the tokening principle on which classical theory depends is no longer in play. Thus, classical theory also fails to fully explain systematicity with regard to recursively definable capacities (Phillips and Wilson 2012).

In recent decades, computer scientists have turned to category theory to develop a systematic treatment of recursive computation (see, e.g., Bird and Moor 1997). We have adapted this theory for an explanation of systematicity in regard to recursive cognitive capacities. Conceptually, recursive capacities are decomposed into an invariant component, the recurrent part, and a variant component, with the capacity-specific computation taking place at each iterative step. The invariant component corresponds to the underlying recursive data structure (e.g., stepping through a deck of cards) underpinning the group of systematically related capacities. The variant component corresponds to the computation at each step (e.g., comparing cards for the smaller or larger card). In category theory terms, every recursive capacity is an algebra, called an *F-algebra*, built using an endofunctor F. Under very general conditions, a category of such F-algebras has a universal construction called an initial F-algebra, and hence provides an explanation for systematicity with regard to recursive capacities.

Definition (F-algebra, initial algebra, catamorphism). For an endofunctor $F : \mathbf{C} \to \mathbf{C}$, an *F-algebra* is a pair (A, α), where A is an object and $\alpha : F(A) \to A$ is a morphism in $\mathbf{C}$.

An *initial algebra* (A, in) is an *initial object* in the category of F-algebras $\mathbf{Alg}(F)$. That is, $in : F(A) \to A$ is a morphism in $\mathbf{C}$, and there exists a unique F-algebra homomorphism from (A, in) to every F-algebra in $\mathbf{Alg}(F)$.

A *catamorphism* $h : (A, in) \to (B, \beta)$ is the *unique* F-algebra homomorphism from initial F-algebra (A, in) to F-algebra $(B,)$. That is, $h \circ in = \beta \circ F(h)$ and the uniquely specified h for each such is denoted *cata* (i.e., $h = cata$). (See figure 9.4.)

The dual constructions: *F-coalgebra*, *final coalgebra*, and *anamorphism* are also used to explain related instances of systematicity (see Phillips and Wilson 2012 for details).

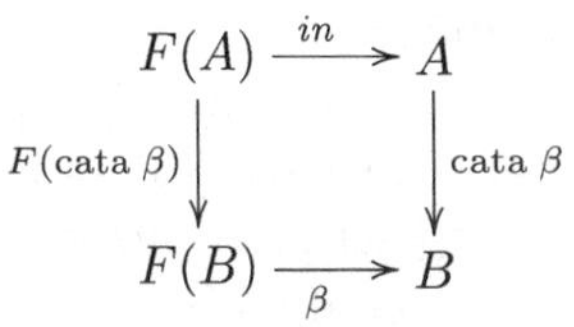

$$F(A) \xrightarrow{\ in\ } A$$

$$F(\text{cata } \beta) \Big\downarrow \qquad\qquad \Big\downarrow \text{cata } \beta$$

$$F(B) \xrightarrow[\ \beta\ ]{} B$$

Figure 9.4
Commutative diagram for catamorphism.

In each case, the recursive capacity depends only on the common arrow $in : F(A) \to A$ and the unique arrow cata$\beta : A \to B$ (see figure 9.4). In outline, searching for the smallest number in a list of numbers is given by $fold(\infty, lower)$ l, where $fold$ is the common recursive part, $(\infty, lower)$ is the task-specific component, ∞ applies to empty lists, $lower$ returns the lower of two numbers, and l is a list of numbers. For example, $fold(0, lower)[3,2,5] = lower(3, lower(2, lower(5, \infty))) = 2$. Searching for the largest number in a list is given by $fold(0, higher)$ l, where $(0, higher)$ is the task-specific component, 0 applies to empty lists, and $higher$ returns the higher of two numbers. For example, $fold(0, higher)[3,2,5] = higher(3, higher(2, higher(5, 0))) = 5$.

4   Category Theory in Context

The abstract and abstruse nature of category theory may make it difficult to see how our explanation relates to other theoretical approaches, and how it should make contact with a neural level of analysis. For all of the theoretical elegance of category theory, constructs must also be realizable by the underlying brain system. The relationship between our category theory explanation and a classical or connectionist one is analogous to the relationship between, say, an abstract and concrete specification of a group in mathematics: particular classical or connectionist architectures may be models of our theory. Here, we sketch some possibilities.

4.1   Classical Models

Our category theory explanation overlaps with the classical one in the sense that the common constituent of a collection of complex cognitive capacities is "tokened" (i.e., imparted or executed) whenever each complex cognitive capacity is. Notice, however, that tokening in the category theory sense is the tokening of arrows, not objects; analogously, tokening is of processes, not symbols. Classical systems also admit symbols as processes.

Nonetheless, our explanation goes significantly beyond the classical one in that we require not just any arrow, but rather, an arrow derived from a universal construction. Thus, no further assumptions are required to guarantee that each and every capacity is uniquely constructed from it. Moreover, the virtue of this arrow-centric perspective is that, contra classicism, our explanation extends to nonsymbolic domains, such as visual cognition, without further adjustment to the theory.

Universal constructions such as adjunctions and F-(co)algebras as the basis for a theory of cognitive architecture are unique to our theory, and go significantly beyond the widespread use of isomorphism (cf. analogy models, including Suppes and Zinnes 1963; Halford and Wilson 1980) in cognitive science generally. From a category theory perspective, two systems that are isomorphic are essentially the "same" up to a change of object and arrow labels. An adjunction is more general and potentially more useful: two systems (involving different sorts of processes) in an adjoint relationship need not be isomorphic, while still being in a systematic relationship with each other.

4.2 Nonclassical Models

For conciseness, we treat connectionist and Bayesian models in the same light, despite some significant advances in Bayesian modeling (see, e.g., Kemp and Tenenbaum 2008). By regarding connectionist and Bayesian networks as graphs (with additional structure), one could consider a category of such graphs as objects and their homomorphisms as morphisms. That additional structure could include "coloring" graph nodes to distinguish corresponding input, output and internal network activation units, and functions corresponding to propagation of network activity. The category **Grph** of graphs and graph homomorphisms has products, suggesting that a suitable category of graphs with additional structure can be devised that also has products. In these cases, systematicity is realized as a functor from the product (as a category) into a category of connectionist/Bayesian networks, thus guaranteeing an implementation of the systematicity property within a connectionist/Bayesian-style framework.

A similar approach also applies to dynamic systems models. In a simple (though not exhaustive) case, one class of dynamic systems can be treated as a finite state machine. A category of such machines and the structure-preserving morphisms also has products (see Arbib and Manes 1975). Thus, again, systematicity is implemented as a functor from the product (as a category) into this category of finite state machines.

5   Testing the Theory

Our theory can be tested using a series of tasks, where each task instance is composed of an invariant component corresponding to the (co)universal arrow and a variant component corresponding to the unique arrow of the underlying universal construction (see Phillips and Wilson 2012, text S2). An example of this kind of design is familiar in the form of so-called *learning set* paradigms (see, e.g., Kendler 1995).

5.1   Nonrecursive Example

A series of simple classification tasks illustrates one kind of experimental design that can be used to test for systematicity in terms of universal constructions. For the first task, suppose participants are given stimuli to be classified into one of two classes. Let S be the set of stimuli, and R the set of (two) responses associated with each class. Hence, the morphism $t : S \rightarrow R$ is the stimulus-response process for the first task instance. Next, suppose the task is modified (for the next task instance), say by changing the responses to each stimulus class from left and right to up and down. Let R' be the set of responses associated with the new task, and $t' : S \rightarrow R'$ the associated stimulus-response process. Since the responses are determined by the classes rather than directly by the stimuli, each task instance t decomposes into the task-invariant classification component $c : S \rightarrow C$, where C is the set of classes, and the task-variant response mapping component $r : C \rightarrow R$ uniquely. That is, c corresponds to Ψ, the couniversal arrow, in figure 9.2 for a universal construction.

A test of systematicity for this example is whether participants can correctly predict the stimulus response classification on new task trials after receiving sufficient trials to determine r, the response mapping. In the general case that there are n possible responses, so n trials are needed to determine the correct mapping (one trial for each possible response), but no more. Thus, systematicity is evident on correct prediction for the remaining $m - n$ stimulus-response trials, assuming there are more stimuli than responses (i.e., $m > n$). (See Phillips and Wilson 2010 for a further example.)

5.2   Recursive Example

Phillips and Wilson (2012) provide an example of systematicity with respect to recursively definable concepts in the form of finding the smallest/largest item in a list. Here, we illustrate how this kind of systematicity can be directly tested. Suppose participants are given pairs of stimuli (e.g., shapes) from which they must predict the "preferred" (or rewarded) shape—essentially, a discrimination task. This preliminary task allows participants to learn the total order associated with the set of stimuli. Upon completion of the discrimination task, participants are presented with a list of stimuli, selected from the set used in the preliminary task, and asked to find the most preferred stimulus in that list. Upon completion of this second task, participants are then required to identify the unpreferred stimulus from a pair of stimuli, and then the least preferred stimulus in a list. Evidence of systematicity in this example is correct determination of the least preferred stimulus without further feedback. This paradigm could be further extended by changing the set of stimuli between task instances. (See Phillips and Wilson 2012 for another example.)

6   Beyond Systematicity

Beyond systematicity are other questions that a general (category) theory of cognitive architecture should address. We round out this chapter by considering how a categorial theory of cognitive architecture may address such issues.

6.1   Systematicity and Nonsystematicity: Integration

Not all cognitive capacities are systematic, as we mentioned in the first section. The classical proposal (Fodor and Pylyshyn 1988), and so far ours too, speak only to the systematic aspects of cognition, while leaving nonsystematic aspects unaddressed. Fodor and Pylyshyn (1988) recognized the possibility of some kind of hybrid theory: say, a classical architecture fused with some nonclassical (e.g., connectionist) architecture to address cognitive properties beyond the scope of (or unaccounted for by) classical theory. Aizawa (2003) warns, however, that hybrid theories require a higher explanatory standard: not only must each component theory account for their respective phenomena, but there must also be a principled account for why and when each component theory is invoked. Here, we sketch how our category theory approach could be extended to incorporate nonsystematic aspects of cognition.

As Phillips and Wilson (2012) suggest, if we regard a category as a model of a cognitive subsystem, then combining two categories by taking a fibered (co)product can be regarded as the integration of two subsystems into a larger combined system. Clark, Coecke, and Sadrzadeh (2008) provide an example of how subsystems can be combined categorically for modeling aspects of language. Their example is a hybrid symbolic-distributional model of grammar as the fibered product of a symbolic and a distributed (vector-based) component. In our case, one category realizes systematicity, another realizes nonsystematicity, and the category derived from their fibered (co)product realizes both.

For our purposes, though, we must also consider the principle dictating which component is to be employed and under what circumstances. We have suggested (Phillips and Wilson 2012) that a cost–benefit trade-off may be the basis of such a principle. For instance, there are at least two ways to add numbers such as 3 and 5. One can employ a systematic counting procedure by counting from the first number (3) the number of increments indicated by the second number (5). This procedure has the benefit of working for any two numbers (systematicity), but at the cost of being slow when the numbers are large. Alternatively, one can simply recall from memory the sum of the given numbers. This second procedure has the benefit of speed, but the cost of unreliability (unsystematic): the sums of some pairs may not have been memorized, and moreover, time and effort are required to memorize each pair.

To accommodate such possibilities, we further suggest here that our category theory approach can be extended by associating a cost with each morphism within the framework of *enriched category theory* (Kelly 2005). Enriched category theory considers categories whose hom-sets (i.e., sets of arrows between pairs of objects) have additional structure. For example, by defining a partial-order over a set of arrows each hom-set becomes a poset (i.e., a set with the extra order structure)—the category is enriched over the category of posets, **Pos**, the category of partially ordered sets and order-preserving functions. In this way, a choice between arrows (alternative cognitive strategies) can be based on an order principle—choose the strategy with the lower associated cost, when the alternatives are comparable.

Note, however, that we are not yet in a position to provide such principles. One could, of course, simply fit data by assigning adjustable parameters to each morphism, akin to a connectionist network. However, this maneuver merely affords compatibility with the data. What we really require is a principle necessitating when a particular subsystem is employed, lest we also succumb to the kinds of ad hoc assumptions that have bedeviled other approaches to the systematicity problem (Aizawa 2003).

6.2 Category Theory, Systematicity, and the Brain

Any theory of cognitive architecture must ultimately be reconcilable with the underlying neural architecture. Ehresmann and Vanbremeersch (2007) have provided a general description of how biological, neural, and cognitive systems may be cast within a category theory framework, though their work was not intended to address the systematicity problem.

Our category theory approach to the systematicity problem suggests an intriguing connection between the implied components of a categorial cognitive architecture (universal constructions) and brain structure. Note that the universal constructions we have employed to address various instances of systematicity involve endofunctors. The composition of adjoint functors is necessarily an endofunctor, and F-(co)algebras are based on endofunctors. An analogue of recurrency in the brain is the reciprocating neural connections within and between brain regions. Thus, one place to look for a correspondence between cognitive and neural architectures, at least in regard to the systematic aspects of cognition, are recurrent neurally connected brain regions. These kinds of connections are prevalent throughout the brain. Conversely, brain regions lacking such connections suggest corresponding cognitive capacities lacking systematicity. Of course, reciprocal neural connections may have other functional roles, and the computational connection to adjunctions is only speculation at this point.

We have begun investigating the relationship between category theory constructs and the brain (Phillips, Takeda, and Singh 2012). A pullback (fibered product), which featured as the kind of universal construction in our explanation of quasi systematicity (Phillips and Wilson 2011), also corresponds to integration of stimulus feature information in visual attention. By varying the "arity" (unary, binary, ternary) of the fibered product matching the number of feature dimensions (color, frequency, orientation) needed to identify a target object, we observed significantly greater EEG synchrony (phase-locking) between frontal and parietal electrodes with increasing arity. These results also provide a category theory window into development, discussed next.

6.3 Development and Learning

Cognitive development, in some cases, can also be seen as instances of systematicity. The capacity for inferential abilities, such as *transitive inference* and *class inclusion* are consistently enabled around the age of five years (see Halford, Wilson, and Phillips, 1998). Children who have the capacity to make transitive inferences typically also have the capacity to make inferences based on class inclusions. Conversely, children who fail at class inclusion also fail at transitive inference. Thus, we can see this equivalence as another instance of the systematicity schemata: capacity c_1 if and only if capacity c_2, all else being equal (McLaughlin 2009).

We have given a category theory explanation for these data (Phillips, Wilson, and Halford 2009) to overcome some difficulties with our earlier *relational complexity* approach (Halford et al. 1998). This common inferential capacity was explained in terms of the arity of the underlying (co) product. Older children (above age five years) have the capacity for binary (co)products, whereas younger children do not. Thus, as an instance of a universal construction, and a special case of our systematicity explanation, having the capacity for transitive inference implies class inclusion because the underlying categorical structures are dual to each other. In the dual case, the constructions are related by reversal of arrow directions via *contravariant* functors, where each object is mapped to itself, and each arrow to an arrow whose domain and codomain are respectively the codomain and domain of the source arrow.

Learning is also of central importance to cognitive science. A universal construction is a kind of optimal solution to a problem: a (co)universal morphism is an arrow that is a factor (in the sense of function composition) of all arrows to/from a particular construction (functor). To the extent that learning (and evolution) is a form of optimization, universal constructions may provide an alternative perspective on this aspect of cognition.

7 Conclusion

We began this chapter with the distribution of deflected charged particles affording an important insight into the structure of the atom as an analogy to the importance of the distribution of cognitive capacities to understanding the nature of cognitive architecture. We end this chapter with the insight that systematicity affords cognitive science: the atomic components of thought include universal constructions (not symbols, connections, probabilities, or dynamical equations, though these things may be part of an implementation) insofar as the systematicity property is evident.

Acknowledgment

This work was supported by a Japanese Society for the Promotion of Science (JSPS) Grant-in-Aid (Grant No. 22300092).

References

Aizawa, K. 2003. *The Systematicity Arguments*. New York: Kluwer Academic.

Arbib, M. A., and E. G. Manes. 1975. *Arrows, Structures, and Functors: The Categorical Imperative*. London: Academic Press.

Awodey, S. 2006. *Category Theory*. New York: Oxford University Press.

Baez, J. C., and M. Stay. 2011. Physics, topology, logic and computation: A Rosetta Stone. In *New Structures in Physics*, ed. B. Coecke, 95–172. New York: Springer.

Barr, M., and C. Wells. 1990. *Category Theory for Computing Science*. New York: Prentice Hall.

Bird, R., and O. de Moor. 1997. *Algebra of Programming*. Harlow: Prentice Hall.

Clark, S., B. Coecke, and M. Sadrzadeh. 2008. A compositional distributional model of meaning. In *Proceedings of the Second Symposium on Quantum Interaction*, ed. P. Bruza, W. Lawless, K. van Rijsbergen, D. Sofge, B. Coecke, and S. Clark, 133–140. Oxford: College Publications.

Coecke, B. 2006. Introducing categories to the practicing physicist. In *What Is Category Theory?* vol. 3, 45–74. Monza, Italy: Polimetrica.

Ehresmann, A. C., and J.-P. Vanbremeersch. 2007. *Memory Evolutive Systems: Hierarchy, Emergence, Cognition*, vol. 4. Oxford: Elsevier.

Eilenberg, S., and S. Mac Lane. 1945. General theory of natural equivalences. *Transactions of the American Mathematical Society* 58:231–294.

Elman, J. L. 1990. Finding structure in time. *Cognitive Science* 14:179–211.

Fodor, J. A., and B. P. McLaughlin. 1990. Connectionism and the problem of systematicity: Why Smolensky's solution doesn't work. *Cognition* 35 (2):183–204.

Fodor, J. A., and Z. W. Pylyshyn. 1988. Connectionism and cognitive architecture: A critical analysis. *Cognition* 28 (1–2):3–71.

Hadley, R. F. 1994. Systematicity in connectionist language learning. *Mind and Language* 9 (3):247–272.

Halford, G. S., and W. H. Wilson. 1980. A category theory approach to cognitive development. *Cognitive Psychology* 12:356–411.

Halford, G. S., W. H. Wilson, and S. Phillips. 1998. Processing capacity defined by relational complexity: Implications for comparative, developmental, and cognitive psychology. *Behavioral and Brain Sciences* 21 (6):803–831.

Hinton, G. E. 1990. Mapping part-whole hierarchies in connectionist networks. *Artificial Intelligence* 46 (1–2):47–76.

Kelly, G. M. 2005. Basic concepts of enriched category theory. *Reprints in Theory and Applications of Categories* 10.

Kemp, C., and J. B. Tenenbaum. 2008. The discovery of structural form. *Proceedings of the National Academy of Sciences of the United States of America* 105 (31): 10687–10692.

Kendler, T. S. 1995. *Levels of Cognitive Development*. Mahwah, NJ: Erlbaum.

Mac Lane, S. 2000. *Categories for the Working Mathematician*, 2nd ed. New York: Springer.

Marcus, G. F. 1998. Rethinking eliminative connectionism. *Cognitive Psychology* 37 (3):243–282.

Marquis, J.-P. 2009. *From a Geometrical Point of View: A Study of the History and Philosophy of Category Theory*. New York: Springer.

McLaughlin, B. P. 2009. Systematicity redux. *Synthese* 170:251–274.

Phillips, S. 1998. Are feedforward and recurrent networks systematic? Analysis and implications for a connectionist cognitive architecture. *Connection Science* 10 (2):137–160.

Phillips, S., Y. Takeda, and A. Singh. 2012. Visual feature integration indicated by phase-locked frontal-parietal EEG signals. *PLoS ONE* 7 (3):e32502.

Phillips, S., and W. H. Wilson. 2010. Categorial compositionality: A category theory explanation for the systematicity of human cognition. *PLoS Computational Biology* 6 (7):e1000858.

Phillips, S., and W. H. Wilson. 2011. Categorial compositionality II: Universal constructions and a general theory of (quasi-)systematicity in human cognition. *PLoS Computational Biology* 7 (8):e1002102.

Phillips, S., and W. H. Wilson. 2012. Categorial compositionality III: F-(co)algebras and the systematicity of recursive capacities in human cognition. *PLoS ONE* 7 (4):e35028.

Phillips, S., W. H. Wilson, and G. S. Halford. 2009. What do Transitive Inference and Class Inclusion have in common? Categorical (co)products and cognitive development. *PLoS Computational Biology* 5 (12):e1000599.

Rosen, R. 1958. The representation of biological systems from the standpoint of the theory of categories. *Bulletin of Mathematical Biophysics* 20:317–341.

Rumelhart, D. E., G. E. Hinton, and R. J. Williams. 1986. Learning representations by back-propagation of error. *Nature* 323:533–536.

Smolensky, P. 1990. Tensor product variable binding and the representation of symbolic structures in connectionist systems. *Artificial Intelligence* 46 (1–2):159–216.

Suppes, P., and J. L. Zinnes. 1963. Basic measurement theory. In *Handbook of Mathematical Psychology*, ed. R. D. Luce, R. R. Bush, and E. Galanter. London: Wiley.

van Gelder, T. 1990. Compositionality: A connectionist variation on a classical theme. *Cognitive Science* 14:355–384.

Wilson, W. H., N. Marcus, and G. S. Halford. 2001. Access to relational knowledge: a comparison of two models. In *Proceedings of the 23rd Annual Meeting of the Cognitive Science Society*, ed. J. D. Moore and K. Stenning, 1142–1147. Austin, TX: Cognitive Science Society.

III

10 Systematicity and Architectural Pluralism

William Ramsey

1 Introduction

In 1987, Jerry Fodor came to the University of California, San Diego to present his now-famous systematicity argument(s) against the connectionist outlook on cognitive processes that was just beginning to blossom there at the time. Somehow I wound up with the task of picking him up at the airport, although it was hardly a task I minded. Fodor had sent an earlier draft of the paper he had cowritten with Zenon Pylyshyn and because I was finishing a dissertation on the philosophical implications of connectionism, I was happy to get a chance to discuss the argument with him before all the shouting began. Meeting him at the airport, Fodor seemed a little on edge, and I joked that he was sort of like a Christian walking into a den of lions. Fodor grinned and said, "Yeah ... or maybe more like a lion walking into a den of Christians."

As it turned out, Fodor's version was probably a little closer to the truth. He forcefully presented his and Pylyshyn's main challenge (Fodor and Pylyshyn 1988)—that connectionist models cannot capture systematicity among representational states without simply implementing a classical architecture—and offered detailed counter-replies to the various rebuttals the pro-connectionist crowd put forth. So began what has turned into one of the more extensive and valuable debates over cognitive architecture. Today, it is hard to find an introductory text or course on cognitive science that doesn't include a segment on Fodor and Pylyshyn's systematicity challenge to connectionism. Besides generating a broad range of responses and counter-responses, the challenge has forced both classicists and connectionists to clarify the fundamental commitments and assumptions that help define their respective theories.

In this chapter, I am going to recommend a line of response that might initially seem odd because it accepts the idea that systematicity is a real aspect of cognition but rejects the idea that connectionists should be deeply worried about it. In fact, I am going to defend the conjunction of three seemingly incompatible propositions:

(1) Systematicity is indeed a real aspect of human cognitive processing.
(2) Connectionism, as such, does not provide a good explanatory framework for understanding systematicity.
(3) (1) and (2) are not necessarily major problems for connectionism.

In other words, I'll suggest that Fodor and Pylyshyn are justified in claiming that representational systematicity is an important aspect of cognitive activity, but they are wrong to suppose that a failure to explain it should substantially undermine a given architecture's promise or credibility. Fodor and Pylyshyn's argument assumes that the mind has a single basic cognitive architecture and representational system; thus, if systematicity is real, it must be a *fundamental* feature of cognition. While this assumption has always been problematic, as various authors have noted, it has now been rejected by a growing crowd of investigators who have adopted different forms of architectural pluralism. With pluralism, there is no reason to think that explaining representational systematicity is necessary for a viable theory of cognitive processes. This is good news for connectionists, because there is reason to think that a full explanation of systematicity would require them to give up many elements of their theory that are distinctive and noteworthy. Indeed, I'll suggest that when properly understood, some forms of connectionist processing should be treated as lacking not just representational states that are related systematically, but rather as lacking representational states altogether.

To show this, I'll first review the systematicity argument as I understand it and consider a few of the related issues that have arisen in discussions of the argument. I will also briefly discuss a couple of responses to the challenge that have been pursued, and explain why I don't find them promising. Then I will emphasize how the challenge depends on an assumption of architectural monism and note how that assumption has been abandoned by many cognitive researchers today. With various options available for some kind of architectural pluralism, such as dual process theories, connectionists now have a straightforward response to the systematicity challenge that allows them to preserve what is distinctive about their outlook on cognition. I will finish by encouraging connectionists to consider moving even further away from the classical framework by embracing an account of cognitive operations that is not only nonsystematic but also nonrepresentational.

2   The Systematicity Challenge

Since a great deal has been written on what the systematicity challenge to connectionism actually is and isn't, I won't dwell on that question here. Instead, I will simply adopt what I take to be the most charitable reading of Fodor and Pylyshyn's argument, as articulated by writers like Aizawa (2003) and McLaughlin (2009). According to this view, the challenge is based on the need to explain an alleged psychological law regarding our mental representational capacities. The psychological law is roughly this: the capacity of cognitive agents to mentally represent[1] certain states of affairs (or propositions) is significantly related to their capacity to mentally represent certain other states of affairs. The states of affairs in question whose representations appear to be related are ones that share key elements. For instance, the state of affairs of John loving Mary is related to the state of affairs of Mary loving John in that these different states involve the same people and the same relation of love. The alleged representational systematicity of cognitive agents is a claim about the mental representations of related states of affairs like these. The claim is that the ability to token or produce the representational vehicle for one of these states of affairs is related to the ability to token the representational vehicle for the other. In other words, the ability of agents to produce the cognitive representation *John loves Mary* is significantly related to the ability to token the representation *Mary loves John*. The challenge is to explain why this is so for all such representations of relevantly related states of affairs.

Just what is meant here by "significantly related to" has been the subject of considerable discussion. In their original article, Fodor and Pylyshyn claim that the ability to represent certain states of affairs is "intrinsically connected to" the ability to represent certain others. But that doesn't provide much clarification, since we aren't told in detail what "intrinsically connected to" means. More recently, it has been suggested that the significant relation is a form of counterfactual dependency or capacity (Aizawa 2003). In particular, if you possess the capacity to represent the condition aRb (e.g., John loving Mary), you thereby must possess the capacity to represent the condition bRa (e.g., Mary loving John). Indeed, in a later paper, Fodor and McLaughlin put things exactly this way: "If you meet the conditions for being able to represent aRb, *you cannot but meet the conditions for being able to represent bRa*" (Fodor and McLaughlin 1990, 202, emphasis in original). For our purposes, we can simply stipulate that representational systematicity is *some* form of significant dependency involving the tokening of different representations of states of affairs that are themselves

related insofar as they share constituent elements. It is thus claimed to be
a law of psychology that the representational capacities of cognitive agents
exhibit this type of interrelatedness—where representational capacities
come in clusters of this sort.

Of course, there is nothing for a *psychological* theory to explain regarding
the interrelatedness of different states of affairs that are represented. If
anything, that is a matter of metaphysics; presumably the relevant states
of affairs are related insofar as they involve the same people, relations,
properties, events, and so on. But while the targets of mental representations may be related in all sorts of ways, the representational vehicles
themselves, it seems, need not be, at least not in principle. In other words,
John's loving Mary might well be represented by one sort of neural or
computational state, while Mary loving John might be represented by an
entirely different and independent neural or computational state. More to
the point, our ability to generate a representation of John loving Mary
could in principle have nothing at all to do with our ability to generate a
representation of Mary loving John. The representation of John loving
Mary could stand in the same relation to the representation of Mary loving
John that it stands to the representation of dogs hating cats. But, goes the
argument, as a fact of psychology, it clearly doesn't. The capacity to represent the first two are mutually dependent in a way that the capacity to
represent the first and third are not. And that difference cries out for an
explanation.

Given that the alleged law is grounded in a psychological disposition
or capacity, the trait in question would presumably be explained through
a more basic feature of the representational system. What needs explaining
is not a particular sort of representational performance or any particular
set of tokenings; rather, it is a dispositional property of the underlying
system. In effect, the question Fodor and Pylyshyn posed is this: what
is it about the way representation happens in the brain that makes it the
case that our ability to represent some things is dependent on our ability
to represent other things, namely, new states of affairs with the same
elements?

Classicists like Fodor and Pylyshyn insist that the classical account of
representation—what is often called a "language of thought" account—
successfully provides an answer to this question, whereas other accounts,
such as the one provided by connectionists, does not. Classicists claim that
mental representations of different states of affairs are molecular structures
made up of atomic representations that are themselves representations
of the state of affairs' elements. In other words, on the classical view, our

mental representation of John loving Mary has a molecular structure composed of more atomic representations (i.e., concepts) representing John, loving, and Mary. Hence, our capacity to produce a mental representation of John loving Mary is related to our capacity to token a mental representation of Mary loving John because these distinct molecular representations share the same conceptual parts and the same basic syntactic form. Where they differ is with regard to which parts are playing which syntactic roles—a difference that is, of course, quite significant. This feature of classical representations is often referred to as a "compositional" or "combinatorial" syntax and semantics (sometimes abbreviated to "compositional semantics"). What is distinctive about a classical account is that the representational vehicle—the representation itself—has a compositional structure that reflects the "structure" of the represented state of affairs. Putting it simply, the meaning of the molecular representation is a consequence of the meaning of the atomic parts and of their role in the molecule.

Thus, according to the classical view, the reason the representations of cognitive agents are systematically related is that, like the relevant states of affairs, they too share the same components. The relatedness of representational capacities is explained mereologically: the underlying representational system provides the building blocks for different molecular representations. If you have the building blocks for constructing a representation of John loving Mary, then you automatically also have the building blocks needed for constructing a representation of Mary loving John (but not for constructing a representation of dogs hating cats, or any other molecule that requires different parts). It is this more basic conceptual representational atomism, along with a compositional syntax and semantics, that explains the systematicity laws. The systematicity of thought is, in the classical framework, explained in the same way we would explain the systematicity of language. The ability to produce the English sentence "John loves Mary" is related to our ability to produce the sentence "Mary loves John" because the two sentences share the same words—their demands on the speaker's vocabulary are the same.

If the classical account can explain systematicity in this way, the next obvious question—the one at the heart of Fodor and Pylyshyn's challenge—is how do connectionists explain it without simply implementing a classical architecture? In connectionism, representations are typically something like an activation pattern of network nodes (Rumelhart, McClelland, and the PDP Research Group 1986). While these distributed representations of states of affairs may have component parts, such as individual activated nodes, those parts do not represent elements of states of affairs.

Generally, connectionist representations do not employ a compositional semantics. Connectionist representations are, after all, supposed to be different from classical symbols in terms of structure and form. Consequently, connectionists must come up with some other way to explain the relevant relatedness of our representational capacity. If representational vehicles are distributed patterns of activation, or something similar, then what is it about the pattern representing John loving Mary that supports the ability to produce the pattern representing Mary loving John (and vice versa) but does not support the ability to produce a pattern representing dogs hating cats? This question is the crux of the systematicity challenge for connectionists.

Before looking at some possible responses to this challenge, a few points are in order. First, it should be noted that Fodor and Pylyshyn raise the representational systematicity challenge alongside a host of other challenges, such as explaining the productivity of thought (our ability to represent a practically infinite array of different states of affairs) and explaining the systematicity of inference (the fact that our ability to infer P from P & Q & R is also intrinsically connected to our ability to infer P from P & Q). While these further challenges are certainly deserving of attention, I am going to restrict my discussion to representational systematicity. As I hope will become clear, much of what I have to say about representational systematicity also applies to these other matters as well.

Second, although the systematicity challenge is widely treated as an argument designed to refute connectionism, it is, as Fodor and Pylyshyn themselves note, based on a more fundamental argument actually designed to *support* one very specific account of mental representation structure against *all* competing accounts. Fodor (1987), for example, had presented a prior and more general argument about the systematic nature of representations in support of the classical framework in the appendix of *Psychosemantics*. Here the target was anyone committed to representationalism (the term he uses is "intentional realism") but who rejects a language of thought account, that is, anyone who embraced a view of representation structure lacking a compositional syntax and semantics.[2] In effect, Fodor and Pylyshyn took an inference-to-the-best/only-explanation argument designed to support an account of representation that includes a compositional semantics, and, seeing that connectionists appear to be offering an account of representation without a compositional semantics, simply reworked it to apply specifically to connectionism. This is worth pointing out because I have on occasion heard people express sympathy for Fodor and Pylyshyn's argument *against* connectionism and, in a completely different context, disagreement with a language of thought account of mental representation. Yet it is hard to see how you can have it both ways. Everything Fodor and Pylyshyn say against connectionism regarding systematicity can also be (and has been) said against *any* theory of mental representation that lacks a compositional semantics—that is, accounts that reject, for the most part, a language of thought framework. That would include accounts of representation that have been very popular among philosophers—like the account put forward by Fred Dretske (1988).[3] If the systematicity challenge is a problem for connectionists, then it is a problem for anyone who rejects classicism.

A further important point concerns the nature of the explanandum that is at the heart of the challenge. As a number of writers have noted, while connectionists may be able to program networks so that whenever they have the capacity to produce one sort of representation they can also produce the relevant other representations, this would not provide a solution to the systematicity challenge. As Fodor and McLaughlin put it, the challenge is "not to show that systematic cognitive capacities are *possible* given the assumptions of a connectionist architecture, but to explain how systematicity could be *necessary*—how it could be a *law* that cognitive capacities are systematic—given those assumptions" (1990, 202). This requires some clarification. Fodor and McLaughlin are not, of course, claiming that minds are necessarily systematic; in fact, one of their central claims is that representational systematicity is a contingent feature that demands an explanation. As we've defined it, representational systematicity is the feature of representational capacities coming in clusters of a certain sort. The capacity to represent a given state of affairs comes with the capacity to represent other states of affairs that share the same elements. With this conception, the necessity is built into the nature of systematicity itself—creatures with representational systematicity are such that if they can represent aRb they can necessarily represent bRa. It is a mystery why this is so, and classicists want to claim it is a natural by-product of the underlying architecture of cognition. In the case of classical architecture, it is indeed a by-product of its compositional semantics. The problem with simply wiring it directly into a connectionist network is not that the resulting network will fail to be necessarily systematic (perhaps it would be), but rather that the resulting capacities would *not* be a natural by-product of the underlying architecture. After all, if we ask, "Why do minds have representational systematicity?" the answer can't be anything like "Because someone or something must have wired them that way."

3   Responses to the Systematicity Challenge

As might be expected, connectionists have offered a wide range of different responses to the challenge. Perhaps the most popular line of response involves various attempts to provide an account of representational systematicity that involves something other than a compositional semantics—an account that explains systematicity in strictly connectionist terms. This rebuttal requires putting forth an account of connectionist processing that (a) provides a satisfactory explanation of real systematicity—of how the relevant representations are related in the relevant way, and (b) does not employ a compositional semantics. Smolensky (1991a,b), for example, has offered various accounts of distributed representations that he suggests provide something like the structure needed to answer the challenge. On one version, nodes representing micro-features combine to generate distributed representations of various combinations. The resulting distributed representations lack classical constituents but nevertheless are decomposable into representations of various elements. However, Fodor and Pylyshyn argue that this arrangement fails to capture representational systematicity because of the context-sensitivity of relevant contributing vectors. In short, the vector representing John in the distributed representation of John loving Mary is different from the vector representing John in the distributed representation of Mary loving John. Consequently, it is far from clear what explains why a capacity to represent the one is dependent on a capacity to represent the other, since the two representations do not involve the same building blocks.

In another version, Smolensky (1991a) modifies his account by introducing superpositional representation structure, whereby a distributed representation includes retrievable "components" that can be recovered through mathematical analyses like tensor division. In this account, the vectors that contribute to the representation of a state of affairs avoid the context-sensitivity issue but are still different from classical symbols. With classical representation constituents, the atomic parts must be tokened whenever the molecular is generated. However, in Smolensky's account, the activation vector representing something like John loving Mary does not strictly speaking have a representation of John as a constituent part. If anything, it is thought to be present tacitly insofar as it can be retrieved through tensor division. This feature prompts once again the classical rebuttal that such an arrangement fails to explain representational systematicity (Fodor and McLaughlin 1990). Since the representations of John loving Mary and Mary loving John do not share any constituent components, it is left a mystery why the capacity to represent one should guarantee a capacity to represent the other (any more than it guarantees the capacity to represent something with Fred as subject). Of course, there are various ways that vector representations could be made more like classical constituents of a larger distributed representation. But then the account starts to look like the implementation of a classical account with a compositional semantics.

There are perhaps further modifications that Smolensky could offer to deal with these concerns. And, of course, his is not the only attempt to provide a connectionist account of systematicity (see, e.g., Chalmers 1990; Hadley and Hayward 1997). However, I'm inclined to think that this is not the most interesting line of response to Fodor and Pylyshyn's argument. Anything sufficiently nonclassical in nature—that is, anything that truly lacks a compositional semantics—is open to being described as failing to account for "real" systematicity. And anything that does explain "real" systematicity in an uncontroversial way is open to being described as merely an intriguing implementation of a classical architecture. As far as I know, no one has ever provided even a hypothetical mechanism that both provides an *uncontroversial* explanation of systematicity and yet does not employ a compositional semantics. In fact, I'm inclined to think that a compositional syntax and semantics is related to representational systematicity in a way that is deeper than that of *an* instantiating mechanism. I'm inclined to think that, fundamentally, the relation is such you can't have latter without the former. But then any attempt to explain systematicity without implementing a classical architecture, and by extension, a language of thought, is a sucker's game, ultimately doomed to failure.[4]

Jonathan Waskan has pointed out that there is some reason to doubt this last suggestion (that anything with a combinatorial semantics that can handle systematicity would qualify as a language of thought architecture). As he has argued at some length (Waskan 2006), a representational architecture based on scale models might involve the sort of compositional structure that could account for systematicity. Certain models of "the cat is on the mat" arguably have a representation of the cat and a representation of the mat as constituent parts in *some* sense, and the arrangement of the parts of the representation is certainly semantically salient. We can at least begin to see how such a representational system would give rise to representational capacities coming in clusters—why the capacity to construct of model of "the cat is on the mat" would be linked to a capacity to construct a model of "the mat is on the cat." Yet, as Waskan notes, scale models are hardly language-like or sentential in nature.

In reply to this point, it should be recalled that Fodor and Pylyshyn's argument was designed primarily to show that our cognitive architecture has a compositional semantics and syntax—that is how "classical architecture" is more or less defined. Thus, insofar as it can be argued that scale models have a compositional semantics and syntax (or something close enough) that can explain systematicity, then such an architecture would qualify as classical in nature. On this interpretation, there are different sorts of classical representational systems, with a language of thought account and scale model account being subtypes. As someone who has argued that a model-based approach to representation is commonly found in various classical computational accounts of cognition (Ramsey 2007), this strikes me as quite sensible (see also Block 1995; Braddon-Mitchell and Jackson 2007; Rescorla 2009). A more extreme approach would be to loosen the notion of a language of thought to include nonsentential representational systems, including scale models, as long as the representational system has a combinatorial structure that can handle the various challenges of representing propositions. I will leave it to the reader to judge the plausibility of this approach and the degree to which it is consistent with Fodor's own perspective. It should be noted, though, that the matter becomes much less interesting if it degrades into a question about nothing more than how we use labels like "language of thought."

If it is a mistake for connectionists to attempt to explain systematicity without implementing a compositional semantics, then what other rebuttals are possible? An alternative strategy is to deny the reality of systematicity altogether—that is, to deny that systematicity is a feature of cognition that requires explanation. This rebuttal comes in both a strong and a weak form. The strong version would entail denying that any aspect of cognition is systematic in the way Fodor and Pylyshyn suggest—the appearance of representational systematicity in cognition is something of an illusion (see, e.g., Aizawa 1997; Clark 1989; Matthew 1994). The weak version admits that some form of cognitive processing may well be representationally systematic, but denies that it is a fundamental feature permeating all of cognition.

As Fodor and Pylyshyn point out, the strong version looks like a nonstarter, at least for language-using cognitive agents like ourselves. Understanding languages like English (that clearly involve a compositional syntax and semantics) entails understanding that they are made up of semantically evaluable parts (words) that can contribute to other sentences. At some level, then, it seems our ability to understand the sentence "John loves Mary" is indeed linked to our ability to understand the sentence "Mary loves John," so linguistic understanding appears to be systematic. If we adopt the highly plausible assumption that we possess mental representations of words like "John," "loves," and "Mary," then it becomes hard to doubt that these representations are involved in our representations of the sentences "John loves Mary" and "Mary loves John." In any event, the question here is not whether the systematicity involved in language processing can be explained with a compositional semantics, but only whether or not it reflects systematicity among the relevant mental representations. Since a very strong case can be made that it does, an across-the-board denial of representational systematicity seems like a mistake.

These are, of course, empirical matters that will depend on what the facts actually reveal about how we process language. At the same time, however, we should not pretend that our intuitive access to conscious thoughts does not provide considerable force to the claim that thought is systematic or that it has a compositional structure. Introspectively, it certainly *seems* like my ability to consciously think a thought about something like John loving Mary shares something with my ability to entertain a thought about Mary loving John—something that it doesn't share with my ability to think that the cat is on the mat. Moreover, while most of us have become appropriately wary about the reliability of introspection regarding the workings of the mind, it is far from obvious that our conscious access to cognitive processes should count for *nothing* at all, or that it can't serve as some sort of qualified data. What about something as innocuous as my ability to form a mental image of a square on top of a circle and then switch these around so I wind up with an image of the circle on top of the square? It is hard to see how the ability to form the former image isn't closely tied to my ability to form the latter image, especially since I am just mentally rearranging the relevant parts. In fact, a good deal of consciously accessible cognition appears to exhibit a degree of representational systematicity that seems hard to deny.

The upshot is that it is reasonable to suppose that some degree of representational systematicity is a feature of at least certain forms of linguistic processing and perhaps other sorts of very explicit or conscious cognition. The strong denial of systematicity looks like a dead end. However, as noted above, there is a weaker version of systematicity denial that is not so obviously wrong and, I'll suggest, is actually quite plausible. The weaker version simply rejects the claim that all of cognition is fundamentally systematic or requires a representational system with compositional semantics. It embraces the view that there are different architectures and underlying

principles that govern different areas of cognition. It is this view to which
I now turn.

4 The Appeal of Architectural Pluralism

As we have noted, the force of the systematicity challenge depends on the
pervasiveness of representational systematicity in cognitive operations.
Fodor and Pylyshyn write as if there is one basic cognitive competence
that is to be explained by either a classical cognitive architecture or a connectionist cognitive architecture or perhaps some other option—all of
which are mutually exclusive. While cognitive subsystems might be functionally diverse and perform different tasks and types of operations, they
assume all of this is nevertheless driven by a single type of computational
architecture and representational system. This is what leads them to
conclusions like: "So, *the* architecture of the mind is not a Connectionist
network" (1988, 27, emphasis added). I'll call this assumption that cognition is driven by one basic underlying computational architecture with its
own representational system and principles "architectural monism."[5]

If we assume that the mind is made up of different systems and processing architectures with very different representational capacities, then the
systematicity argument against connectionism loses its force as a refutation
of connectionism. We can call this alternative assumption "architectural
pluralism." According to architectural pluralism, Fodor and Pylyshyn are
perhaps right to think that an important area of cognitive processing
involves representational systematicity, but they are wrong to assume that
it is a pervasive feature of all of cognition.

The closest Fodor and Pylyshyn come to considering architectural pluralism is in their discussion of what they refer to as "infraverbal" cognitive
agents, such as nonhuman animals. Here they briefly entertain the idea
that connectionism's failure to account for systematic language processing
would not be devastating because connectionists could account for other
areas of nonsystematic cognition:

A connectionist might then claim that he can do everything "up to language" on the
assumption that mental representations lack a combinatorial syntax and semantic
structure. Everything up to language may not be everything, but it's a lot. (1988, 40)

The authors then go on to reject this option by insisting that it is simply
not plausible that animal minds are not thoroughly systematic. As they
note, this would imply that animal minds exhibit punctate capacities:
"such animals would be able to respond selectively to *aRb* situations but

quite unable to learn to respond selectively to *bRa* situations ... (though you could teach the creature to choose the picture with the square larger than the triangle, you couldn't for the life of you teach it to choose the picture with the triangle larger than the square)" (1988, 41). While they admit that this is an empirical question, they think the evidence strongly favors ubiquitous representational systematicity.

But is the alternative—the denying of pervasive representational systematicity in animal minds—so implausible? Their argument is not terribly convincing. It may indeed be true that we are not likely to find creatures who can learn to selectively respond to square-larger-than-triangle but who cannot learn to selectively respond to triangle-larger-than-square. But it is far from clear that we need to appeal to a compositional representational system to explain this. After all, it would be equally surprising to discover that there are creatures who can learn to selectively respond to square-larger-than-triangle but could not selectively respond to circle-over-diamond. For that matter, it would be equally surprising to learn that there are creatures who can learn to selectively respond to a circle but cannot learn to selectively respond to a square or a triangle or a wide array of other similarly simple yet unrelated figures. Of course, in these cases we can't explain the agent's broader representational capacities by appealing to a compositional representation system because, presumably, the representations of these figures do not share semantically evaluable parts. So it seems we already know that a nonsystematic representational capacity—the capacity to represent items and properties of similar complexity but that share no common elements—will need to be explained in animals without appealing to a compositional semantics. But then it might well turn out that a representational capacity involving representations of items that *do* share elements will simply be an extension of this more basic capacity.[6] Even on a classical account it must be conceded that some dimension of our representational capacity is not systematic.[7]

The appeal of architectural pluralism as a response to the systematicity challenge has been noted by several authors (see, e.g., Clark 1989; Sloman 1996; Waskan and Bechtel 1997; Waskan 2006). Yet I believe the force of this rebuttal has not been fully appreciated, especially in light of various developments that have occurred in cognitive science over the last twenty-five years. Certainly one of the more significant changes in cognitive science has been the growing popularity of different forms of architectural pluralism. These provide a good reason to reject the architectural monism that Fodor and Pylyshyn assume, and thus a reason for thinking that the systematicity argument yields at most the conclusion that connectionism cannot account for *all* of cognition. While architectural pluralism has come about in a variety of forms, I will discuss just one version here: the dual process approach. First, however, it will help to reconsider the cognitive relevance of implementational architectures in light of current attitudes.

Fodor and Pylyshyn famously suggest that if connectionism cannot account for systematicity, then it could perhaps help provide a story about how the cognitive architecture is implemented in the neural wiring of the brain. While they regard such a role as important, they certainly do not believe it should be treated as an aspect of psychological or cognitive theorizing. Psychology would be concerned with the mechanisms, relations, and capacities described at a higher level of analysis—at the level of representations and symbol manipulation. Connectionism, by contrast, would help explain how the activity of networks of neurons brings about the classical computations and symbol manipulations that, as such, comprise the cognitive realm.

Today, the idea that implementation details are largely irrelevant to psychology will strike many as flat-out wrong. Relatively few writers now accept the notion that cognitive theorizing can ignore neurological facts, or even think that there are clean and tidy divisions between psychological operations and functions, on the one hand, and the ways they are implemented, on the other (Bechtel and Abrahamsen 2002; Standage and Trappenberg 2012). Moreover, we now have quite sophisticated accounts of how a connectionist architecture could implement something like a classical symbol system (see, e.g., Smolensky and Legendre 2011), and these reveal processes of cognitive significance at many levels below classical processes. Even if we assume that the mind has a uniform multitiered structure—that, say, all cognitive subsystems involve something like a connectionist architecture implementing a classical architecture—we have compelling reasons for thinking that the lower implementing levels should absolutely belong as part of cognitive theorizing.

But the real problem for the systematicity challenge today, as I see it, is the growing group of architectural pluralists who do not think that the mind has the same basic cognitive architecture throughout. One popular version of this pluralism is what has come to be known as the "dual-process" or "dual-system" picture of cognitive architecture (Evans 2010; Evans and Frankish 2009; Evans and Over 1996; Frankish 2004; Osman 2004; Sloman 1996; Stanovich and West 2000). As the label implies, these writers have developed the idea that the mind is composed of two types of systems that operate in fundamentally different ways. One system, S1, is generally thought to process information in a way that is fast, automatic,

System 1 (S1)	System 2 (S2)
– Evolutionarily old	– Evolutionarily new
– Fast	– Slow
– Parallel	– Sequential
– Automatic	– Controlled
– Unconscious	– Conscious
– Associative	– Syntactic
– Implicit knowledge	– Explicit knowledge

Figure 10.1
Typical feature listings for the two systems.

and nonconscious. It is assumed to be evolutionarily older and underlies such capacities as pattern recognition and associative reasoning. A second system, S2, is assumed to process information in a way that is slower, more deliberate, and more rule-governed. It is thought to be much newer and supports our conscious thought processes. Figure 10.1 presents a common sort of table contrasting the commonly assumed different properties of the two systems (Evans and Frankish 2009).

There are now several different versions of the dual-system account (for a nice discussion, see Evans and Frankish 2009). On some versions, distinct cognitive capacities like procedural knowledge or logical inference are thought to be housed in either S1 or S2, whereas in other accounts, both systems are thought to contribute something to basic cognitive processes like reasoning or memory. Space does not allow a detailed examination here, but one aspect of this picture bears directly upon our discussion. Given the kinds of features unique to the two systems, as presented in figure 10.1, the dual-system framework provides a natural opportunity for a kind of hybrid theory, with connectionist-type models explaining the processing in S1 and more classical accounts explaining the processing of S2. Many dual-system proponents have endorsed just such an outlook (see, e.g., Sloman 1996; Sun 2002). Moreover, it is perfectly reasonable to suppose that the sort of consciously available, explicitly entertained thoughts that make the systematicity challenge so compelling (e.g., thoughts like *John loves Mary*), are largely the domain of one system, namely S2. And if systematicity is primarily a feature of S2, then we are provided with a fairly straightforward strategy for diffusing Fodor and Pylyshyn's challenge. This type of architectural pluralism implies that, for the most part, the sort of cognitive operations that connectionism best explains need not involve systematic processing. As Sloman puts it, "[Fodor and Pylyshyn's] argument is only relevant to one form of reasoning" (1996, 5). Given the division of labor suggested by the dual-architecture model, with a connectionist-style architecture explaining S1 and a classicist account explaining S2, then even if connectionist representations fail to accommodate representational systematicity, this doesn't undermine the theory because representational systematicity need not be an aspect of cognition that connectionist architectures are required to explain.

For example, cognitive operations in S1 are often described as being heavily context-dependent. Often this is in reference to the ways in which situational factors influence the different operations and generate biases or specific outcomes. However, in some models, context is thought to influence the nature of the conceptual representations invoked as well (see, e.g., Smith and Conrey 2007). This is regarded as a virtue since it accommodates a large body of psychological literature on the way contextual factors influence the way we represent different items (Barsalou 1987). In such models, the way John is represented when an agent thinks about John loving Mary may very well differ in various ways from the way he is represented in S1 thoughts about Mary loving John. Indeed, the computational models of this are quite similar to the account of weak compositionality put forth by Smolensky (1991b). As we saw above, Fodor and Pylyshyn would insist that such a representational system would lack real representational systematicity. Yet many contemporary dual-system theorists would merely reply, "And the problem with that is ... ?" In other words, the proper response to the claim that context-dependent conceptual representations can't produce representational systematicity is not to fiddle with them so that they do. The proper response is to broaden our conception of how information is represented in the brain and to insist that, as the psychological literature suggests, significant areas of cognition almost certainly lack full-blown systematicity.

This is just one of the ways in which the dual-system outlook promotes the kind of architectural pluralism that substantially diminishes the systematicity challenge. Of course, this approach brings with it its own issues. For example, some reject the idea that a connectionist architecture should explain S1 whereas a classical architecture accounts for S2. As Samuels puts it, "It is, to put it mildly, obscure why connectionist processes should exhibit the S1 cluster whilst classical ones exhibit the S2 cluster" (2009, 142). Samuels's point is perhaps true of some of the features, especially those for which we currently have no idea about how they might be implemented, such as conscious access. Yet some of the features associated with S1 line up exactly with the traditional properties normally ascribed to connectionist modeling. For example, the processing of S1 has been traditionally described as being largely associative in nature. Insofar as associative processing is thought to be a hallmark of connectionist networks (Rumelhart, McClelland, and the PDP Research Group 1986), it is perfectly reasonable to suppose that S1 is a natural home for connectionist frameworks. The same is true of a variety of other features, such as parallel processing and non-explicit knowledge storage.

However, it is important to recognize that the sort of architectural pluralism needed for a connectionist answer to Fodor and Pylyshyn's challenge does not require things to line up in neat and tidy ways. It does not require that S1 be exclusively the explanatory domain of connectionist modeling and S2 be exclusively the domain of classical modeling. Nor does it require that there be only two fundamental types of systems or processes—another aspect of dual-system theory over which there has been considerable debate (Evans and Frankish 2009). In fact, it does not even require that representational systematicity appear in only one of the two (or more) basic computational architectures. All it requires is the following very weak claim: there are significant areas of cognition in which the psychological processes and mechanisms do not include a representational framework in which the representations are related systematically. Given the strength and the breadth of support that architectural pluralism is receiving from different areas of cognitive science, that claim is extremely plausible. But then a core assumption of the systematicity argument that was, at least initially, taken for granted is now something we have good reason to reject.

An important virtue of this way of responding to the systematicity challenge is that it allows connectionists to preserve much of what is distinctive about their theoretical framework. Fodor and Pylyshyn's argument demands connectionists and others to abandon novel, nonclassical ways of thinking about the mind and instead explain cognition in traditional terms. This is analogous to Lamarck attacking Darwin because evolution does not explain certain basic features of Lamarckian theory, such as adaptive change within a given individual over its lifetime in response to environmental challenges. The proper Darwinian response was not to attempt to explain this aspect of organismic change, but instead to insist that although real, it was not something their theory treated as central to the explanation of how adaptive traits are passed along. Evolutionary theory went in a new direction and demanded a shift in the way we think about the inheritance of adaptive traits. Similarly, instead of trying to mold their explanatory framework to accommodate traditional aspects of the classical account of cognition, connectionists are better served by demanding a significant shift in the way we think about the mind. Rather than trying to accommodate representational systematicity without implementing a classical framework, connectionists should instead insist on the sort architectural pluralism that significantly restricts representational systematicity to only certain processes. They can then go about the business of providing their own account of cognitive activity, with their own principles and theoretical posits.

I would like to close by encouraging connectionists to push things a little further in this direction. Connectionists widely and universally embrace the general assumption that cognitive processes require representations of *some* sort. As Fodor and Pylyshyn make clear, the systematicity debate is not over the existence of representations, but over whether or not connectionist-style representations are of the *right* sort.[8] In fact, the standard view is that if connectionist accounts failed to posit representations of some form in their descriptions of how networks perform various tasks, then they would cease to offer anything that could qualify as a *cognitive* theory. As Fodor and Pylyshyn put it: "It's the architecture of representational states and processes that discussions of *cognitive architecture* are about. Put differently, the architecture of the cognitive system consists of the set of basic operations, resources, functions, principles, etc. ... whose domain and range are the *representational states* of the organism" (1988, 10).[9]

This nearly universal outlook suggests that the "representational theory of mind" has lost its status as an empirical hypothesis and has instead become something like a conceptual truth. For any theory to qualify as a psychological theory or cognitive model, it is necessary that it invoke representational states of some form. If there aren't representations, then by definition the processing isn't cognitive. Think about that for a minute. It wasn't that long ago that the majority of psychologists regarded the positing of inner representations in psychological theorizing as a deep conceptual mistake. Now, the opposite view is held—anyone who tries to explain cognitive processes without representations is conceptually confused because the former is defined through the latter. Of course, the specifics of how all this is supposed to work are a bit unclear. Even in computational operations that are saturated in symbols, lots of other processes and supporting operations involve entities that are not representations. Are these thereby noncognitive? Is there a percentage of representations required for an operation to count as cognitive? What if some subsystem takes representational structures as input and generates different representational structures as outputs, but nothing internal to its processing is representational in nature. Is that a cognitive system or not?

My view is that all of this is fundamentally wrongheaded. It is wrong to treat representational posits as a necessary condition for cognitive theorizing or to suggest that they serve as some sort of demarcating criterion for what counts as psychology. Instead, we should return to what I believe was the correct view all along: that representationalism is an empirical hypothesis about the inner workings of the mind. While there are very good reasons for thinking that much of what we want cognitive scientists to explain involves different sorts of representations, not all of it must. Thus, we should treat the relation between representation and cognition in much the same way we currently treat the relation between consciousness and cognition. For many, it was once self-contradictory to speak of unconscious or nonconscious mentation. Descartes, for example, believed that all thought must be conscious.[10] Today, of course, very few people hold this view, as we have come to appreciate just how much of what we regard as psychologically salient occurs in the nonconscious realm. Acknowledging the realm of nonconscious cognition has marked an essential shift in our overall theorizing about the mind. We should do the same with regard to nonrepresentational cognition.

There are, I believe, a variety of ways in which this sort of shift away from representationalism would enhance (or at least broaden) our understating of the sort of processing taking place in the faster, automatic, and nonconscious regions of the mind commonly ascribed to S1. Take, for example, the procedural memory associated with an acquired skill through practice. On many connectionist-style accounts, this know-how comes about through alterations in the connection weights as a consequence of repeated trials or some other type of extended learning process. Consequently, the system's know-how is generally regarded as residing within these connection weights. Information is claimed to be encoded superpositionally, with separate items of information superimposed on top of one another in the resulting weight configuration of the connections, and with no particular entity or structure encoding any specific content (Rumelhart, McClelland, and the PDP Research Group 1986; Smolensky 1988). This picture is thus an increasingly popular way of thinking about tacit or implicit representation, whereby information is stored holistically and there is no one-to-one mapping between discrete items and representational content.

As I have argued at length elsewhere (Ramsey 2007, chap. 5), this popular way of thinking about tacit representation actually makes little sense and is completely devoid of any theoretical value. It seems to be based on a conflation of representations with dispositional properties of the acquired weight configuration. To qualify as a representation, something needs to play a certain functional role in the system's processing—it needs to be doing something recognizably representational in nature. At the very least, it should have an identifiable content that in some way, at some level of analysis, is relevant to the job it performs. I do not see how the weight configuration of a network does any of this. It is true that the resulting weight configuration is crucial to the way the network transforms input patterns into specific outputs. However, the fact that a structure is merely *causally relevant* to the processing, or that it is the part of the system that is modified so that a cognitive task can be performed—these considerations do not provide adequate justification for treating the connection weights as representations.

I do not regard this critique of this notion of connectionist representation as a reason for doubting that the modification of connection weights is a potentially valuable and plausible account of how cognitive skills are acquired and stored. The problem is not with equating cognitive skill acquisition with weight modification; the problem is with then describing that process as the development of representations. Connectionists should not only reject the idea that all cognitive operations are systematic; they should also reject the idea that *all* cognitive processes require inner representations.

So here is one area where I believe connectionists should go further in staking their independence from the classical framework and, in the process, expand our conception of cognitive processes. While S1 is often treated as lacking a classical, highly symbolic, rule-governed computational architecture, it is nevertheless thought to be a representational architecture, perhaps with distributed, tacit, or some other sort of nonsymbolic or subsymbolic representations. And yet, if our cognitive ability to, say, recognize faces comes about through something like a connectionist weight modification of a connectionist-style network situated in the cognitive architecture of S1, and if that cognitive capacity resides in the dispositional properties of the network, then connectionists would be providing an account of an important *cognitive* skill that, at least in this regard, is not only unsystematic, but is also nonrepresentational. Fodor and Pylyshyn's challenge was an extremely strong attempt to define cognition on classical terms, appealing to processes that are most naturally and easily explained with the traditional computational framework. Connectionists should strive to avoid this framework, and in the process help us see things differently with a new understanding of cognitive processes and the sort of entities and structures that are essential to them.

Acknowledgments

A version of this essay was presented at the 2012 Annual Summer Interdisciplinary Conference, Cala Gonone, Sardinia, June 7–12. Feedback from this audience was extremely useful. I am also grateful to Jonathan Waskan for extremely helpful feedback on an earlier draft.

Notes

1. Mental representations are typically characterized in the form of propositional attitudes—e.g., beliefs, desires, hopes, fears. A common locution used in the literature to capture the relevant capacity is the ability to "think the thought," though the argument is not simply about beliefs—systematicity includes the ability to, say, desire that John love Mary and how it is (allegedly) counterfactually dependent on the ability to desire that Mary love John.

2. In other works, Fodor uses the term "monadic" to refer to one sort of account of representation structure that lacks a compositional semantics (Fodor 1985).

3. Dretske (1988) explicitly rejects classical computationalism and strongly suggests that at least with regard to simple representations of states of affairs (like a fly currently buzzing in front of the cognitive agent), the representation lacks semantically evaluable components.

4. Some writers have wondered if a close conceptual link between systematicity and a compositional semantics reveals that Fodor and Pylyshyn merely beg the question against connectionism by tacitly presupposing a compositional semantics in their stipulation of the systematicity explanandum (Aizawa 2003). Yet even if X is necessary for Y, it is not begging any questions to argue for X on the basis of Y. Even if it should turn out that the only way in which real systematicity could arise is if we have a representational system with a compositional semantics, it would not be begging the question to argue from systematicity to a compositional semantics.

5. Given Fodor's defense of a fairly strong modularity thesis involving several different encapsulated subsystems, this assumption might seem quite odd. Yet Fodor's modularity thesis is completely compatible with the view that individual modules more or less process information in the same way, employing the same computational principles and representational capacities. For Fodor, while the specific computational routines and information employed may differ greatly from one module

to the next, the underlying computational architecture is always assumed to be classical in nature.

6. At one point, Fodor and Pylyshyn note that "it would be *very* surprising if being able to learn square/triangle facts implied being able to learn that quarks are made of gluons or that Washington was the first President of America" (1988, 44). Yes, that would be quite surprising, but not because gluon/quark facts fail to share elements with square/triangle facts. The surprise would be due to the very different levels of conceptual sophistication required for representing circles and triangles versus gluons and quarks. Note that it would not be all that surprising if the ability to learn square/triangle facts implied the ability to learn circle/diamond facts.

7. After all, there is nothing systematic about our capacity to acquire and represent basic concepts that serve as representational atoms.

8. It is far from clear who Fodor and Pylyshyn have in mind when they refer to nonrepresentational connectionists. Insofar as some supporters of connectionism, like the Churchlands, might be viewed as eliminativists, they are not antirepresentationalists. Connectionist-style eliminativism of the sort endorsed by the Churchlands (or in Ramsey, Stich, and Garon 1990) is not the denial of mental representations but rather the denial of a certain *sort* of mental representation—namely, propositional attitudes like beliefs (see, e.g., P. S. Churchland 1986; P. M. Churchland 1989).

9. Similarly, Aydede (2010) tells us that, for any or all non-representational accounts, "it is hard to see how they could be models of *cognitive* phenomena."

10. The analogy is not perfect since consciousness belongs as part of the explanandum of much of psychology, whereas representations are, to a greater degree, part of the explanation.

References

Aydede, M. 2010. The language of thought hypothesis. In *The Stanford Encyclopedia of Philosophy*, ed. Edward N. Zalta, http://plato.stanford.edu/archives/fall2010/entries/language-thought/.

Aizawa, K. 1997. Explaining systematicity. *Mind and Language* 12 (2):115–136.

Aizawa, K. 2003. *The Systematicity Arguments*. Norwell, MA: Kluwer Academic.

Barsalou, L. W. 1987. The instability of graded structure: Implications for the nature of concepts. In *Concepts and Conceptual Development*, ed. U. Neisser. Cambridge: Cambridge University Press.

Bechtel, W., and A. Abrahamsen. 2002. *Connectionism and the Mind*, 2nd ed. Malden, MA: Blackwell.

Block, N. 1995. The mind as the software of the brain. In *An Invitation to Cognitive Science*, vol. 3, ed. E. Smith and D. Osherson, 377–425. Cambridge, MA: MIT Press.

Braddon-Mitchell, D., and F. Jackson. 2007. *Philosophy of Mind and Cognition*. Cambridge, MA: Blackwell.

Chalmers, D. J. 1990. Syntactic transformations on distributed representations. *Connection Science* 2: 53–62.

Churchland, P. M. 1989. *A Neurocomputational Perspective: The Nature of Mind and the Structure of Science*. Cambridge, MA: MIT Press.

Churchland, P. S. 1986. *Neurophilosophy*. Cambridge, MA: MIT Press.

Clark, A. 1989. *Microcognition: Philosophy, Cognitive Science, and Parallel Distributed Processing*. Cambridge, MA: MIT Press.

Dretske, F. 1988. *Explaining Behavior*. Cambridge, MA: MIT Press.

Evans, J. S. B. T. 2010. *Thinking Twice: Two Minds in One Brain*. Oxford: Oxford University Press.

Evans, J. S. B. T., and K. Frankish, eds. 2009. *In Two Minds: Dual Processes and Beyond*. Oxford: Oxford University Press.

Evans, J. S. B. T., and D. Over. 1996. *Rationality and Reasoning*. Hove, UK: Psychology Press.

Fodor, J. 1985. Fodor's guide to mental representation. *Mind* 94:76–100.

Fodor, J. 1987. *Psychosemantics*. Cambridge, MA: MIT Press.

Fodor, J., and B. McLaughlin. 1990. Connectionism and the problem of systematicity: Why Smolensky's solution doesn't work. *Cognition* 35:183–204.

Fodor, J., and Z. Pylyshyn. 1988. Connectionism and cognitive architecture: A critical analysis. *Cognition* 28:3–71.

Frankish, K. 2004. *Mind and Supermind*. Cambridge: Cambridge University Press.

Hadley, R. F., and M. B. Hayward. 1997. Strong semantic systematicity from Hebbian connectionist learning. *Minds and Machines* 7 (1):1–37.

Matthew, R. 1994. Three-concept monte: Explanation, implementation, and systematicity. *Synthese* 101 (3):347–363.

McLaughlin, B. 2009. Systematicity redux. *Synthese* 170 (2):251–274.

Osman, M. 2004. An evaluation of dual-process theories of reasoning. *Psychonomic Bulletin & Review* 11:988–1010.

Ramsey, W. 2007. *Representation Reconsidered*. Cambridge: Cambridge University Press.

Ramsey, W., S. Stich, and J. Garon. 1990. Connectionism, eliminativism, and the future of folk psychology. In *Philosophical Perspectives 4*, ed. J. Tomberlin, 499–533. Atascadero, CA: Ridgeview Publishing.

Rescorla, M. 2009. Cognitive maps and the language of thought. *British Journal for the Philosophy of Science* 60 (2):377–407.

Rumelhart, D., J. McClelland, and the PDP Research Group. 1986. *Parallel Distributed Processing*, vol. 1: *Explorations in the Microstructure of Cognition: Foundations*. Cambridge, MA: MIT Press.

Samuels, R. 2009. The magical number two, plus or minus: Dual-process theory as a theory of cognitive kinds. In *In Two Minds: Dual Processes and Beyond*, ed. J. S. B. T. Evans and K. Frankish. Oxford: Oxford University Press.

Sloman, S. 1996. The empirical case for two systems of reasoning. *Psychological Bulletin* 119 (1):3–22.

Smith, E. R., and F. Conrey. 2007. Mental representations as states not things: Implications for implicit and explicit measurement. In *Implicit Measures of Attitudes*, ed. B. Wittenbrink and N. Schwarz, 247–264. New York: Guilford.

Smolensky, P. 1988. On the proper treatment of connectionism. *Behavioral and Brain Sciences* 11:1–23.

Smolensky, P. 1991a. Connectionism, constituency, and the language of thought. In *Meaning in Mind: Fodor and His Critics*, ed. B. Loewer and G. Rey, 201–227. Oxford: Basil Blackwell.

Smolensky, P. 1991b. The constituent structure of connectionist mental states: A reply to Fodor and Pylyshyn. In *Connectionism and the Philosophy of Mind*, ed. T. Horgan and J. Tienson, 281–308. Dordrecht: Kluwer Academic.

Smolensky, P., and G. Legendre. 2011. *The Harmonic Mind*, vols. 1 and 2. Cambridge, MA: MIT Press.

Standage, D., and T. Trappenberg. 2012. Cognitive neuroscience. In *The Cambridge Handbook of Cognitive Science*, ed. K. Frankish and W. Ramsey. Cambridge: Cambridge University Press.

Stanovich, K., and R. West. 2000. Individual differences in reasoning: Implications for the rationality debate. *Behavioral and Brain Sciences* 23:645–665.

Sun, R. 2002. *Duality of the Mind*. Mahwah, NJ: Erlbaum.

Waskan, J. 2006. *Models and Cognition*. Cambridge, MA: MIT Press.

Waskan, J., and W. Bechtel. 1997. Directions in connectionist research: Tractable computations without syntactically structured representations. *Metaphilosophy* 28: 31–62.

11 Systematicity Laws and Explanatory Structures in the Extended Mind

Alicia Coram

The systematicity argument championed by Jerry Fodor, Zenon Pylyshyn, and Brian McLaughlin (Fodor and Pylyshyn 1998; Fodor and McLaughlin 1990) became a magnet for debates over the cognitive architecture, providing an empirically based challenge to those who did not accept classical computationalism. In general, the alternative explanations of systematic patterns that were proposed by connectionist models did not directly challenge the assumption that mental representations are both syntactically and semantically internal, focusing instead on alternative means of achieving compositionality without classical vehicles of content (see, e.g., Van Gelder 1990; Chalmers 1993; Smolensky 1995). However, some recent discussions locate the explanatory structures for such patterns outside the brain. While not always couched in terms of the extended mind theory, proposals by authors such as Menary (2007) and Symons (2001) reconsider the explanations for such patterns to encompass aspects of the linguistic environment.

In what follows, I will outline one variety of such an explanation as part of a more general review of the ways in which the extended mind theorist might address the systematicity challenge. The attempt here is to meet the systematicity challenge as directly as possible, accepting most of the definitions made in the original argument insofar as they can be accommodated in this framework while questioning some underlying assumptions about the nature of representation. The proposed explanation considers the relevant explanatory structures that determine the nature of instances of systematicity laws to be extended, though allowing that the mechanisms that directly support them may be internal to the cognitive agent.

This shift of the explanatory burden away from properties that are possessed intrinsically by intracranial mental states can be compared to a similar explanatory strategy employed by Pylyshyn in relation to imagistic phenomenon, and his work provides a useful framework for both understanding and potentially empirically verifying this method of explaining evidence for the systematicity of thought. A brief review of some of the empirical evidence that exists for the systematicity of thought suggests that the patterns that do exist in such cognitive abilities can more easily be accommodated by this explanation.

1   The Systematicity Challenge and the Extended Mind

1.1   The Challenge

Fodor's original construction of the systematicity argument begins with the observation of certain lawlike patterns in cognitive abilities that he claims have the status of psychological laws. His toy examples of the systematicity of thought are by now familiar—the ability to think thoughts such as *John loves Mary* coexists (ceteris paribus) with the ability to think thoughts such as *Mary loves John*, and so on for any clusters of similar thoughts. This is captured in the general schema:

Ceteris paribus, a cognizer is able to think the thought that aRb if and only if he is able to think the thought that bRa.

McLaughlin has clarified that "think the thought" is intended to mean "mentally represent," and furthermore that mental representations have propositional contents, so the relevant schema can be restated as: "Ceteris paribus, a cognizer is able to mentally represent that aRb if and only if the cogniser is able to mentally represent that bRa" (McLaughlin 2009, 272). This claim is also extended to inferential abilities; however, the systematicity of thought will be the focus of this chapter.

Several authors (see, e.g., Schiffer 1991; Cummins 1996; Johnson 2004) have argued that such definitions are uninformative and risk either begging the question regarding the nature and structure of mental states or biasing the identification of systematic patterns toward those most naturally explained by classical computational architectures. However, as McLaughlin notes, the general schema does not commit the advocate of the systematicity argument to anything beyond the claim that the relevant contents "will be related, albeit non-equivalent" (McLaughlin 2009, 254) mental contents. He also rejects the possibility of providing a list of necessary and sufficient conditions identifying the clusters of thoughts that are in fact related by systematicity laws, and he does not believe that such conditions are required to pose a challenge to nonclassicists (ibid., 253). In part, this is because of the controversial nature of theories about propositional content, which he argues means that it is "indeed best to identify the thought abilities in question other than by appeal to any hypothesis about what the similarities in propositional-content consist in, for just about any such hypothesis will be contentious" (ibid., 270).

While these are good reasons for being reticent about providing further detail, evaluating the empirical claims made in the service of the argument for classicism requires a clearer working definition of what they are taken to be, especially in the context of considering alternative means of identifying such laws where there might be disagreement about the contrast cases that patterns of systematicity stand out against. Any further specification can be considered as a kind of prediction of what systematicity laws would look like, given a certain theoretical framework and cognitive architecture. A more generic statement of classical systematicity laws that incorporates the assumed relationship between thought and language in classicism can be stated in the following way:

Cognizers exhibit classical systematicity if, whenever they can mentally represent a proposition p they can also mentally represent propositions similar to p (ceteris paribus), where similarity is understood in terms of permuting semantic constitutents of the same semantic category within p, or substituting any semantic constituents in p with others of the same semantic category that they possess. As a first approximation, contents expressed using the same constituents in grammatical natural language sentences should be taken as indicative of this similarity, unless there is a principled reason for them to diverge.

This description is clearly not theory neutral; however, it allows that these patterns can be redescribed in terms that do not take mental representational abilities to consist in the kind of internal, compositionally structured states advocated by this theory. It also leaves open the possibility that the thoughts that are actually related by systematicity laws (and so fulfill the general schema) are not all or only those that fulfill the definition of classical systematicity laws. The attempt to provide further specification of the content of the relevant mental states is intended to provide more substance to the claim and a clearer burden of proof in relation to proposed exceptions to evidence marshaled in support of classicism.

The challenge presented in the original argument was for the proponent of connectionist systems to explain these patterns without invoking a language of thought architecture. Furthermore, Fodor and McLaughlin argued that this explanation must fulfill the criteria of explaining these patterns as psychological laws, and hence as nomologically assured by the central tenets of the theory (Fodor and McLaughlin 1990), and to do so by providing a mechanistic explanation. I will return to these explanatory conditions after a general overview of the different ways in which the proponent of the extended mind theory might take up the systematicity challenge and define patterns of systematicity.

1.2 Extended Cognition

The central claim of the extended theory of mind is that the cognitive system is not bounded by the skin or skull, but needs to be understood as a system that encompasses aspects of the world that have often been taken to be external to the mind (see Clark and Chalmers 1998; Clark 2000, 2001, 2003, 2010b). The extended mind theory can be understood as a form of active externalism whereby "some cognitive processing is constituted by active features of the environment" (Menary 2010, 2)—processes and vehicles in the extracranial environment are coupled with internal resources to form the cognitive system. Sometimes this claim is motivated by considerations of functional parsimony, whereby the similarity of the role played by "external" and internal resources in cognitive tasks is taken to give them equal claim to the cognitive mantle, as in the position most often adopted by Andy Clark and David Chalmers. However, such functionalist considerations are not the only means of arriving at this position. For example, Menary (2007) argues that it is the high degree of coordination and integration exhibited by the extra- and intracranial aspects of the world in certain cognitive tasks that warrants their consideration as a whole system in some instances.[1]

Within this broad framework, different commitments can be made on many points. For example: what conditions are required for some part of the extracranial world to count as part of the cognitive system (such as the conditions of "glue and trust" described by Clark and Chalmers); the extent to which the brain and nervous system are central to cognitive activities; how to divide the extended cognitive system into subsystems; whether properties like consciousness can be extended (Noë 2010); the symmetry of the causal relationships between intra- and extracranial states; and so on. However, all versions of the extended mind approach will allow that, at least to some degree, specifying the vehicles (and/or contents) of mental states as well as cognitive processes can involve looking beyond the brain of the cognitive agent.

In terms of cognitive modeling, this approach is particularly compatible with the explanatory framework of dynamic systems theory, which allows for intra- and extracranial aspects of cognitive processes to be easily modeled as one system (see, e.g., van Gelder 1995; Port and van Gelder 1995). There are various kinds of dynamic models, including those that provide interpretations of connectionist systems (see Zednik 2008 for discussion), and the manner in which the underlying cognitive architecture is conceived bears on questions concerning the role of mechanistic explanation for psychological generalizations, as will be discussed further below.

The very possibility that lawlike psychological generalizations could be explained in an extended framework has been questioned by some authors such as Adams and Aizawa (2001) and Rupert (2004), who argue that the diversity of physical states and causal processes that can be involved in cognition from this perspective makes such explanations untenable. There is concern that the empirical generalizations found in an unextended understanding of mind will not be visible in an extended framework, and that other law-like generalizations will not emerge in this framework (Adams and Aizawa 2001, 61). However, as Clark (1997) has noted, lawlike generalizations can also be made in thoroughgoing dynamic systems models—for example, the lawlike regularities in finger oscillation patterns described by Kelso (1995).[2] While an extended cognitive science may not so readily highlight certain kinds of generalizations associated with intracranial processing, it would appear question-begging to assume that these are the only, or even the most important, generalizations about cognition that can be made.

Within the general extended mind framework, the systematicity challenge could be taken up in several ways, in terms of both the definition and explanation of these patterns. These can be considered in regard to a related group of claims about cognitive states and explanation made by classicism: (1) the definition of cognitive abilities as representational; (2) the claim that mental representations are propositional; (3) the claim that mental representation is intracranial (both semantically and syntactically); (4) the corresponding identification of psychological resources as intracranial; and (5) the claim that the best explanation of such patterns will be one that identifies a mechanism in virtue of which they exist. It is (3), (4), and to a lesser extent (5) that will be the focus of this chapter; however, before discussing these in further detail I will briefly discuss the possibilities raised by a consideration of (1) and (2).

Although there is no universally accepted position regarding the nature of cognitive states in the extended theory of mind, dynamic systems models of the cognitive architecture are sometimes touted as nonrepresentational (see, e.g., van Gelder 1995; Clark 1997). One means of addressing the systematicity challenge would be to reject the assumption that the relevant cognitive abilities exhibiting these patterns are representational, and instead take up a modified challenge of explaining the behavioral, empirical evidence (as Clark [1990, 149] suggests: "What stands in need of empirical explanation is not the systematicity of thought but the systematicity of behaviour, which grounds thought ascription"). If this approach were adopted, the kinds of patterns identified in classical systematicity laws could be made at least partially visible by stating them in terms of the semantic relations between the representational objects of nonrepresentational states—so, for instance, the ability to mentally represent *John loves Mary* could be reconstrued as the ability to be intentionally related to a public representation of *John loves Mary*.[3] Although it is possible to understand the explanation to be discussed in the following section in this way, I will retain the terminology of mental representation for two main reasons: first, because avoiding it risks amplifying the distinction between intra- and extracranial states in a way that does not reflect the central tenets of the extended mind theory; and second, because a suitably adapted, extended account of mental representation is available that takes on the systematicity challenge at face value and addresses some of the concerns of proponents of nonrepresentationalism.

In regard to (2), the claim that mental representation involves propositional content could also be challenged, especially if this is taken to entail a rejection of imagistic or other possible models of mental content. I will chiefly be concerned with explaining instances of systematicity laws that do fulfill this further specification, but it should be noted that rising to this challenge does not require a commitment to the claim that all cognitive activity involves propositional representation.

From this perspective, patterns of classical and other propositional, representational systematicity may be part of a family of patterns in cognitive processes.

1.3 Defining Extended Systematicity

Within the systematicity challenge as given, there is clearly no restriction to an internalist conception of mental representations of the kind advocated by classicism, or to any particular structure of the vehicles of such representational states. It is here that the extended framework provides the most obvious point of challenge to the identification and explanation of these regularities. Although the empirical generalizations about mental representational abilities that Fodor, McLaughlin, and others have in mind may not yield a precise mapping from classicism to an extended framework, the initial strategy will be to reconceive some of the central evidence provided for classical systematicity. This is not intended to exhaust the intended scope of classical systematicity, and this point will be explored later.

The focus of Fodor and McLaughlin's original challenge and many subsequent discussions has been the systematicity of language comprehension, which is taken to be strongly indicative of patterns in underlying mental representational abilities due to the conception of these abilities as involving the translation of sentences into content-matching intracranial mental representations. However, these abilities (along with other representational abilities) can instead be reconceived as representation-using abilities that do not involve the tokening and manipulation of intracranial states that are possessed of certain intrinsic syntactic and semantic properties.

The act of mental representation understood in this way can involve extracranial representations and the norms governing their use in a number of different ways. In general terms, Clark and Chalmers have discussed the way in which manipulating physical symbols can aid in the restructuring of an agent's epistemic environment—for example, shuffling Scrabble tiles to suggest new combinations or manipulating falling Tetris blocks (Clark and Chalmers 1998). In his consideration of experiments that are suggestive of how "material tokens" (written words, numerals, graphs, and so on) can aid cognitive aptitude, Clark (2006) focuses on three main uses of linguistic and other representational tokens: as a source of additional targets for attention and learning; as a resource for directing and maintaining attention on complex conjoined cues; and as providing proper parts of hybrid thoughts. For example, in one experiment, primates were trained to apply tags to pairings of objects to indicate if they were same-object or different-object pairings. They were subsequently able to complete tasks involving higher-order sameness and difference that were beyond their non-tag-trained counterparts.[4] Clark suggests that such interaction with physical symbols creates new perceptual targets (either real or reimagined) that can serve to lessen the burden of cognitive tasks. In other cases, agents can exploit the "modality-transcending" properties of symbols to combine information from numerous perceptual sources. Finally, Clark considers cases where words in a public code can supplement more basic cognitive capacities as part of hybrid thoughts. For example, he considers the work of Dehaene and colleagues, who suggest that the use of number words results in a new capacity to think about an "unlimited set of exact quantities" that does not depend on a content-matching representation of those numbers: "Instead, the presence of actual number words in a public code (and of shallow, imagistic, internal representations of those very public items) is itself part of the coordinated representational medley that constitutes many kinds of arithmetical knowing" (Clark 2006, 9, discussing the work of Dehaene 1997; Dehaene, Spelke, Pinel, Stanescu, and Tviskin 1999).

There are several (non-mutually exclusive) means of conceiving of the relationship between brains and representational symbols that shed light on these experiments. For example, Dennett sees a central role for linguistic systems in particular as a kind of virtual machine, as "a temporary set of highly structured regularities imposed on the underlying hardware by a program" (Dennett 1991, 216). Clark suggests the relationship is one of "complementarity," whereby languages and other symbol systems aid cognition without being replicated by "the more basic modes of operation and representation endemic to the biological brain" (Clark 2006, 1). In his consideration of dynamic explanations, Symons presents the following "rough initial hypothesis" of the relationship between language and the brain: "I suggest that language-like structures, as well as patterns of intelligent behavior can be understood as providing complex multi-dimensional targets for the organism. ... This approach can recognize the indispensability of the intentional or representational idiom in psychological explanation, without locating representations in the brains of intelligent agents" (Symons 2001, 522).

Menary proposes an "integrationist solution" involving both "internal, non-classical vehicles and processes and external classical vehicles and processes." The solution he proposes is "to take two systems and put them together. The classical system of structured linguistic sentences is an external and autonomous system. The connectionist system of learning algorithms and pattern recognition techniques is coupled to the external linguistic system of spoken and written sentences. The external system provides the forms of grammatical structure and, thereby, the rules by which sentences are structured" (Menary 2010, 150).

For purposes of explaining the phenomenon of systematicity, I will offer a characterization of the relationship between the intra- and extracranial states found in representation-using activities in terms similar to those employed by Pylyshyn (2003) when he considers the relationship that holds between thought and imagery. In this understanding, the degree of interconnection between the intra- and extracranial aspects of the cognitive system leads to a range of derivative structural effects on internal cognitive processing that should not be considered as resulting from intrinsic structural properties of intracranial states. This rests on distinguishing between the properties that are intrinsically possessed by the intracranial states of the cognitive system, and those properties that may affect such processing but are intrinsically possessed by the extracranial vehicles or contents of cognitive activities. Of particular interest for the systematicity argument will be those structural properties that are attributed to internal mental states by classicism—namely, semantic and syntactic compositionality.

Although intended as a defense of a classical architecture, Pylyshyn's work on mental imagery here offers a model for not only understanding the role this distinction plays in explaining patterns in cognitive abilities, but potentially for empirically validating the claim that such properties are not intrinsic to internal states. The key point of difference between the account offered here and Pylyshyn's position is that it is not unidirectional: language and other representational systems can affect intracranial states just as intracranial states can affect representational systems. This accommodates an understanding of the structural properties of such systems as mutually influencing parameters of an extended cognitive system, in a similar vein to Symons's understanding of languages as "both sources of and subject to selective pressure" (Symons 2001, 522, in discussion of Deacon 1997).

Representation using activities can then be understood as the ability to manipulate these predominantly extracranial representations according to certain norms of use. Menary sees the manipulation of external representations, the interpretation of symbols, and the structuring and correcting of the activities of cognizers in a problem-solving task as normative "in the sense that we learn or acquire a practice that is an established method of manipulating notations to produce an end" (Menary 2011, 143). This account can be applied to all of those uses of representations considered above.

From within this rough framework, we can generate a definition of patterns of systematicity as the coexistence of abilities to use representational systems, as governed by cognitive norms, wherein relations between those abilities can be understood in terms of semantic relations in the extracranial representational schemes. As noted earlier, it is not clear that representation-using abilities will exhaust the kinds of propositional abilities that systematicity laws are intended to range over, although I will suggest in the final section that there are problems with extending systematicity laws beyond cognitive abilities that employ representations. It may also be possible to understand similarity here in terms other than relations between extracranial representations.

It should also be noted that viewing the cognitive system as extended means that the formulation of ceteris paribus exceptions to such laws will differ from the classical model. The "all else" that is assumed to be held equal in the identification of such laws is intended to limit the claim that thought is systematic to those cases where the operant idealizations of psychology are realized, thus ruling out exceptions where there is some disruption at the implementational level of the system or influence from another cognitive mechanism at the psychological level (see, e.g., Fodor 1991). Such ceteris paribus restrictions can be recast to encompass the idealizations of an extended psychological system, depending on the model that is used.

2 Explanation, Structure, and Psychological Resources

The general explanation to be offered here appeals to a principle of cognitive efficiency coupled with the way that structural properties of extracranial representational schemes can have derivative effects on cognitive processing.

Fodor and McLaughlin have claimed that both classicists and connectionists accept the same general model of explanation for the systematicity of mind, citing four points of accord: "i) that cognitive capacities are generally systematic in this sense, both in humans and in many infrahuman organisms; ii) it is nomologically necessary (hence counterfactual supporting) that this is so; iii) that there must therefore be some psychological mechanism in virtue of the functioning of which cognitive capacities are systematic; iv) and that an adequate theory of the cognitive architecture should exhibit this mechanism" (Fodor and McLaughlin 1990, 185).

In regard to (iii), the psychological mechanisms that Fodor and McLaughlin argue must be appealed to are the operations, functions, resources, principles, and so on specified in the theory of the cognitive architecture, as distinct from the folk psychological or implementational level of the system. Such mechanisms are taken to be defined over the microstructure of the things that satisfy the law in question—in this case, the microstructure of representational states.

In this regard, the principal difference from classicism for the extended mind theorist is the rejection of the supervenience of mind on brain in favor of a kind of restricted global supervenience. This means that extracranial resources, and especially structural properties of extracranial representational vehicles, can be among the psychological resources that are called upon to explain such patterns. Although this may incorporate a variety of physical types and causal processes, as Clark has argued, even a radical diversity does not prevent them being considered part of a unified system: "It is part of the job of a special science to establish a framework in which superficially different phenomenon can be brought under a unifying explanatory umbrella. ... Moreover, it is by no means clear that acceptable forms of unification require that all the systematic elements behave according to the same laws. As long as there is an intelligible domain of convergence, there may be many subregularities of many different kinds involved" (Clark 2010b, 51).

I will first consider the general nature of the explanatory strategy to be offered in the extended framework, before providing more detail of the role played by extracranial representational structure.

2.1 Explanation and Mechanism

It is clear that offering a mechanistic account of psychological phenomenon is a generally accepted goal of cognitive science. Leaving aside for the moment the contentious claim that these lawlike regularities need to be explained as a matter of nomological necessity, providing such an account would establish systematicity laws as something more than accidental generalizations, illuminate them as explicable by common underlying causes, and provide a satisfying response to questions regarding how the cognitive system sustains such regularities.

However, some proponents of the dynamic models take them to provide nonmechanistic, covering law explanations for cognitive phenomenon whereby they are explained by deduction from a law together with statements of the antecedent conditions of the system (see Walmsley 2008), as exemplified in van Gelder and Port's statement that "dynamical modeling ... involves finding ... a mathematical rule, such that the phenomena of interest unfold in exactly the way described by the rule" (van Gelder and Port 1995, 14). One option would be to offer a covering law explanation of patterns of systematicity. This would not necessarily be problematic; covering law explanations are perfectly acceptable for things like planetary motion, and such explanations can be counterfactual supporting (see Walmsley 2008 for discussion). In this case, the structures of extracranial representations might be incorporated as constraints on the system.

However, as noted by Wright and Bechtel (2007) and Zednik (2008), there does not seem to be anything preventing a dynamical, extended framework from providing a mechanistic explanation. In general, mechanistic explanations operate from the assumption that "many target phenomena and their associated regularities are the functioning of composite hierarchical systems ... composed of component parts and their properties. Each component part performs some operation and interacts with other parts of the mechanism (often by acting on products of the operation of those parts or producing products that they will act on), such that the coordinated operations of parts is what constitutes or comprises the systemic activity of the mechanism" (Wright and Bechtel 2007, 45).

As Wright and Bechtel have noted, this leaves open many questions, the last of which is directly relevant for the extended mind account: "What are the necessary and sufficient conditions for a given sequence of causal interactions to be constitutive of a mechanism? How important is endurance, and when do gaps in temporal continuity cease being mere interruptions? To what extent does repair or particulate change in organization involve the creation of a new mechanism? Are there determinate limits on how much spatial or structural disconnectedness a mechanism can exhibit?" (Wright and Bechtel 2007, 46).

A mechanistic account does not require that the relevant components are located intracranially, and Zednik describes the way that variables, parameters, differential equations, and "the representational currency of dynamical analysis" can serve to describe mechanistic parts, operations, and relations (Zednik 2008, 247). This seems to leave open the possibility of an extended mechanistic account, whereby phenomena like systematicity (which I will argue in the final section only seem to arise in a wider symbolic environment) are only fully explained by considering mechanisms that extend beyond the brain.

2.2 Derivative Structure and Explanation

The explanation given of systematicity from within a classical framework comes in two parts, both of which rely on the semantic and syntactic compositionality of mental states and the sensitivity of these architectures to them. Semantic compositionality here means that there are semantically simple and complex representations. The semantic value and mode of combination of the constituent simple representations determine the semantic value of complex representations, and the simple representations pass on all of their semantic value to the complex representations. (See, e.g., Fodor and Lepore 2002.) By definition, those representations that contain the same semantic parts in classical architectures will contain the same syntactic parts. It is this, together with the fact that cognitive processes are sensitive only to the formal and syntactic properties of mental representations, that is taken to explain systematicity as a psychological law.[5]

Unlike classicism, the principles of compositionality that are here taken to determine the nature of these patterns apply primarily to extracranial representations. It is the distinction between the intrinsic and derived properties of intracranial states that provides the basis for this alternative explanation. Pylyshyn proposes a litmus test for deciding whether some property is best understood as intrinsic to mental states themselves (that is, is a property of the vehicle of mental states) or the contents of those states. Echoing Fodor's claim that "one cannot, in general, infer from what is represented to the nature of the vehicle of representation" (1975, 177), when considering imagistic phenomenon he claims that "it is the things we think about, not the patterns in our brain we think with" that possess imagistic properties (Pylyshyn 2003, 380). He claims that intrinsic properties of mental states are cognitively impenetrable—that is, they are not influenced by the cognizer's other mental states, nor do they differ from cognizer to cogniser.[6] By contrast, properties that are intrinsic only to the contents of mental states but not their intracranial vehicles are cognitively penetrable. For instance, one can imagine any color one likes when asked what color results when light is shone through a red and a blue transparent gel, and evidence suggests that people's beliefs about how colored light mixes informs their answers to such questions (Pylyshyn 2003, 298). That is, "colored" mental states do not literally obey the laws covering the mixing of colored things. Pylyshyn argues that all imagistic properties (such as spatial properties) are cognitively penetrable in this way, and that the kind of "functional" space we sometimes think in is a set of externally imposed restrictions on information and information processing that are not bound by the laws that restrict physical space.

For Pylyshyn, this argument is underpinned by an acceptance that there is an intrinsic, language-like structure to the mental states we think with. Properties like systematicity (along with other evidence regarding language learning and use, semantic coherence, and Principle P[7]) are offered as support for this claim. However, some authors including Clark, Dennett, and Garfield have argued that this does not warrant the ascription of language-like properties to thought (see discussion in Clapin 2002). For example, Garfield argues that "the kernel of truth in the language of thought hypothesis is the intuition that representation must have determinate, and indeed, compositional, content, and that only language can provide that. It does not follow, however, that thought is in language, only that it is of language" (Garfield 2005).

This suggests that phenomena such as systematicity could be considered as cognitively penetrable effects of our interactions with extracranial representational vehicles. This way of framing the role of compositional structure is compatible with many of the interpretations of the relation between brain and symbols noted previously, and it presents the possibility that an explanation exploiting extracranial properties could be to some extent empirically examined.[8]

Explaining systematicity does not require the commitment that such structures are intrinsically possessed by intracranial mental states: an explanation of sufficient strength to provide an answer as to how and why cognitive systems exhibit the particular patterns of systematicity that they do can be provided by an account that does not view these structural properties as intrinsically intracranial.

This does not mean that there is no relation between intra- and extracranial structures, or that intracranial structures and mechanisms are not vital to explaining phenomena such as systematicity. As Deacon notes, the selection pressures on languages and brains could be expected to go two ways (Deacon 1997, discussed in Symons 2001). For example, Martin Sereno (2005) has proposed that linguistic systems developed to piggyback on nonsemantic features of the brain that were responsible for audio recognition and production in early "talking song." The deeply ingrained nature of representation-using activities could be expected to influence our plastic brains in a myriad of ways, but there is reason to doubt that this involves the creation of entrenched classical structures and processes (see, e.g., Garson 1998). Furthermore, some connectionist models suggest that intracranial vehicles do not have to possess such a structure in order to exhibit a kind of compositionality (see, e.g., Smolensky 1995 and the discussion of functionally compositional systems in van Gelder 1990).

The explanation proposed here is that patterns of systematicity be understood as a product of the extracranial linguistic environment, with these patterns supported in individual cognitive agents by intracranial mechanisms of the kind "endemic to the biological brain" (Clark 2006). This can be appreciated by understanding how a similar explanation could apply to simple intentional states like perception. It seems unproblematic to claim that certain patterns in sensory capacities, such as vision or hearing, can be adequately accounted for by the structure found in the objects of perception (together with a general principle of cognitive efficiency, to be discussed further below). The same basic perceptual resources (such as edge-detection neurons, color rods and cones, and so on) and any necessary mechanisms (for example, those that may be required for binding different perceptual elements from the same object together) would be used in both cases. The explanation of why it is these patterns rather than other patterns that exist is given by the structure of the extracranial world. This explanation can be expanded to incorporate the skills necessary to employ cognitive norms and practices to effectively use representations. For the purposes of this explanation, these are not necessarily qualitatively different from sensorimotor states and processes. To illustrate a similar point, Margaret Wilson (2002) employs the analogy of using fingers to aid with counting. Someone might initially use very obvious physical actions, such as holding out each finger in turn and counting it off. Moving more "inward," they might use imperceptible muscle twitches in the relevant fingers. They might then only imagine twitching their muscles or counting off their fingers, perhaps using the same kind of motor resources as those accompanying the actual action. Finally, this imagining of muscle-twitching or finger-counting could become so much a part of their cognitive routine that certain groups of neurons become dedicated to this task, until the kinds of cognitive resources that are used are far removed from those used in the original motor actions.

This general explanation inverts the claims about the source of compositionality in thought and language found in Fodor's work. However, Fodor (2001) has argued that languages are not, as a matter of empirical fact, compositional, and he explicitly rules out the possibility that minds inherit their compositionality from natural languages. Aside from the controversial nature of this claim, as Robbins (2002) has argued, languages need not be entirely compositional to play this explanatory role. The indefinite compositionality of natural languages is enough here. As Symons argues, we should take these patterns to be "the manifestations of a relatively stable social and cultural landscape that the organism must negotiate in something like the way our limbs and muscles negotiate geographical landscapes" (Symons 2001, 539).

There are stable enough patterns of compositionality in natural languages to account for this, and the lack of complete compositionality of language would go some way to explaining cases that seem problematic for the systematicity argument, as will be discussed in the following section.

3   Evaluating the Evidence

In the preceding discussion, the claim that mental capacities exhibit classical systematicity has been left unchallenged, and I have focused on representation-using abilities such as sentence comprehension without considering the range of other abilities that the original argument was intended to implicate. McLaughlin has explicitly resisted a discussion of other empirical evidence. Although he states that "having propositional attitudes such as belief, desire, intention, and the like by no means requires the ability to use a natural language," he also claims:

Since systematicity laws specify what the thought abilities they cite are abilities to think (rather than quantifying over things thought) and since we will state the laws in our language, a tangle of controversial issues arises for the view that non-language using cognizers can non-vacuously satisfy them. I think that pre-verbal children, deaf people that have never learned to sign, as well as primates, for instance, will non-vacuously satisfy various systematicity laws. But I cannot here even begin to try to untangle the controversies that claim raises. So, instead, I will follow Fodor and Pylyshyn's (1988) lead and note that even if it were the case that all of the thought abilities cited in systematicity laws require the ability to use a natural language, the laws require explanation. (McLaughlin 2009, 259–260)

However, it is clear that patterns in the cognitive abilities of non-language-users provide a better and less potentially biased arena for comparing the extended and classical explanations. While recognizing that the difficulties that attend their consideration make it impossible for us to make any definite claims about them, there are several apparent exceptions to classical systematicity laws in these areas that are illuminating for the current discussion: given that systematicity is used as evidence for a cognitively impenetrable, language-like structure of thought, a lack of classical systematicity in these areas could be evidence for the cognitive penetrability of these structures.

3.1 Sentence Comprehension

The systematicity of language comprehension is claimed to provide evidence for classical systematicity to the extent that natural language sentences reflect the structure of the underlying thoughts in classical architectures. This was explicitly built into the definition of classical systematicity laws previously, suggesting some burden of proof on the classicist to give principled reasons for cases where these patterns diverge. Such reasons can clearly be offered in cases such as idiomatic expressions. However, there are other examples where the lack of systematicity in language comprehension (often because of a lack of semantic compositionality) is not so easily accounted for (for a full discussion of issues arising with compositionality, and possible replies, see Szabó 2000). For example, the meaning of complex nominals like "air guitar" are determined by their constituents, but not fully so (see Braisby 1998 for a discussion of complex nominals and compositionality), and cases of "category mismatch" resulting in apparently nonsensical yet grammatical sentences (such as "three is happy") have been offered as counterexamples to the closely related generality constraint (see, e.g., Camp 2004).

There are several possible means by which to exclude these examples from classical systematicity laws. In relation to cases where the thoughts seem to be semantically underdetermined, it could either be claimed that there is ambiguity in a polysemous meaning, or that there is further semantically irrelevant information (as Fodor claims in the case of adjectives—so, for example, "red watermelon" just means "red for a watermelon" [Fodor 1998]). However, these are not uncontroversial: the former relies on the existence of a mental lexicon that can accommodate such constituents without trivializing the claims of systematicity, and it is unclear whether the latter captures what it is to understand the relevant constituents (see, e.g., Siebel 2000).

Many cases of category mismatch are intelligible in some sense—as Camp (2004) notes, we can often grasp the inferential roles that such sentences can figure in—and so might be included among potential propositional contents. McLaughlin also seems to support this in the case of another proposed counterexample: "One mentally represents that {Tom} is the sole member of Tom when one disbelieves that {Tom} is the sole member of Tom (perhaps taking it to be necessarily false that Tom even has members in the sense in question). Disbelief differs of course from non-belief. One disbelieves that p if and only if one believes that not-p" (McLaughlin 2009, 255). But it is not clear that this kind of comprehension will exhibit systematicity—it is at least controversial whether the way in which we understand such sentences uses the same psychological resources as "ordinary" sentences.[9]

In general, although it may be possible to provide a principled reason for the lack of classical systematicity in these cases, and they leave untouched the indefinitely many cases of this pattern that the argument rests on, the systematicity of linguistic comprehension cannot be unquestionably accepted. An explanation based on the semantic properties of extracranial representations provides a much simpler reading of these patterns. As Menary notes:

Classical formulations of linguistic systematicity are too simplistic because they focus on the syntactic properties of sentences and ignore their pragmatic and semantic features. These features are important because they place restrictions on the combination of linguistic constituents. These features are pre-eminently features of public language and not an internal language of thought. Therefore, linguistic systematicity is enforced, through learning, by features of external public language and not syntactic properties of the language of thought. (Menary 2007, 157)

If the explanatory structures are extracranial, then systematic patterns are predicted only to the extent that the cognitive norms governing the use of these schemes are compositional. Compositionality is a cognitively penetrable property—it depends on the beliefs and knowledge of the representation-users for its causal role in generating systematic patterns.

3.2 Systematicity in Non-Language-Users

The explanation of systematicity as largely a derivative effect of the structure found in extracranial representations is further supported by the general pattern that cognitive abilities become more systematic as cognizers begin to use representations. This is, of course, an untidy causal story. However, in general, the minds of non-language-users appear to exhibit a restricted kind of systematicity that does not fulfill the definition of classical systematicity laws. Two examples that can be used to illustrate this are the cognitive abilities of preverbal infants before they possess the concept of object permanence and the context-sensitivity of animal thought.

As Ronald Chrisley (1993) has argued, infants at a particular stage of cognitive development will seemingly be able to entertain thoughts like "Mother is behind me and the ball is in front of me" (when their mother is audible or partially visible) but not "Mother is in front of me and the ball is behind me." These infants have not reached the stage Piaget (1972) refers to as object permanence, and so lack the ability to conceptualize the existence of objects when they are unperceived. It is difficult to rule out such states as noncognitive under many conceptions. For example, they reliably track the object in question when it is perceptible; demonstrate certain expectations associated with its behavior (such as expressing surprise if the object suddenly changes shape); and the infant's behavior in respect to the object is not rigid.

There are examples of animal cognition that pose similar problems for the classical explanation of systematicity, including the capacity to perform transitive inferences described by Susan Hurley (2003), and other cases of context sensitivity in animal thought. Camp (2009) has argued that animal thought is often recombinable only in principle. In practice, the expression of many cognitive abilities is tied to the presentation of the right kind of stimulus, or in some cases all but prohibited because the scenario suggested by the proposition is naturalistically untenable. A similar point is made by Dennett, who uses it to draw attention to the language-based intuitions that the systematicity argument rests on: "There are organisms of which one would say with little hesitation that they think a lion wants to eat them, but where there is no reason at all to think they could 'frame the thought' that they want to eat the lion! The sort of systematicity that Fodor and McLaughlin draw our attention to is in fact a pre-eminently language based artefact, not anything one should expect to discover governing the operations in the machine room of cognition" (Dennett 1991, 27).

While these examples do not rest on any strong empirical basis, they do suggest a pattern of context-dependence that is difficult to accommodate within the methodological solipsism of classical architectures.

3.3   Capacities and Abilities

There are several means of accounting for these cases of limited systematicity in infant and animal thought. One is to argue that the cognizers in question possess a representation with a propositional content that differs from the standard content attributed by our linguistic expression (e.g., "cup*") in a way that explains its behavior in respect to systematicity laws. In this case, the representation "cup*" could be taken to refer to a cuplike object that ceases to exist when unperceived. While a mental sentence like "Mother is behind me and the cup* is in front of me" might be well formed, the mental sentence "The cup* is behind me and mother is in front of me" might not be. However, it is difficult to see how such indexically determined content sits within a classical architecture, and partially systematic patterns remain that still require explanation.

The classicist could also note that the claim that thought is systematic is a claim about the capacity to think thoughts with similar propositional content, rather than a claim about the expression of this ability. Furthermore, these capacities may be environmentally or context dependent in a way that can be part of the formal properties that the classical cognitive system is sensitive to. As Aizawa (2003) has noted, the fact that the classical cognitive architectures are insensitive to the content of cognitive states does not rule out the influence of other properties that are not part of the local syntax of representations—for example, the position of representation on a search list may be related to how frequently it is tokened—and Fodor (2000) has also noted various relational but nonsemantic properties that classical architectures can be sensitive to. This leaves room for the claim that some capacities may require training to be expressed by appealing to a performance–competence distinction. Similarly, Camp (2004) argues that training chimps on a compositional symbol system (coupled with giving them some pragmatic reason for exercising particular abilities) may provide the right kind of environmental support to recombine existing thoughts in ways that would not otherwise be possible. However, she notes that it is difficult to adjudicate between the possibility that the relevant ability arises in conjunction with symbol use and the possibility that they have a latent ability or competence that is triggered by symbol use.

If these examples are cases where the performance of a latent ability requires some additional support, the apparent lack of systematicity might be accounted for by the ceteris paribus nature of these laws. In particular, if there is some aspect of the unrepresentationally mediated world that inhibits an existent ability by triggering a competing mechanism that is overcome by employing symbols (see Hurley and Nudds 2006 for discussion), then it could be argued that the idealizations of psychology are not held equal in these cases. This interpretation can be illustrated by an experiment conducted by Sarah Boysen and colleagues (Boysen and Bernston 1995, discussed in Clark 2006) involving a primate, Sheba. In one experimental setup, Sheba was offered two plates with different quantities of candy and asked to indicate which plate she would share with the researcher. Sheba would then be given the plate that she did not point to, but she seemingly could not bring herself to point to the plate with less candy. However, when the quantities were instead represented by corresponding numerals written on cards (which Sheba had been trained to recognize), she quickly learned to point to the card with the smaller of the two numerals out of any given pair in order to receive the larger plate of food. In this case, it seems the symbol use was a means to overcome distracting or competing cognitive mechanisms that inhibited performance.

However, it is not clear that this is always the case. For example, in considering the evidence that symbol comprehension in primates and dolphins indicates some form of systematicity, Prince and Berkely (2000) also note that recent work suggests that training on symbol systems alters the animal's psychology (Tomasello, Savage-Rumbaugh, and Krueger, 1993).

3.4 Propositional Thought and Stimulus Dependence

Another possible strategy would be to remove these examples from the domain of abilities that systematicity laws cover because they do not meet the criteria of being propositions. For example, Bermúdez makes the following demand for stimulus independence in the context of ascribing propositional thought: "A ... key element of propositional thinking is that propositions should be independent of the particular context of thinking. That is to say, it should be possible to grasp a proposition both without knowing its truth value and without any contact with the state of affairs that proposition is about" (Bermudez 2003, 39, quoted in Camp 2009).

While adopting such a means of ruling out these cases of animal cognition runs counter to Fodor, Pylyshyn, and McLaughlin's intended scope of systematicity laws, it could always be argued that they are simply mistaken on this point.

However, as Camp (2009) has argued, there are many instances in which animals capable of only "basic" cognition nonetheless engage in what looks like representation without direct contact with the state of affairs the mental state is about. Honeybees are a case in point: their "waggle-dance" stands in for the location of nectar in distant locations, but we surely would not want to claim that they mentally represent whereas a preverbal infant or a chimp does not.

While such cases are far from providing a definitive argument against the classical explanation of systematicity, the explanation proposed in this chapter—that such patterns should be considered as cognitively penetrable phenomena explained primarily by the structure of extracranial representations—does not encounter these difficulties. We would expect thought to become more classically systematic the more that such extended cognitive systems operated in an environment containing compositional representational schemes because it is the structure of these schemes that provides the explanation for why we see these particular patterns. Weaker patterns would be consistent with the mechanisms of more basic intracranial processes that such linguistic skills have evolved in tandem with.

4 Conclusion

The extended theory of mind challenges some of the most basic assumptions made in the classical framework by shifting the explanatory burden from the intrinsic structure of internal mental states to the semantic rules, structures, and mechanisms of extended mental states. While there are many possible means of addressing the systematicity challenge from this framework, one promising type of explanation casts these patterns as the derivative effect of this extended cognitive microstructure. In this light, the syntactic and semantic structural properties that underlie Fodor's explanation of systematicity can be considered as cognitively penetrable properties of intracranial states that can nonetheless figure in mechanistic or covering law explanations.

This strategy has some benefits over its classical counterpart when the actual patterns that exist in language comprehension and other cognitive abilities are considered. Although they are far from definitive, some general patterns in the empirical evidence suggest that an explanation that is contingent upon the semantic and syntactic structures of extracranial representational schemes provides a better fit for the patterns we witness in representational abilities. While there are good reasons for Fodor and McLaughlin to be wary of placing too much emphasis on evidence from nonlinguistic systematicity, it is this area that will prove one of the most fertile for investigating the nature of the architecture that underlies such patterns in mind.

I have offered here a loose approach to the definition and explanation of patterns of systematicity rather than a specification of a particular architecture or interpretation of the exact nature of the relation between brains and material symbols. The central question is whether compositional semantic and syntactic structures need to be considered as intrinsic to mental states in order to offer a satisfactory explanation of systematicity, or whether the derivative effect of structures in the linguistic environment are adequate. I have suggested that an acceptable explanation of why these patterns exist rather than others, and how the extended cognitive system generates and sustains them, is available without this step.

Notes

1. It should be noted that Menary distinguishes his integrationist approach from the extended mind theory. However, this distinction is not directly relevant for the current argument.

2. Both "in-phase" (left to right) and "anti-phase" (symmetrical mirroring) patterns of index-finger movement are stable at low frequencies, but only the in-phase pattern remains stable as the frequency is increased past a critical point.

3. For instance, Jay Garfield (2005) has suggested eschewing representational terminology in favor of the terminology of intentionality, as he argues that *only* public representational schemes can stand in a norm-governed relation with the world of the kind required to be considered representational.

4. For instance, they may tag a pairing of a cup and a shoe with a blue circle, and a pairing of two shoes with a red triangle. The higher-order task would require them

to decide whether pairings-of-pairings such as "cup-shoe" and "cup-shoe" were instances of higher-order difference or higher-order sameness (as they are in this example, where both are pairs of different objects) (Thompson, Oden, and Boysen 1997).

5. McLaughlin adds that "the grammar of the symbol system generates all and only the mental sentences whose meanings are the contents of propositional attitudes that the cognizer has the ability to have" (2009, 258).

6. An example of an impenetrable effect would be the kinds of things Fodor attributes to modular processes—for example, in the case of perceptual illusions, knowing that there is an illusion does not change the way the illusion appears.

7. "Principle P" appears in an appendix to *Psychosemantics* (Fodor 1987), and infers the structure of effects from the structure of behavior. Clapin interprets it in the following way: "If c1 events are implicated in the causal history of e1 events, then they are implicated in the causal history of complex e1 and e2 events" (Clapin 1997, 263).

8. To my knowledge, there is no research specifically intended to investigate the cognitive penetrability of structural properties like compositionality. However, research such as that conducted by Johnson-Laird et al. indicates the kind of evidence that might be bought to bear on the issue. Citing experiments on the kinds of mistakes people make in reasoning, Johnson-Laird et al. (2000) argue that patterns in the way that rules of inference are breached, and in particular subjects' susceptibility to "illusions of consistency," cannot be accommodated by a theory of classically structured mental states and structure-sensitive processing. Johnson-Laird uses this to argue that such processes cannot occur because of blindly applied syntactic rules over classically structured states; however, it also suggests that the states and processes underlying logical reasoning are influenced by the beliefs people have about how these rules are applied.

9. An interesting study that is relevant to this question involves the use of "functional shift" in Shakespeare that suggests different brain processes are used for syntactically anomolous but sensible sentences; grammatical but not sensible sentences; and grammatical, sensible sentences (see Davis n.d.).

References

Adams, F., and K. Aizawa. 2001. The bounds of cognition. *Philosophical Psychology* 14 (1):43–64.

Aizawa, K. 2003. *The Systematicity Arguments*. Norwell, MA: Kluwer Academic.

Bermúdez, J. L. 2003. *Thinking without Words*. Oxford: Oxford University Press.

Boysen, S. T., and G. G. Berntson. 1995. Responses to quantity: Perceptual versus cognitive mechanisms in chimpanzees (Pan troglodytes). *Journal of Experimental Psychology: Animal Behavior Processes* 21:82–86.

Braisby, N. 1998. Compositionality and the modelling of complex concepts. *Minds and Machines* 8:479–508.

Camp, E. 2004. The generality constraint and categorical restrictions. *Philosophical Quarterly* 54 (215):209–231.

Camp, E. 2009. Putting thoughts to work: Concepts, systematicity, and stimulus-independence. *Philosophy and Phenomenological Research* 78 (2):275–311.

Chalmers, D. 1993. Connectionism and compositionality. *Philosophical Psychology* 6 (3):305–319.

Chrisley, R. 1993. Connectionism, cognitive maps, and the development of objectivity. *Artificial Intelligence Review* 7:329–354.

Clapin, H. 1997. Problems with Principle P. *Pacific Philosophical Quarterly* 78 (3):261–277.

Clapin, H., ed. 2002. *Philosophy of Mental Representation*. New York: Oxford University Press.

Clark, A. 1990. *Microcognition*. Cambridge, MA: MIT Press.

Clark, A. 1997. The dynamical challenge. *Cognitive Science* 21 (4):461–481.

Clark, A. 2000. Minds, brains, and tools. In *Philosophy of Mental Representation*, ed. H. Clapin, 66–90. New York: Oxford University Press.

Clark, A. 2001. Reasons, robots, and the extended mind. *Mind & Language* 16 (2):121–145.

Clark, A. 2003. *Natural Born Cyborgs*. Oxford: Oxford University Press.

Clark, A. 2006. Material symbols. *Philosophical Psychology* 19 (3):291–307.

Clark, A. 2010a. Coupling, constitution, and the cognitive kind: A reply to Adams and Aizawa. In *The Extended Mind*, ed. R. Menary, 81–100. Cambridge, MA: MIT Press.

Clark, A. 2010b. *Memento*'s revenge: The extended mind, extended. In *The Extended Mind*, ed. R. Menary, 43–66. Cambridge, MA: MIT Press.

Clark, A., and D. Chalmers. 1998. The extended mind. *Analysis* 58:10–23.

Cummins, R. 1996. Systematicity. *Journal of Philosophy* 93 (12):591–614.

Davis, P. n.d. The Shakespeared brain. *Reader* 23:39–43. http://thereaderonline.co.uk/features/the-shakespeared-brain/.

Deacon, T. 1997. *The Symbolic Species: The Co-evolution of Language and the Brain.* London: Penguin.

Dehaene, S. 1997. *The Number Sense.* New York: Oxford University Press.

Dehaene, S., E. Spelke, P. Pinel, R. Stanescu, and S. Tsivkin. 1999. Sources of mathematical thinking: Behavioral and brain-imaging evidence. *Science* 284:970–974.

Dennett, D. 1991. *Consciousness Explained.* Boston: Little, Brown.

Fodor, J. 1975. *The Language of Thought.* Cambridge, MA: Harvard University Press.

Fodor, J. 1987. *Psychosemantics.* Cambridge, MA: MIT Press.

Fodor, J. 1991. You can fool some of the people all of the time, everything else being equal: Hedged laws and psychological explanations. *Mind* 100 (397):19–34.

Fodor, J. 1998. There are no recognitional concept; not even red. *Philosophical Issues* 9:1–14.

Fodor, J. 2000. *The Mind Doesn't Work That Way: The Scope and Limits of the Computational Mind.* Cambridge, MA: MIT Press.

Fodor, J. 2001. Language, thought, and compositionality. *Mind and Language* 16 (1):1–15.

Fodor, J., and E. Lepore. 2002. *The Compositionality Papers.* Oxford: Clarendon Press.

Fodor, J., and B. P. McLaughlin. 1990. Connectionism and the problem of systematicity: Why Smolensky's solution doesn't work. *Cognition* 35:183–204.

Fodor, J., and Z. Pylyshyn. 1988. Connectionism and the cognitive architecture. *Cognition* 28:3–71.

Garson, J. W. 1998. Chaotic emergence and the language of thought. *Philosophical Psychology* 11 (3):303–315.

Garfield, J. 2005. Intention: Doing away with mental representation. http://www.smith.edu/philosophy/jgarfieldintention.htm.

Hurley, S. 2003. Animal action in the space of reasons. *Mind and Language* 18 (3):231–256.

Hurley, S., and M. Nudds. 2006. *Rational Animals?* Oxford: Oxford University Press.

Johnson, K. 2004. On the systematicity of language and thought. *Journal of Philosophy* 101 (3):111–140.

Johnson-Laird, P. N., P. Legrenzi, V. Girotto, and M. S. Legrenzi. 2000. Illusions in reasoning about consistency. *Science* 288 (531):531–532.

Kelso, J. A. S. 1995. *Dynamic Patterns: The Self-Organization of Brain and Behavior.* Cambridge, MA: MIT Press.

McLaughlin, B. P. 2009. Systematicity redux. *Synthese* 170:251–274.

Menary, R. 2007. *Cognitive Integration: Mind and Cognition Unbounded*. London: Palgrave Macmillan.

Menary, R., ed. 2010. *The Extended Mind*. Cambridge, MA: MIT Press.

Noë, A. 2009. *Out of Our Heads*. New York: Hill & Wang.

Piaget, J. 1972. *The Psychology of the Child*. New York: Basic Books.

Port, R. F., and T. van Gelder, eds. 1995. *Mind as Motion: Explorations in the Dynamics of Cognition*. Cambridge, MA: MIT Press.

Pylyshyn, Z. 2003. *Seeing and Visualizing: It's Not What You Think*. Cambridge, MA: MIT Press.

Robbins, P. 2002. How to blunt the sword of compositionality. *Noûs* 36 (2):313–334.

Rupert, R. 2004. Challenges to the hypothesis of extended cognition. *Journal of Philosophy* 101:389–428.

Schiffer, S. 1991. Mentalese—Compositional semantics? In *Meaning and Mind: Fodor and His Critics*, ed. B. Loewer and G. Rey. Oxford: Blackwell.

Sereno, M. L. 2005. Language origins without the semantic urge. *Cognitive Science Online* 3 (1):1–12.

Siebel, M. 2000. Red watermelons and large elephants: A case against compositionality? *Theoria* 15 (2):263–280.

Smolensky, P. 1995. Connectionism, constituency, and the language of thought. In *Connectionism: Debates on Psychological Explanation*, ed. G. Macdonald, 164–198. Cambridge: Blackwell.

Symons, J. 2001. Explanation, representation, and the dynamical hypothesis. *Minds and Machines* 11:521–541.

Szabó, Z. 2000. *Problems of Compositionality*. London: Routledge.

Thompson, R. K., D. L. Oden, and S. T. Boysen. 1997. Language-naive chimpanzees (Pan troglodytes) judge relations between relations in a conceptual matching-to-sample task. *Journal of Experimental Psychology. Animal Behavior Processes* 23: 31–43.

van Gelder, T. 1990. Compositionality: A connectionist variation on a classical theme. *Cognitive Science* 14:335–384.

van Gelder, T. 1995. What might cognition be, if not computation? *Journal of Philosophy* 7:345–381.

Walmsley, J. 2008. Explanation in dynamical cognitive science. *Minds and Machines* 18:331–348.

Wilson, M. 2002. Six views of embodied cognition. *Psychonomic Bulletin & Review* 9 (4):625–636.

Wright, C., and W. Bechtel. 2007. Mechanisms and psychological explanation. In *Philosophy of Psychology and Cognitive Science*, ed. Paul Thagard, 31–79. New York: Elsevier.

Zednik, C. 2008. Dynamical models and mechanistic explanations. In *Proceedings of the 30th Annual Conference of the Cognitive Science Society*, ed. B. C. Love, K. McRae and V. M. Sloutsky. Austin, TX: Cognitive Science Society.

12 Systematicity and Conceptual Pluralism

Fernando Martínez-Manrique

1 Introduction

The systematicity argument (henceforth SA), offered by Fodor and Pyly-shyn (1988) against the plausibility of connectionism as an alternative theory of cognition, can be characterized in terms of three claims—an empirical claim, an explanatory claim, and a definitional claim—from which a dilemma arises for connectionism. Let me present the four elements in outline before saying a little more about each of them:

SA
(i) Empirical claim: Systematicity is a pervasive property of cognition.
(ii) Explanatory claim: The only plausible explanation for systematicity is to posit a compositional system of representations.
(iii) Definitional claim: Compositionality is a defining property of classical representational systems.
(iv) Dilemma: If connectionism is not compositional, then it cannot account for systematicity and so it does not provide a full account of cognition (from (i) and (ii)); if connectionism can account for systematicity, then it is actually implementing a classical system (from (ii) and iii)).

SA has been haunting connectionist approaches ever since, and main responses to it can be classified depending on whether they focus on (i), (ii), or (iii).[1] Much can be said about the relative success of each such response, yet one important common point is that Fodor and Pylyshyn's argument would work as a global refutation of connectionist explanations only if systematicity were regarded as a property of cognition in general. However, SA per se does not include the latter commitment. Truly, Fodor and Pyly-shyn stated that "there's every reason to believe that systematicity is a thoroughly pervasive feature of human and infrahuman mentation" (1988, 37). Yet, unless further arguments for the universality of systematicity are provided, the statement can be read simply as claiming that it is an important phenomenon that needs explanation, and practically everybody agrees that much. This leaves open the issue of whether some cognitive domains or processes are not systematic in the way intended by SA, and one may conjecture, as many connectionist authors do, that some nonclassicist model could just account for them.

Still, the fact that some cognitive processes were not systematic in the way intended by SA would not be enough for nonclassical models to carry the day. To this end, they not only must show that their models can deal with such cognitive processes but that they are in a better position than their classical competitors to do so. In other words, they need something like an SA for themselves—let me call it the nonclassical systematicity argument—that would run roughly as follows:

NSA
(i′) Empirical claim: X is a pervasive property of cognition.
(ii′) Explanatory claim: The only plausible explanation for X is property Y.
(iii′) Definitional claim: Y is a defining property of such and such nonclassical systems.
(iv′) Dilemma: If classicism cannot account for property Y, then it does not provide a full account of cognition (from (i′) and (ii′)); if classicism can account for Y, then it is actually implementing a nonclassical system (from (ii′) and (iii′)).

My aim in this chapter is to provide a path to construct such an argument. I want to stress that my main focus is not NSA itself, but the elements that may allow us to get at NSA. First, I offer an overlook of the connectionist answers to SA, classified as focusing on (i), (ii), or (iii), followed by a quick assessment of the debate. This assessment is negative for the connectionist side, in the sense that it never managed to substantiate an alternative explanation of the phenomenon pointed out by Fodor and Pylyshyn. Of course, I lack the space to go into details, so connectionist fans of this or that particular reply may think that I am being unfair to it. Yet, apart from the general considerations that I will provide to back my negative assessment, it seems to me that it is reinforced by the sheer fact that there is no agreement with respect to which reply to SA works best. My aim in this section, thus, is just to motivate the view that classical models still stand as the *most plausible* explanation for classical systematicity. Second, I will deal with the question of whether systematicity is actually a general property of cognition. I will argue that the best chances to support such a view come from regarding Evans's well-known Generality Constraint as a constraint on the architecture of conceptual creatures—a constraint that only concepts that exhibit classical systematicity seem to satisfy. Then I will show a different way of understanding the constraint, in terms of *attributions of belief*, that opens the door to architectures with concepts that do not exhibit classical systematicity. Third, I will present and motivate the thesis of conceptual pluralism, arguing that concepts split into subkinds that share two fundamental properties: they are central and they grant belief-attributions. I will draw on Camp's (2009) analysis to make the case that there are actually two kinds of concepts. Finally, I will rely on dual systems theory and on Penn et al.'s (2008) recent review of differences between animal and human cognition to motivate a plausible scenario of two different processing systems that work on different kinds of concepts with properties that give rise to two different sorts of systematicity. I will then sketch a way in which NSA could be filled, but my goal is not to endorse a particular nonclassical approach as a filler for the argument. To repeat, my aim is not to try to reply to SA for the umpteenth time, but simply to show that although nonclassical approaches lack the resources to meet SA, the elements for an alternative NSA argument can be provided.

2   The Elements of the Systematicity Argument

2.1   The Empirical Claim

Fodor and Pylyshyn plainly took their claim about the systematicity of cognition as an empirical one. Systematicity can be characterized as the property of having the ability to think systematically related thoughts. It is a matter of fact that creatures that have the ability to think *aRb* have also the ability to think *bRa*. Apparently, some critics failed to see this point. For instance, early in the debate Clark (1989) argued that the relation between the abilities to think *aRb* and *bRa* is not an empirical but a conceptual fact. It is not that we cannot find organisms with punctate thoughts but that the fact that they are punctate is enough to deny that they are thoughts. It is part of our concept of what it takes to have thoughts that they be systematically related. McLaughlin (1993) replied that if systematicity is a conceptual property then the challenge posed by Fodor and Pylyshyn would be strengthened, given that we would get an a priori constraint for the constitutive basis of cognition. More recently, Chemero (2009) also complained (1) that SA is a conceptual argument (or, as he calls it, a Hegelian one) against an empirical claim, and (2) that Fodor and Pylyshyn provide almost no empirical evidence to support premise (i). Actually, Chemero is wrong about both complaints. First, having poor empirical evidence for one's argument does not make it a conceptual argument—it makes it a poor argument. Second, their empirical evidence is not so poor as Chemero intends us to believe. It is based on a parallelism with language understanding, the most famous example being that just as you do not find anyone who can understand "John loves Mary" but cannot understand "Mary loves John," you do not find anyone who can think that John loves Mary but cannot think that Mary loves John. Fodor and Pylyshyn think that examples like this come on the cheap, so it is no wonder that they do not feel the need to provide plenty of them. In other words, they assume that the extent of their empirical evidence is as large as the extent of language itself.

Other critics accepted the claim as an empirical one but rejected it as false. Some of them focused on the idea of systematicity as "a thoroughly pervasive feature of human and *infrahuman* mentation" (Fodor and Pylyshyn 1988, 37, emphasis added), and alleged that nonhuman animals do not exhibit the sort of systematicity exemplified by the *J loves M* case (Sterelny 1990, 182–183; Dennett 1991; Kaye 1995). More recently, Gomila et al. (2012) reject the claim that systematicity is a general property of cognition. In their view, it is only related to those cognitive abilities that are possible by the acquisition of language, and it is derived precisely from the systematicity of linguistic structure. As I will argue later, I concur with Gomila et al. that there are grounds to deny that systematicity is a general property of cognition. Yet, this does not entail a rejection of the classical explanation. On the one hand, even if SA only applied to human cognition, or to language-related cognition, it would still be a *significant* property. On the other hand, the best explanation of this property is still classical. For instance, even if the explanation of systematicity lay in the properties of language, as Gomila et al. (2012) contend, the way of fleshing out such an explanation is still by regarding language as a classical system itself—that is, systematicity is still explained in terms of language's alleged compositional structures and processes that are sensitive to those structures.[2] So inasmuch as connectionism could not avail itself of this explanation, it would be in trouble to account for cognition, and this is how many authors viewed the issue. In other words, connectionist attempts at rejecting systematicity as a *general* property of cognition would not entail, even if they were successful, rejecting classicism as the architecture of at least *part* of cognition.

2.2   The Explanatory Claim

The second claim in SA is an instance of a "best explanation" argument. The idea is that a straightforward and plausible way of explaining systematic relations of the *J loves M* type is to posit a compositional semantics, that is, a system of context-free, recombinable semantic pieces in which the semantics of the composed whole depends in a systematic way on the semantic values of the pieces. Many critics focused on this explanatory relation. Some of them complained that the *explanandum*—that is, systematicity—had been poorly characterized and consequently devoted their efforts to reformulate it in a way that could be explained by nonclassical systems. For instance, Clark (1989, 149) insisted that what has to be explained "is not the systematicity of thoughts but the systematicity of the behavior, which grants thought ascription"; Goschke and Koppelberg (1991) and Bechtel (1994) regarded systematicity not as a property of thoughts but of an external symbolic language; Niklasson and van Gelder (1994) and Cummins (1996; Cummins et al. 2001) examined forms of systematicity different from the language-based cases; Johnson (2004), on the other hand, addressed systematicity from the linguistic perspective and provided a definition of systematicity so as to contend that language is not systematic after all.[3]

Other critics focused instead on the *explanans*—that is, compositionality—and tried to offer distinctions that helped connectionism to meet the explanatory challenge. The most notable of them was due to van Gelder (1990), who made a distinction between concatenative and functional compositionality.[4] In concatenative composition, tokenings of constituents of an expression (and the sequential relation between them) are preserved in the expression itself. In functional compositionality, general, reliable processes decompose an expression in their constituents and produce it again from them, but it is not necessary that the expressions contain their constituents. Van Gelder argued that even if connectionist networks only exhibit the latter kind of compositionality, this is enough to account for systematicity.

The trouble with reformulations of compositionality is that they failed to provide a global alternative explanation of systematicity, for example, one that relied on functional compositionality as a fundamental property of nonclassical cognition, in the same sense as compositionality plays the central role in classical conceptions. Even though many people, myself included, acknowledge the relevance of the distinction, I know of no overarching connectionism conception in which it plays that pivotal role. So regarding van Gelder's (1991) prediction that functional compositionality would be one of the central aspects for connectionism to become a truly alternative paradigm, one must say that it is a prediction yet to be fulfilled. Indeed, we will see later that recent approaches that dwell on van Gelder's distinction use it to characterize the features of two different systems, so functional compositionality could be seen as playing an explanatory role only in part of cognition.

The trouble with reformulations of systematicity, on the other hand, is that they easily change the subject matter. The facts that behavior or language are also systematic, or that there are nonlinguistic instances of systematicity, do not deny the systematicity of thought that is the basis for SA. It is good to say that there are other things to explain apart from the systematicity in SA, but unless one wants to say that the latter property is unreal, SA itself remains untouched. Indeed, the line that I am going to follow in this chapter is an instance of the "change subject matter" strategy—but not to defeat SA. Instead, I will create a different argument that leaves room for nonclassical systems as an account of part of cognition.

2.3 The Definitional Claim

Having a combinatorial syntax and semantics for mental representations, and having processes that are sensitive to the structure of the representations so constructed are defining properties of classical models, according to Fodor and Pylyshyn (1988, 13). There are two sides to this claim: one is that given the principles of classical computationalism, explaining systematicity comes as a necessary consequence, that is, it is not possible to have a classical system that is not systematic in the demanded sense. The challenge can be thus reformulated as a demand that the opponent should provide models based on different principles, in which systematicity appears as a consequence of those principles (Fodor and McLaughlin 1990). In terms of Aizawa (1997, 2003), the challenge is not to exhibit systematicity—that is, to show that it is possible to have a systematic connectionist model—but to explain it—that is, to show that systematicity follows necessarily from the principles of the theory. It is the latter challenge that connectionists fail to meet. My view is that even if systematicity is not strictly entailed by the principles of classical models, as Aizawa contends,[5] it is still the case that these models have a much more robust explanation of the phenomenon than their connectionist counterparts.

The second side of the definitional claim is that if compositionality and structure-sensitive processes are defining properties of classical systems, then any system that resorts to them will count ipso facto as a classical one. The early debate between Smolensky (1988, 1991a,b) and Fodor et al.

(Fodor and McLaughlin 1990; Fodor 1997) can be understood in those terms, and the gist of the dilemma posed by Fodor et al. comes to this: if Smolensky is capable of showing that his models do have a constituent structure, then they are implementations of a classical system, given that they are based on the same relevant explanatory principles; if they do not have a constituent structure, then they cannot account for systematicity. The countless subsequent connectionist attempts of proving that this or that network has systematic capabilities—I will save space referring to Hadley (1994) for a review and criticism of early attempts, and to Frank et al. (2009) for later ones—are subject, despite their differences, to basically the same sort of objection.

2.4 Quick Assessment of the Debate

I think that connectionist attempts never provided a satisfactory answer to SA, and this applies both to those that tried to reformulate systematicity or compositionality and to those that tried to provide practical refutations of the classicist challenge. The problem with the former, as I said above, is that they easily changed the subject matter without really meeting the challenge. The problem with the latter is that they easily fell prey to the classicist dilemma.

Someone could object that this assessment is too quick and unfair to some of the connectionist contenders, and it is possible to point toward this or that particular model to argue that it offers better chances for dealing with the classical challenge.[6] I do not deny that some models work better than others and that the process of trying to cope with SA has unraveled many interesting aspects of the properties of both classical and non-classical systems. What I deny is that there is, as of today, an answer that satisfies most authors on the connectionist side, and this is enough to be at least suspicious that the challenge has been met. To put but one recent example, Frank et al. (2009) review previous connectionist attempts to provide a model with semantic systematicity (Hadley 1994) without implementing a classical system. They find all of them wanting only to propose their own model that, allegedly, succeeds in the task. One gets the impression that it is only a matter of time before someone comes up with a similar criticism of their model and makes a similar optimistic claim.

Indeed, I think that the problem with connectionist attempts can be put in different terms: what Fodor et al. were demanding was not a new family of computational models but a new family of explanatory principles. Even though connectionists claimed to be providing just this when they talked about vector representations, learning algorithms, activation propagation, and the like, the thing is that they did not have an easy day when it came to explain how those principles connected with explaining the relation between the ability to think *J loves M* and *M loves J*. It seemed that in order to do so it was necessary to appeal to how those relations emerged from the network's behavior. Yet all the explanatory load seemed to remain *on what emerged*—the elements *J*, *M*, and *love* and their relations—and not on the goings-on of the system from which it emerged. The latter was, to use the classical parlance, implementation detail. To put it bluntly, what connectionism had to provide, but failed to do so, is a new theory of mind.

3   How to View Systematicity as a General Property of Cognition

As I said, SA rests simply on the claim that a lot of cognition is systematic, not necessarily all of it. However, two claims, when taken jointly, may sustain the view that systematicity is a general property of cognition. These claims, which are part and parcel of Fodor's view of mind, are:

(1) *Cognition as concept involving.* As Fodor says (1998, vii), "The heart of a cognitive science is its theory of concepts." What distinguishes cognition from, say, perception is that cognitive processes work on concepts. Hence processes that work on nonconceptual representations are of relatively little interest for the central claims about the nature of cognition.

(2) *Compositionality as a nonnegotiable property of concepts.*[7] Whatever concepts are, they are compositional, that is, they can be combined with other concepts to form larger conceptual structures in such a way that the content of the compound is a function of the contents of the concepts it contains and their mode of combination.

Taken together, (1) and (2) entail that the constitutive elements of cognition—concepts—have a fundamental property—compositionality—that is the source of systematicity—that is, a conceptual system is ipso facto a systematic system. In other words, systematicity is a general property of cognition that derives from the nature of the cognitive elements.

Do we have good grounds to maintain (1) and (2)? I am going to assume that (1) is right and I will take issue with (2). I am not going to provide an argument for (1), but let me say briefly that it is an assumption that, tacitly or explicitly, is widely endorsed in cognitive science. Even in those accounts that try to blur the distinction between cognition and perception, such as Prinz's neo-empiricist theory of concepts (Prinz 2002), there is something that distinguishes concepts from other mental representations and, therefore, that distinguishes cognition from perception. For instance, in Prinz's view, even if concepts are copies of percepts the former have the distinctive property of being under internal control.

Let me thus focus on (2). The question of compositionality has been on the agenda for years, especially due to Fodor's insistence on using it against non-atomistic theories of concepts (Fodor 1998; Fodor and Lepore 2002). His argument, in a nutshell, goes like this: Concepts are the basic elements of thought; compositionality is a "nonnegotiable" property of concepts; but non-atomistic theories of concepts—that is, those that contend that concepts are structured representations such as prototypes—are incapable to meet compositionality demands; hence non-atomic concepts are ill suited to figure as the basic elements of thought.[8] The argument is also relevant for connectionist approaches because many of them endorse, in one way or another, the idea that concepts are structured entities.

There is a recent defense against the compositionality argument— endorsed by Prinz (2002, 2012), Robbins (2002), and Weiskopf (2009a)— that is relevant for the issue of the nonnegotiability of compositional properties of concepts. The defense relies on the notion that compositionality is a modal property, that is, the idea is that concepts *can* combine compositionally but they do not necessarily do so all the time. Prinz (2012) contends that this weaker requirement allows us to regard prototypes as compositional given that there are cases in which they behave compositionally (i.e., the semantics of the compound is fully determined by the semantics of its parts), and there are others in which the compositional mechanism may not be used, or may be regularly supplemented with other combination mechanisms. The modal defense could then be easily extended to other cases of structured concepts.

I think that this defense is weak. First, notice that the "can" involved in it demands that there is something in the nature of concepts that allows them to be compositionally combined. So the defense assumes that compositionality is a general *constitutive* property of concepts, and it seems to demand that there are general compositional mechanisms that can work on concepts, even if sometimes they are not used. If this is the case, then it still follows that systematicity is a general property of cognition, even if sometimes it does not show up. Second, to show that prototypes are compositional, the relevant thing is to show that they are combined *as prototypes*. Yet it seems that instances of prototype combination are compositional inasmuch as their prototypical features are simply dropped away.

Although I do not wish to address the debate on compositionality in the limited space of this chapter, I dare to say that Fodor's criticisms have never been properly rebutted. Compositionality is still a problem for prototypes and other structured concepts. However, the compositionality of concepts cannot be used to support the view that systematicity is a general property of cognition. The reason is that Fodor's argument for the compositionality of concepts hinges precisely on the systematicity of cognition—that is, if cognition is systematic, the better explanation is a compositional system—so the extent to which concepts are compositional will be given by the extent to which cognition is systematic. But you still need an argument to show that cognition is *generally* systematic in the way classicism demands. Otherwise, one can hypothesize that a part of cognition is systematic in the required sense—hence works on compositional concepts, hence poses a problem for prototype-like explanations—and another is not—hence does not work on compositional concepts, hence might be accounted for by prototypes or other structured concepts. This hypothesis entails defending a version of conceptual pluralism, which I will provide in the next section. Before doing so, I want to consider a different (though related) argument that may offer independent reasons to hold that cognition is systematic and compositional.

The argument arises from Evans's well-known Generality Constraint. The constraint can be succinctly put thus:

If a subject can be credited with the thought that a is F, then he must have the conceptual resources for entertaining the thought that a is G, for every property of being G of which he has a conception. (Evans 1982, 104)

Weiskopf (2010) argues that the constraint can be understood as an *architectural* constraint, that is, "as a constraint on the sorts of representation combining capacities a creature must have in order to possess concepts" (2010, 109, n. 1). The constraint acts as a closure principle for the conceptual system so that "nothing could be a concept unless it was capable of entering into this kind of system of relations, and nothing could count as possessing a conceptual system unless it had a system of representations that were organized in such a way" (2010, 109). Notice that this is the sort of claim that turns systematicity into a non-empirical property, in the sense I referred to in section 2.1. In other words, systematicity would be a demand on mental architecture derived not from our theories on how concepts actually are but from deep intuitions on what concepts *have to be*.

I agree that the Generality Constraint arises from deep intuitions about thought. However, I contend that it is possible to interpret it in a way that does not pose the strong architectural constraint that Weiskopf suggests.

If one looks closely to the formulation by Evans, the constraint can be seen primarily in terms of *how to credit* a subject with a thought. In other words, it is a constraint on how to attribute beliefs: it is not possible to attribute a creature the belief that *a is F* and the belief that *b is G* without allowing the *possibility* of attributing it the belief that *a is G* and the belief that *b is F*. We need an extra assumption to turn the Generality Constraint into an architectural constraint that demands full combinability of concepts in the creature's internal system of representation. This is the assumption that concepts are components of thoughts that have to be combinable in ways that mirror the structure of the beliefs attributable to the creature. Yet, as I am going to argue, there is room to resist this view as a general relation between beliefs and concepts. There will be cases in which concepts will combine in complexes whose structure mirrors the structure of the corresponding beliefs but there will also be others in which there is no such mirroring. In the latter case, a creature can be credited with the belief, and the credit is grounded in its representational abilities, but the elements in its representations will not correspond part-to-part to the elements in the attributed belief.

In short, what I am going to defend is a version of *conceptual pluralism* that allows us to resist the line of reasoning that leads from the intuitions of the Generality Constraint to the conclusion that systematicity is a general property of cognition. The point is that the conclusion is warranted only for concepts that have the property of being combinable in ways that mirror the structure of the beliefs. If there are other elements of cognition that can be still regarded as concepts but which do not have such a property, then they will not be systematic in the way required by SA. Thus I must do two things to support this line of defense: the first is to show that conceptual pluralism is a cogent notion, that is, that it is possible to find elements in cognition that share fundamental properties that characterize them as concepts yet split into different subkinds; the second is to show that there are subkinds that differ precisely with respect to the property that is the source of systematicity, namely, compositionality.[9]

4   Conceptual Pluralism and Compositionality

Conceptual pluralism is the thesis that concepts constitute a kind that splits into a number of different subkinds. The notion appeared in the context of the debate against Machery's claim that concepts are not a genuine natural kind, and hence they are not fit to figure in psychological theories (Machery 2009). The basis for this eliminativist claim is that what psychologists call concept is served by an assorted collection of representations, such as prototypes, exemplars, or mini-theories, that have very little in common, either in terms of their structure or of the processes that operate on them. So Machery contends that there are not many useful generalizations that can be made about them.

In contrast, pluralistic approaches to concepts (Weiskopf 2009ab) hold that there are different kinds of mental representations that can be rightfully regarded as concepts. Psychological literature shows, indeed, that prototypes, exemplars, or theorylike structures appear to have a role to play in dissimilar cognitive tasks.[10] Yet the conclusion to draw is that minds have the different kinds of representational structures at their disposal, and they make a selective use of each of them depending on the type of task in which they are engaged. Still, those different kinds of representations have enough in common to be regarded as subkinds of a more inclusive, superordinate kind—the kind of concepts.

What are those common properties that unify concepts as a kind? They have to be properties picked at a different level from those that unify each subkind of concepts. In other words, in order to show that concepts are a kind one cannot use criteria that split themselves, that is, criteria that are applied differently to the different hypothesized subkinds. What is needed is some middle point at which one can find common high-level properties that are robust enough to block the eliminativist conclusion but still permit a plurality of kinds that possess them. In other words, one needs to show, first, that there are properties that qualify concepts as a class and, second, that there are different subkinds that share those properties and yet differ in other significant properties. Among the properties of concepts suggested in the literature, there are two that stand out as the most prominent ones: their *centrality* and their role in *attributions of belief*. Consequently, I contend that they pose the minimal common requirements that qualify concepts as a class. On the other hand, I will argue that a significant property in which subkinds of concepts differ is compositionality and, hence, systematicity. Let me elaborate a little on the common properties of concepts in the next subsection and leave the question of the differences in compositionality and systematicity for the following ones.

4.1   Common Properties of Concepts: Centrality and Belief-Attribution

Centrality is the idea that concepts are *central* mental representations, as opposed to *peripheral* ones. By "peripheral" I mean mental representations that are closer to the stimuli or input. This distinction has been used in different ways in theories of concepts. For instance, to point out a couple of recent examples, Camp (2009) singles out stimulus independence as one of the crucial factors that mark conceptuality, whereas Prinz (2002) appeals to internal control as the distinctive property between concepts and percepts—which in his view are undistinguishable with respect to its modality-specific constitution. The distinction between central and peripheral also plays a pivotal role in classical modularist views of mind (Fodor 1983), where peripheral representations correspond to the proprietary bases of input modules, and central representations are typically the concepts handled by the central processor. Indeed, even massive modularist views of mind (Carruthers 2006) make a distinction between conceptual and perceptual modules, which depends on architectural considerations regarding the distance to the input.

The second prominent property of concepts is that they are the representations whose possession allows the possibility of *attributing belieflike states* (as well as other kinds of propositional attitude states) to an individual. I intend this property to be neutral between those theories that hold that beliefs must be actually composed of concepts (Fodor 1998) and more instrumentally inclined theories that hold that beliefs can be ascribed to creatures with representational capabilities without necessarily holding that the tokened representational structures are literally composed of parts that correspond to those of the attributed belief (Dennett 1987). The point I want to make is that it is possible to make compatible, on the one hand, the rejection of the notion of beliefs as actually composed of concepts as smaller pieces with, on the other, a representationalist stance on concepts. Concepts would be the sort of mental representations whose possession allows an organism the possibility of exhibiting behaviors that grant attributions of belief.

Let me illustrate this with a toy example from the literature on animal cognition. Consider birds, such as jays (Clayton et al. 2003), that are capable of remembering the location where they stored food some time ago. One can describe the bird's performance by saying that the jay remembered where it stored the food, which involves attributing it the belief *that there is food at location l*. I think that there are two claims about this description that it is necessary to reconcile. One is the claim that it is a genuinely explanatory statement: it provides a description that allows one to make generalizations that are useful, perspicuous, and predictive. The other is the claim that it possibly strains the capability of birds (more on this later) to say that they are capable of combining concepts such as FOOD and LOCATION so as to form beliefs like the one I mentioned. Following the first claim, someone would contend that the bird does literally possess the structured belief that is composed of those concepts. Following the second claim, someone would contend that belief attribution is a wholly pragmatic affair that does not reflect the innards of the creature. However, there is a middle ground between both contentions: given the bird's food-tracking abilities, it is possible that it deploys actual mental representations for the attributed concepts FOOD and LOCATION, without deploying anything like a structured representation for the attributed belief *there is food at location l*. In other words, one can be (approximately) a *realist* about concepts and, at the same time, be (approximately) an *instrumentalist* about beliefs. Belief attributions like this would not be merely instrumental and observer-dependent but would be supported by certain representational abilities that some organisms possess and others do not. Concepts would thus be *those mental representations that it is necessary to possess so as to be the kind of organism to which one can attribute beliefs.*

Nothing prohibits, however, that in certain cases the structure of the attributed belief could be actually mirrored by the structure of the representational structure that grants the attribution. Yet this does not split the notion of belief into two different kinds—one for beliefs that are representationally mirrored and another for beliefs that are not so. Attributing beliefs has principally to do with the possibility of making predictions and generalizations regarding the organism's behavior, and this possibility can be served whether the representational states that underlie the behavior mirror those beliefs or not. This opens the door to the possibility of having two kinds of concepts, managed by two kinds of mechanisms, that underlie attributions of belief.

The point to consider now is whether there are elements that can be rightfully regarded as concepts, inasmuch as they exhibit the properties of being central and being the representations that underlie attributions of belief, and yet which split into subkinds that differ with respect to properties that are the source of systematicity. The relevant property in this respect, of course, is compositionality.

4.2 Compositional and Noncompositional Concepts

Let me take stock: I said above that the systematicity argument works on the premise that systematicity is a significant property of cognition, yet SA does not contain itself the stronger notion that systematicity is a general property of cognition. To support this latter notion, one may appeal to the claim that concepts are nonnegotiably compositional and back this claim with intuitions from Evans's Generality Constraint. I tried to debunk the idea that the constraint mandates a certain architecture, so as to show that there may be different kinds of representations that possess the minimal requirements for concepthood and which satisfy the constraint. Now it is time to argue that those kinds of concepts differ in some respect that does not allow us to regard systematicity as one of their general properties. I want to claim that there are mental representations that qualify as concepts, in terms of being central and involved in belief attributions, and yet are not compositional, and hence systematic, in the way SA contends. The upshot is that we would have two distinct kinds of concepts that differ in their compositional properties.

Let me back that claim by adapting some ideas from Camp (2009), who provides a careful analysis of the concept of "concept" that takes into account evidence from animal abilities. She begins by noting that notions of "concept" typically oscillate between two extremes: *concept minimalism*, in which for a cognitive ability to be regarded as conceptual it simply has to be systematically recombinable; and *concept intellectualism*, which links conceptuality to linguistic abilities, so that language, or some capacity that is only possible by means of language—for example, the capacity for thinking about one's thoughts—becomes necessary for conceptual thought. Both extremes would delimit a continuum in which Camp thinks it is possible to distinguish three notions of concept:

a minimalist "$concept_1$," denoting cognitive, representational abilities that are causally counterfactually recombinable; a moderate "$concept_2$," denoting cognitive, representational abilities that are systematically recombinable in an actively self-generated, stimulus-independent way; and an intellectualist "$concept_3$," denoting $concept_2$-type representational abilities whose epistemic status the thinker can reflect upon, where we assume that this latter ability is possible only in the context of language. (Camp 2009, 302)

$Concept_1$ is involved in activities that demand little more than passive triggering and marks the lower limit of the notion. $Concept_2$ is typically associated with cognitive abilities engaged in instrumental reasoning. This cognitive activity, which we find in a number of nonhuman animals, demands from the creature the capacity to represent states of affairs that are not directly provided by the environment, namely, the goal-states that the creature wants to achieve and the means-states that bring it closer to that goal in a number of stages. Finally, $concept_3$ marks the notion's upper limit, and it is here, Camp contends, where Evans's Generality Constraint can be met, because only $concept_3$ grants full recombinability, that is, the capacity to combine arbitrarily any a and b with any F or G of which the creature has a conception. $Concept_2$ cannot grant this capacity because, even if its representational power is removed from the immediate environmental stimulation, its deployment is still tied to the creature's immediate needs. To put it in Camp's terms, a chimpanzee would never entertain any of the potential thoughts that Evans's constraint refers to "because they are utterly useless for solving any problems that it actually confronts" (2009, 297). In contrast, creatures with language and the ability for epistemic reflection—the requirements for $concept_3$—can find some use for the most arbitrary combinations once they have certain epistemic drives, such as curiosity and imagination.

Appealing as I find this analysis, there are two important points that I find unconvincing. First, Camp states that $concept_1$ is less theoretically useful to provide an account of conceptual thought. In fact, I think that it is doubtful that this notion even meets the minimal requirements for concepthood. Camp relies on some capacity for recombination as a minimal requirement to count as conceptual. However, the fact that this capacity can be found in systems that are directly triggered by perceptual stimulation ought to make one suspicious of the proposal. As I pointed out in section 3, one wants an account of cognition as concept-involving in a way that lets one distinguish it from perception. Centrality, I argued in section 4.1, is a way to mark such a distinction. Yet the notion of $concept_1$ is clearly tied to noncentral capacities, so it does not meet the minimal criteria for concepthood. Recombinability is a red herring because one can find it in nonconceptual structures.

The second unconvincing point is Camp's treatment of the Generality Constraint. Camp endorses the view, which I resisted above, that it is an architectural constraint. At the same time, she contends that it works as an ideal rather than as a necessary constraint to grant conceptual thought. To meet the constraint one needs the fully systematic recombinability that permits arbitrary combinations to occur. Yet, in her view, even creatures with $concept_3$ capacities would often not meet the constraint given that many times they would be reluctant to form the arbitrary thoughts that, according to the constraint, they must be capable of forming.[11] This way, she makes room for a way to accept the Generality Constraint that, at the same time, allows one to regard conceptuality as a matter of degree. In other words, the constraint is OK, but it is too strong to be met in full for most practical concerns. Now, the reasoning behind this conclusion seems to me to be close to the reasoning behind the modal defense of compositionality that I discussed in section 3, and thus it commits the same sort of mistake but in the opposite direction. Let me explain.

Recall that the reasoning of the modal defense was that representations that can *sometimes* combine compositionally count as compositional, even if at other times they do not so combine. This was used to support the compositionality of representations such as prototypes. Camp's reasoning is that creatures with concept$_3$ capacities *sometimes* are not capable of entertaining certain combinations for practical purposes. This is used to deny that they meet the Generality Constraint "in full." The mistake in both cases is the same: both overlook the fact that what it takes for representations to count as compositional, and to meet the Generality Constraint, is that they are capable of being arbitrarily recombined *as a matter of how they are constituted* (and given certain processes sensitive to this constitution). It is irrelevant whether as a matter of fact they sometimes do or do not combine. The upshot is that, despite what Camp contends, her notion of concept$_3$ *does* meet the Generality Constraint. But if this is the case, and one still wants to maintain that the constraint restricts the suitable conceptual architecture, now one may object to her pluralist gradable analysis of the notion of concept. One could say that, as we have only a class of representations that meet the constraint—concept$_3$—we'd better regard this class as the *genuine* notion of concept, and the other two notions as varieties of nonconceptual representations.

However, notice that in section 3 I offered an alternative reading of Evans's constraint that poses much more lax restrictions on the representations that a creature must possess in order to satisfy it. So it does not matter much whether a class of concepts includes representations with limited compositionality. What is crucial is that they are central representations whose possession is required in order to grant systematic attributions of belief. In this respect, the notion of concept$_2$ appears to be a suitable candidate for concepthood, unlike the peripheral, perceptually bound representations in concept$_1$.

Where do these considerations leave us? I think that Camp's analysis allows for the existence of just two kinds of concepts. One of them, roughly corresponding to her concept$_2$, has the minimal common requirements to be regarded as conceptual but does not appear to be compositional in the classical sense;[12] hence, it is incapable of giving rise to the sort of systematicity referred in SA. The other kind, roughly corresponding to concept$_3$, is compositional and supports systematicity in SA.

There are two final related issues that I wish to address to finish paving the way to an alternative nonclassical SA. One is: even if it were possible to tell two notions of concept apart, systematicity could still be a general property of cognition. The reason is that each notion could be applicable to a different type of creature. For instance, Camp's analysis suggests a scenario in which $concept_2$ is simply the basis of nonhuman animal thought, while human thought is exclusively constituted by $concept_3$. If this is the case, then one may contend that systematicity is a general property of *human* cognition, which is still a strong claim. To debunk this claim, one ought to show that both kinds of concepts have a place in human cognition.

The second issue is that even if humans possess both kinds of concept, it still may be the case that classical systems can account for them. In other words, one must show not only that SA applies just to a part of cognition—the one that deals with $concept_3$—but that it does not apply to the other part—the one that deals with $concept_2$. I address these two issues in the next and final section.

5 Two Kinds of Systematicity

The aim of this section, then, is to motivate the view that there are two kinds of systematicity in human minds, each of them related to a different kind of concept. It is more than mere wordplay to say that a kind of systematicity involves a kind of system. It is because classical symbol systems have the defining properties that they have *as systems* that they exhibit the sort of systematicity of SA. So to argue for two kinds of systematicity, one must search for reasons that back the existence of two kinds of systems, each of them working on a conceptual kind. Dual systems theory (DST) is an obvious candidate to provide the backbone of such an approach.

DST is the view that human minds are constituted by two distinct kinds of cognitive processing systems (Evans and Frankish 2009). Although their detailed characterizations and properties vary depending on the specific theory, in general terms one is typically characterized as fast, automatic, holistic, inflexible, difficult to verbalize, evolutionarily old, and nonconscious, and the other as slow, controlled, analytic, flexible, more easy to verbalize, evolutionarily recent, and conscious. Following standard usage, I will refer to those systems as S1 and S2, respectively. Even though there are different views about how to articulate this general approach (Evans 2009; Stanovich 2009), I will not take them into account. DST has been mainly applied to explain reasoning and social cognition, but in its most ambitious forms it purports to provide a general vision of mental life, in which the basic distinction between two kinds of systems is the fundamental architectural design of human cognition that helps to account for a range of mental phenomena (Carruthers 2006; Samuels 2009).

The first thing to note is that both S1 and S2 have to be conceptual systems: they are involved in paradigmatic central cognitive processes, such as reasoning, not in perception or other input-controlled processes; and the sorts of behaviors that any of them controls, such as decision taking, give rise to belief-attributions. The question now is whether each system can be conceived of as working on a different conceptual kind. In other words, the question is whether creatures with a dual system architecture are endowed with *both* concept$_2$ and concept$_3$ abilities. The scenario to be considered in terms of DST would be one in which the first kind of concept is handled by S1 and the second by S2. The reason is that the properties exhibited by S1 resemble those of the conceptual capabilities associated with instrumental reasoning that, as we saw above, are arguably present in some nonhuman animals. This also fits the idea the S1 is evolutionarily older than S2, and that S2 is likely to be exclusive to humans. Both systems would be capable of performing typical conceptual functions, such as categorization, reasoning, and meaning extraction, yet in different ways and with different limits on the kinds of thoughts that they are capable of delivering. In particular, S2 would be capable of satisfying the Generality Constraint understood as an architectural constraint but S1 would not. [13]

Is this a plausible scenario? Support for a positive answer can be found in the recent extensive review by Penn et al. (2008) comparing human and animal cognition. Their aim is to show that there is a "profound functional discontinuity between human and nonhuman mind" (2008, 110). The discontinuity is revealed in a wide range of domains, such as the ability to cope with relational (as opposed to perceptual) similarity, to make analogical relations, to generalize novel rules, to make transitive inferences, to handle hierarchical or causal relations, or to develop a theory of mind. Penn et al.'s point is basically to show that the discontinuity between human and nonhuman minds can be cashed out in terms of the presence of a capacity for systematically reinterpreting first-order perceptual relations in terms of higher-order relational structures akin to those found in a physical symbol system (PSS)—the archetypal classical system. [14] This is the sort of capacity that, in Camp's analysis, requires something like concept$_3$.

Nonlinguistic creatures do not exhibit this kind of systematicity. Instead, they manifest a different kind of systematicity that "is limited to perceptually based relations in which the values that each argument can take on in the relation are constrained only by observable features of the constituents in question" (Penn et al. 2008, 127). Borrowing Bermúdez's (2008) term, I will call it *featural systematicity*. Penn et al. think that this systematicity would be accounted for by compositional properties[15] different from those that characterize a PSS. Penn et al. resort to van Gelder's (1990) notion of functional compositionality to account for the kind of compositionality present in animals. Unlike van Gelder, however, Penn et al. do not regard functional compositionality as capable of underlying the sort of systematicity exhibited by humans—that is, as capable of satisfying SA. Animal compositional capacities would be limited to "some generally reliable and productive mechanism for encoding the relation between particular constituents" that would account for "the well-documented ability of nonhuman animals to keep tracks of means-ends contingencies and predicate argument relationships in a combinatorial fashion" (Penn et al. 2008, 125). The animal abilities referred to are basically of the same kind as those that, according to Camp, grant the attribution of concept$_2$.

As in DST—a theory that Penn et al. regard as related to their view—the thesis is that both kinds of systematicity appear in humans, so it is necessary to explain how. Penn et al. propose that the representational system unique to humans "has been grafted onto the cognitive architecture we inherited from our nonhuman ancestors" (2008, 111). In search of an explanation of how such "grafting" might be possible, they resort to computational models. Nonclassical connectionist models might explain the kind of systematicity that we find in animals, whereas recent connectionist-symbolic models might account for the grafting of new human representational abilities to the preexisting representational machinery. Even though they back their proposal with computational models of their own (e.g., Hummel and Holyoak 1997, 2003) one might object that we still lack strong evidence for it. However, I want to consider a different kind of objection that is more relevant for the purposes of this essay: accepting that there are two different processing systems, why could not one resort to a classical explanation for *both* of them? In other words, one could insist on the possibility that animal systematic capabilities were brought about by a classical compositional symbolic system, perhaps limited with respect to the range of represented contents that it can deal with but still working on the same principles of concatenative recombination. If this were the case, one could contend that the difference between both systems—or between the concepts on which they operate, or between the systematicity they exhibit—is not one of *kind*. Classical systems could then still constitute the keystone of cognition in general, just as SA contends.

I think that there are good grounds to reject this possibility. Its problem, in a nutshell, is that symbol systems are *too strong* for that. Recall that from the classicist perspective it is impossible to have a classical system that is not systematic in the sense posited by SA. So if animal minds included in some way a classical system, *then they would ipso facto be endowed with standard full systematicity*. The extensive evidence reviewed by Penn et al. shows precisely that this is not the case. As they contend, if there were no differences in kind, then one could expect that the observed discontinuities would be erased under appropriate conditions. For instance, animals under a "special training regime," which let them access a larger range of contents and relations, would at least approximate human behavior. Yet the evidence shows that even those animals perform poorly.

It seems to me that we have finally reached the elements that would allow us to construct a nonclassical systematicity argument. Recall the general form that such an argument would have:

NSA

(i′) Empirical claim: X is a pervasive property of cognition.

(ii′) Explanatory claim: The only plausible explanation for X is property Y.

(iii′) Definitional claim: Y is a defining property of such and such nonclassical systems.

(iv′) Dilemma: If classicism cannot account for property Y, then it does not provide a full account of cognition (from (i′) and (ii′)); if classicism can account for Y, then it is actually implementing a nonclassical system (from (ii′) and (iii′))

Now we have ways of seeing how the different claims in the argument could be substantiated. First, the X that we have to explain is the kind of systematicity exhibited by nonhuman animals in terms of their limited recombination abilities—limited by their perceptual repertoire even if not bound to the immediate environment, and limited in the kinds of relations that they allow. Moreover, it is also a pervasive property of human cognition, given that it belongs to the inherited part of our cognitive machinery.

Second, the Y that constitutes the best explanation of this systematicity is some property of nonclassical systems. It cannot be a product of classical systems because, as I have just argued, this would endow animals with human systematic capabilities. A plausible candidate for Y comes from the set of properties characteristic of *distributed representations*. Perhaps, as Penn et al. observe, distributed systems as we currently envision them may need to be supplemented to account for animal minds. Yet it would suffice for NSA that distributed representations are essentially involved in the explanation of featural systematicity, and that whatever supplement they require *cannot be* classical.

This would also satisfy the definitional claim, given that it simply says that whatever property Y is, it is constitutive and characteristic of some nonclassical system. In this respect, distribution is a defining property of distributed nonclassical systems, from which it follows that it is simply not possible to be such a system and not to have distributed representations. To conclude, the dilemma for the classicist position comes to this: if it cannot account for the sort of systematicity exhibited by animals and by part of human cognition, then it does not provide a full account of cognition; and if it offers a model that exhibits nonclassical property Y—for instance, distributed representation—then given that Y is definitive of nonclassical systems, the model would count immediately as an implementation of a nonclassical system.

6 Conclusion

After all these years the systematicity argument still poses a powerful challenge to any attempt at explaining cognition. Part of its force resides in its simplicity: "Here is this notorious property of cognition; here is a conspicuous explanation of this property; does anyone have an explanation that does not collapse into ours?" In this chapter, I claimed that the answer to the latter question is negative. Despite the attempts, nobody has come up with an explanation for the sort of systematicity that the argument alludes to that is better than a compositional system of representations. And nobody has a complete account of cognition unless one is able to explain properties of that sort.[16] However, I also contend that this is not the end of the story: there are other cognitive properties to explain, and classicism is not in a better position to do so. Just because one has a powerful explanation of an important mental property, it does not mean that one can transfer this explanation to every other mental property. If, as the evidence increasingly supports, the human mind includes two fundamentally different kinds of systems, and each system exhibits a different way of being systematic, then classical symbol systems cannot account for both of them.

The bottom line can be put thus: while nonclassical systems are *too weak* to account for humanlike compositionality-based systematicity, classical systems are *too strong* to account for noncompositionality-based systematicity. The reason is precisely that any system that has a classical computational-representational architecture *necessarily* exhibits compositional systematicity as a consequence of its architectural design. Yet I have argued that the evidence suggests that, in both animals and humans, there are genuinely cognitive processes that fail to exhibit this kind of systematicity. They are genuinely cognitive because they are concept involving: they are not tied to immediate perceptual stimuli, and they control behaviors that are complex enough so as to merit attributions of belief. If nonclassical approaches are able to explain such processes—and not only, as their critics often complain, early perceptual processing—then they will have an account of part of our mental life, even if not of all of it.

To sum up, the picture of cognition that I tried to motivate in this chapter comes to this: an architecture that supports at least two distinct subkinds of concepts with different kinds of systematicity, neither of which is assimilable to the other. This picture sets a whole new agenda of problems to solve, particularly regarding the relation between both systems. In particular, one may wonder whether nonclassical systematicity is exactly the same in humans and in those animals that exhibit analogous properties, or whether perhaps it is affected by its coexistence with compositional systematicity; one may wonder whether it is possible to integrate both kinds of concepts in some respect, perhaps to form a sort of hybrid structure; one may wonder whether compositional systematicity is exclusively related to linguistic cognition. These are the sorts of questions that I think it will be interesting to address in future research.

Acknowledgments

This essay was finished as part of research project FFI2011-30074-C02-01, funded by the Spanish Ministerio de Ciencia e Innovación. Some of the initial ideas were presented at the First Workshop on Concepts and Perception, University of Córdoba, Argentina, and further developed at the workshop on Systematicity and the Post-connectionist Era, San José, Spain. The author wishes to thank the comments from audiences at both meetings, and particularly Patricia Brunstein, Agustín Vicente, and an anonymous referee.

Notes

1. See McLaughlin 1993 for an earlier—and consequently less complete—classification of connectionist replies to the argument.

2. On the other hand, there are reasons to doubt that language is fully compositional in the required sense. (See Vicente and Martínez-Manrique 2005 for a rejection of the claim that semantics can provide fully determined compositional
thoughts and its consequences for the views that regard language as a cognitive
vehicle). Language may be simply a combinatorial system, and thus the picture
presented by Gomila et al. would be one of a classical compositional system getting
installed thanks to the combinatorial properties of language. But notice that SA is
neutral about how systematicity is acquired, given that its claim is about how it
is explained, and its explanation in such an acquisition model is still a classical
one.

3. See McLaughlin 2009 for an extensive analysis of Cummins's and Johnson's
claims, in which he contends that they miss the point about what has to be
accounted for—which, in his view, are the lawful psychological patterns revealed in
systematic relations between thoughts.

4. Van Gelder and Port (1994) extended the analysis by proposing six parameters—
properties of primitive tokens, and properties of modes of combination—in terms
of which to distinguish varieties of compositionality. However, concatenative versus
functional still seems to be the crucial dimension.

5. Incidentally, Aizawa (1997) thinks that neither connectionist nor classical models
can explain systematicity without the aid of further additional hypotheses.

6. I owe this objection to a referee who wanted to know what was wrong with a
specific model. Obviously, answering questions like this exceeds the limits and goals
of this essay.

7. See Fodor 1998, chap. 2. Fodor's idea of a nonnegotiable condition for a theory
of concepts is that the condition is fallible but abandoning it entails abandoning
the representational theory of mind itself.

8. The problem of compositionality was already detected by early proponents of
prototype theory (Osherson and Smith 1981), and some technical solutions have
been attempted (e.g., Kamp and Partee 1995).

9. A referee has complained that this looks like an unnecessarily circuitous route.
Should it not be enough for the purposes of the essay to show the second, i.e., that
there are elements in cognition that are not systematic in the way required by SA?
I don't think so. The point is that one has to motivate first the view that they are
precisely elements *in cognition*, that is, conceptual elements. Otherwise, one might
brush aside the suggestion that there is a different kind of systematicity by saying
that it has to do with perceptual or other less-than-cognitive elements. The point
of the next section, thus, is to show that there is a general way of characterizing
concepts so that they comply with the Generality Constraint, understood as a

constraint on belief-granting capabilities, while at the same time they split into subkinds that differ in important respects.

10. Although I do not wish to enter the debate on conceptual atomism, I would like to point out that conceptual pluralism allows for the possibility of atomic concepts as one more among the subkinds of concepts. Weiskopf (2009a) seems to forget this possibility when he opposes atomism to pluralism. As I have pointed out (Martínez-Manrique 2010), the relevant opposition is between pluralism and monism, and the former can admit atoms in the repertoire as long as they are not mistaken to be the whole class of concepts.

11. To put it in Camp's words: "We also fall short of full generality: precisely because certain potential thoughts are so absurd, it's unlikely that anyone would ever think them or utter sentences expressing them in any practical context" (2009, 306).

12. As we will see in the next section, there remains the question of whether they are compositional in a different sense. Now, I do not wish to fight for the term "compositional." I am ready to leave it as the property that characterizes the class of concepts present in classical systems (concept$_3$) and accept that the other class of concepts is just noncompositional. What matters for this essay is that these concepts give rise to different systematic properties not accounted for by classical systems.

13. The approach by Gomila et al. (2012) has elements that are congenial to the proposal I am making in this essay. For instance, they also resort to dual systems theory as the overarching architecture of mind. Yet I do not agree with their claim that "this duality also corresponds to the divide between non-systematic and systematic processes" (2012, 112). There is much systematicity in S1 and it is of a kind that demands conceptual processing, even if not the kind of concepts that are processed by S2. So I doubt that dynamic, embodied approaches, as the one they endorse for S1, provide a good account for this system either, at least if they are couched in nonrepresentational terms.

14. To cope with critics of the PSS hypothesis, Penn et al. borrow a milder version from Smolensky (1999), the "symbolic approximation" hypothesis. The distinction is irrelevant for the purposes of this essay, given that the point is still that symbolic approximators require an architecture that is different from the one that supports animal capacities.

15. As I said in note 12, it is irrelevant whether one does not want to call them "compositional," preferring to reserve that term for classical compositionality. What matter is that there is a different kind of systematicity accounted for by different properties of the system.

16. So, attempts at providing a whole alternative framework to computational-representational cognitive science, such as the attempt by Chemero (2009), seem to

be flawed inasmuch as they simply ignore those properties. For instance, there is no single clue in his book about how radical embodied cognitive science would deal with language comprehension or with reasoning processes, just to mention two paradigmatic domains where resort to classical representations is more natural.

References

Aizawa, K. 1997. Explaining systematicity. *Mind and Language* 12:115–136.

Aizawa, K. 2003. *The Systematicity Arguments*. Dordrecht: Kluwer.

Bechtel, W. 1994. Natural deduction in connectionist systems. *Synthese* 101 (3):433–463.

Bermúdez, J. L. 2008. The reinterpretation hypothesis: Explanation or redescription? *Behavioral and Brain Sciences* 31 (2):131–132.

Camp, E. 2009. Putting thoughts to work. *Philosophy and Phenomenological Research* 78 (2):275–311.

Carruthers, P. 2006. *The Architecture of Mind*. Oxford: Oxford University Press.

Chemero, T. 2009. *Radical Embodied Cognitive Science*. Cambridge, MA: MIT Press.

Clark, A. 1989. *Microcognition*. Cambridge, MA: MIT Press.

Clayton, N. S., K. S. Yu, and A. Dickinson. 2003. Interacting cache memories: Evidence for flexible memory use by Western scrub-jays (*Aphelocoma californica*). *Journal of Experimental Psychology: Animal Behavior Processes* 29 (1):14–22.

Cummins, R. 1996. Systematicity. *Journal of Philosophy* 93 (12):591–614.

Cummins, R., J. Blackmon, D. Byrd, P. Poirier, M. Roth, and G. Schwarz. 2001. Systematicity and the cognition of structured domains. *Journal of Philosophy* 98 (4):1–19.

Dennett, D. C. 1987. *The Intentional Stance*. Cambridge, MA: MIT Press.

Dennett, D. C. 1991. Mother nature versus the walking encyclopedia: A Western drama. In *Philosophy and Connectionist Theory*, ed. W. Ramsey, S. P. Stich, and D. E. Rumelhart, 21–30. Hillsdale, NJ: Erlbaum.

Evans, G. 1982. *The Varieties of Reference*. Oxford: Clarendon Press.

Evans, J. St. B. T. 2009. How many dual-process theories do we need? One, two, or many? In *In Two Minds: Dual Processes and Beyond*, ed. J. St. B. T. Evans and K. Frankish, 33–54. Oxford: Oxford University Press.

Evans, J. St. B. T., and K. Frankish, eds. 2009. *In Two Minds: Dual Processes and Beyond*. Oxford: Oxford University Press.

Fodor, J. 1983. *The Modularity of Mind*. Cambridge, MA: MIT Press.

Fodor, J. A. 1997. Connectionism and the problem of systematicity (continued): Why Smolensky's solution still doesn't work. *Cognition* 62 (1):109–119.

Fodor, J. A. 1998. *Concepts*. Cambridge, MA: MIT Press.

Fodor, J. A., and E. Lepore. 2002. *The Compositionality Papers*. Oxford: Oxford University Press.

Fodor, J. A., and B. P. McLaughlin. 1990. Connectionism and the problem of systematicity: Why Smolensky's solution doesn't work. *Cognition* 35 (3):183–204.

Fodor, J. A., and Z. W. Pylyshyn. 1988. Connectionism and cognitive architecture: A critical analysis. *Cognition* 28 (1–2):3–71.

Frank, S. L., W. F. G. Haselager, and I. van Rooij. 2009. Connectionist semantic systematicity. *Cognition* 110 (3):358–379.

Gomila, A., D. Travieso, and L. Lobo. 2012. Wherein is human cognition systematic? *Minds and Machines* 22 (2):101–115.

Goschke, T., and D. Koppelberg. 1991. The concept of representation and the representation of concepts in connectionist models. In *Philosophy and Connectionist Theory*, ed. W. Ramsey, S. P. Stich, and D. E. Rumelhart, 129–162. Hillsdale, NJ: Erlbaum.

Hadley, R. F. 1994. Systematicity in connectionist language learning. *Mind and Language* 9 (3):247–272.

Hummel, J. E., and K. J. Holyoak. 1997. Distributed representations of structure: A theory of analogical access and mapping. *Psychological Review* 104 (3):427–466.

Hummel, J. E., and K. J. Holyoak. 2003. A symbolic-connectionist theory of relational inference and generalization. *Psychological Review* 110 (2):220–264.

Johnson, K. 2004. On the systematicity of language and thought. *Journal of Philosophy* 101 (3):111–139.

Kaye, L. J. 1995. The languages of thought. *Philosophy of Science* 62 (1):92–110.

Kamp, H., and B. Partee. 1995. Prototype theory and compositionality. *Cognition* 57 (2):129–191.

Machery, E. 2009. *Doing without Concepts*. Oxford: Oxford University Press.

Martínez-Manrique, F. 2010. On the distinction between semantic and conceptual representation. *Dialectica* 64 (1):57–78.

McLaughlin, B. 1993. Systematicity, conceptual truth, and evolution. *Philosophy* (Supplement 34):217–234.

McLaughlin, B. 2009. Systematicity redux. *Synthese* 170 (2):251–274.

Niklasson, L. F., and T. van Gelder. 1994. On being systematically connectionist. *Mind and Language* 9 (3):288–302.

Osherson, D. N., and E. E. Smith. 1981. On the adequacy of prototype theory as a theory of concepts. *Cognition* 9 (1):35–58.

Penn, D. C., K. J. Holyoak, and D. J. Povinelli. 2008. Darwin's mistake: Explaining the discontinuity between human and nonhuman minds. *Behavioral and Brain Sciences* 31 (2):109–178.

Prinz, J. 2002. *Furnishing the Mind*. Cambridge, MA: MIT Press.

Prinz, J. 2012. Regaining composure: A defense of prototype compositionality. In *The Oxford Handbook of Compositionality*, ed. M. Werning, W. Hinzen, and E. Machery, 437–453. Oxford: Oxford University Press.

Robbins, P. 2002. How to blunt the sword of compositionality. *Noûs* 36 (2): 313–334.

Samuels, R. 2009. The magical number two, plus or minus: Dual process theory as a theory of cognitive kinds. In *In Two Minds: Dual Processes and Beyond*, ed. J. St. B. T. Evans and K. Frankish, 129–146. Oxford: Oxford University Press.

Smolensky, P. 1988. On the proper treatment of connectionism. *Behavioral and Brain Sciences* 11 (1):1–23.

Smolensky, P. 1991a. The constituent structure of connectionist mental states: A reply to Fodor and Pylyshyn. In *Connectionism and the Philosophy of Mind*, ed. T. Horgan and J. Tienson, 281–308. Dordrecht: Kluwer.

Smolensky, P. 1991b. Connectionism, constituency, and the language of thought. In *Fodor and His Critics*, ed. G. Rey, 201–227. Oxford: Blackwell.

Smolensky, P. 1999. Grammar-based connectionist approaches to language. *Cognitive Science* 23 (4):589–613.

Stanovich, K. E. 2009. Distinguishing the reflective, algorithmic, and autonomous minds: Is it time for a tri-process theory? In *In Two Minds: Dual Processes and Beyond*, ed. J. St. B. T. Evans and K. Frankish, 55–88. Oxford: Oxford University Press.

Sterelny, K. 1990. *The Representational Theory of Mind*. Oxford: Blackwell.

van Gelder, T. 1990. Compositionality: A connectionist variation on a classical theme. *Cognitive Science* 14 (3):355–384.

van Gelder, T. 1991. What is the "D" in "PDP"? A survey of the concept of distribution. In *Philosophy and Connectionist Theory*, ed. W. Ramsey, S. P. Stich, and D. E. Rumelhart, 33–60. Hillsdale, NJ: Erlbaum.

van Gelder, T., and R. Port. 1994. Beyond symbolic: Toward a Kama-Sutra of com-positionality. In *Artificial Intelligence and Neural Networks: Steps toward Principled Integration*, ed. V. Honavar and L. Uhr, 107–126. London: Academic Press.

Vicente, A., and F. Martínez-Manrique. 2005. Semantic underdetermination and the cognitive uses of language. *Mind and Language* 20 (5):537–558.

Weiskopf, D. 2009a. Atomism, pluralism and conceptual content. *Philosophy and Phenomenological Research* 79 (1):131–163.

Weiskopf, D. 2009b. The plurality of concepts. *Synthese* 169 (1):145–173.

Weiskopf, D. 2010. Concepts and the modularity of thought. *Dialectica* 64 (1):107–130.

13 Neo-Empiricism and the Structure of Thoughts

Edouard Machery

Neo-empiricism is one of the most exciting theories of concepts developed in the last twenty years in philosophy and in psychology. According to this theory, the vehicles of tokened concepts are not different in kind from the vehicles of perceptual representations. Proponents of neo-empiricism have touted the virtues of their theory (Barsalou 1999, 2010; Prinz 2002; Gallese and Lakoff 2005), including its alleged empirical support, while critics have raised various concerns (e.g., Markman and Stilwell 2004; Machery 2006, 2007; Mahon and Caramazza 2008; Dove 2009, 2011; McCaffrey and Machery 2012). In this chapter, I will criticize neo-empiricists' views about the nature of thoughts, that is, about those representations that express propositions. In substance, if neo-empiricism were right, our capacity to think—to move from thought to thought—would be either mysterious or a matter of a contingent history of learning. Fodor and Pylyshyn's classic article against connectionism will be useful to develop this criticism, since neo-empiricists' views about the nature of thoughts suffer from problems similar to those Fodor and Pylyshyn diagnosed twenty-five years ago.[1]

Here is how I will proceed. In section 1, I will review the central tenets of neo-empiricism. In section 2, I will review how neo-empiricists characterize occurrent and non-occurrent thoughts. In section 3, I will argue that amodal symbols seem to be needed to have thoughts in long-term memory (i.e., non-occurrent thoughts). In section 4, I will argue that neo-empiricists' views about occurrent thoughts are unacceptable.

1 Neo-Empiricism

Neo-empiricism in contemporary philosophy and psychology is not a reductionist *semantic* theory: it does not state that the content of thoughts can be reduced to perceptual properties such as *red*, *square*, *loud*, and so on.

Prinz (2002) argues at length that the types of representations posited by neo-empiricists (in his terminology, "proxytypes," or in Barsalou's terminology, "perceptual symbols") can be about three-dimensional objects and their physical properties. Nor is neo-empiricism a *developmental* theory about concepts: it does not state that all concepts are learned. Prinz (2002) clearly argues that neo-empiricism is consistent with the claim that some conceptual representations are innate, whereas Barsalou (1999) holds that some conceptual representations are in fact innate.

Instead, neo-empiricism is a theory about the *vehicles* of concepts (Prinz 2002, 109; Machery 2007; McCaffrey and Machery 2012), that is, about the nonsemantic properties of the physical (presumably neural) states that realize concepts. Although neo-empiricists disagree about various points, all of them accept the following two theses about the vehicles of concepts:

(1) The knowledge that is stored in a concept is encoded in several perceptual representational systems.

(2) Conceptual processing involves essentially reenacting some perceptual states and manipulating these perceptual states.

The first thesis describes what it is to token, or entertain, a concept: according to Thesis 1, to entertain a concept is to entertain some perceptual representations. Thus, to think about dogs consists in entertaining some perceptual (visual, auditory, olfactory, proprioceptive, etc.) representations of dogs. These perceptual representations are of the same kind as the perceptual representations one would entertain if one were to perceive dogs. Thus, thinking consists in reenacting, or simulating, the perception of the objects of one's thoughts. The reenacted perceptual representations need not be conscious: in line with modern cognitive science, neo-empiricists hold that one can entertain a perceptual representation unconsciously. This first thesis denies the view held by amodal theorists that the vehicles of thoughts are distinct in kind from the vehicles of perceptual representations (Fodor 1975, 2008; Pylyshyn 1984).

Neo-empiricists have been crystal clear in their embrace of Thesis 1. Barsalou writes (1999, 577–578):

Once a perceptual state arises, a subset of it is extracted via selective attention and stored permanently in long-term memory. On later retrievals, this perceptual memory can function symbolically, standing for referents in the world, and entering into symbol manipulation. As collections of perceptual symbols develop, they constitute the representations that underlie cognition.

Similarly, Prinz proposes that "concepts are couched in representational codes that are specific to our perceptual systems" (2002, 119), and that "tokening a proxytype is generally tantamount to entering a perceptual state of the kind one would be in if one were to experience the thing it represents" (2002, 150).

Neo-empiricists do not seem to agree about the nature of the representations stored in long-term memory. At the very least, they describe them differently. Barsalou (1999) postulates the existence of simulators, which are psychological structures or entities whose function is to create the perceptual representations that are the vehicles of occurrent thoughts. By contrast, Prinz (2002, 146) holds that the perceptual representations of an entity form a network in long-term memory, and that people retrieve different subsets of this network depending on the context in which concept retrieval takes place.

Thesis 1 does not specify the nature of perceptual representations: it does not explain what makes a representation perceptual. Unfortunately, there is no consensus about this issue in the neo-empiricist literature. Some, like Barsalou (1999, 578), hold that perceptual representations are analogical, while others, like Prinz (2002, chap. 5), hold that perceptual representations are simply the representations that occur in perceptual systems. Furthermore, none of the proposals put forward is fully satisfactory. First, it is unclear why only perceptual representations would be analogical since critics of neo-empiricism have argued that analogical representations of quantity are amodal, and they view the research on quantity representation as a counterexample to neo-empiricism (Machery 2007; Dove 2009, 2011). Second, it is difficult to delineate perceptual systems. Where, in neurobiological terms, do perceptual systems stop? Are the temporal poles part of the visual system? Does the frontal cortex belong to any perceptual system?

Because the nature of perceptual representations has not been specified to full satisfaction, it can be unclear whether empirical findings genuinely bear on the debate between neo-empiricism and amodal theories of concepts. In particular, patterns of brain activation in tasks involving retrieving and manipulating concepts can be ambiguous, since it may be unclear whether the brain areas that are activated belong to a perceptual system. But that's not to say that there is no relevant evidence. For instance, McCaffrey and Machery (2012) have argued that the (domain-specific, modality-general) pattern of semantic loss found in semantic dementia is evidence against neo-empiricism.

Thesis 2 is about *conceptual processing*. According to neo-empiricists, thinking consists in manipulating the conscious or unconscious perceptual representations that one entertains when one is thinking. While neo-empiricists have written at length about the nature of the representations that make up our thoughts (Thesis 1), they have not said much about the nature of conceptual processing. Still, in substance, categorization, for instance, would work like this: when one has to decide whether something is a dog, one simulates perceiving a dog, matches one's current perception and one's simulation, and, if the match is close enough, one decides that the object to be categorized is a dog (e.g., Barsalou 1999, 576).

2   Thoughts and Neo-Empiricism

2.1   Occurrent Thoughts

According to neo-empiricists, to entertain the concept of dog—that is, to have occurrent thoughts about dogs as such—consists in simulating perceiving a dog. But what is it to think that Fido is a dog, that dogs bark, that the big dog seems eager to cross the street, or that the dog that barked scared the cat? That is, what is it to entertain a proposition?

Barsalou acknowledges that any theory about the vehicles of concepts has to be consistent with our capacity to entertain propositions:

Another important lesson that we have learned from amodal symbol systems is that a viable theory of knowledge must implement propositions that describe and interpret situations. (1999, 595)

Unsurprisingly, he holds that perceptual symbols can express propositions, and, using the example of a perceived jet in the left part of the visual field, he describes recognitional thoughts that apply concepts to perceived individuals as follows:

As visual information is picked up from the individual [e.g., a jet], it projects in parallels onto simulators in memory. A simulator becomes increasingly active if (1) its frame contains an existing simulation of the individual, or if (2) it can produce a novel simulation that provides a good fit. ... The simulation that best fits the individual eventually controls processing. Because the simulation and the perception are represented in a common perceptual system, the final representation is a fusion of the two. ... Binding a simulator successfully with a perceived individual via a simulation constitutes a type-token mapping. ... Most importantly, this type-token mapping implicitly constitutes a proposition, namely, the one that underlies "it is true that the perceived individual is a jet." In this manner, perceptual symbol systems establish simple propositions. (1999, 596)

Prinz describes entertaining propositional thoughts as follows:

Suppose that Boris forms the desire to hunt a fat gnu, and this desire consists of a perceptual simulation of a person with a rifle pursuing a gnu-shaped animal with an oversized belly. Is there a component of this simulation that counts as the GNU proxytype? Can it be distinguished from the FAT proxytype? Or the HUNTING proxytype?

In some cases, one may be able to distinguish parts of a simulation by intrinsic features. Different bound shape representations, for example, may constitute distinct proxytypes. Perhaps the gnu-shaped representation can be distinguished from the person-shaped representation and the rifle-shaped representation in that way. But there is no separate representation for hunting or for fatness. They are built up from or into the bound shape representations. If there is no distinct proxytype for hunting or for fatness in this representation, how can we say it is a token of a desire to hunt a fat gnu? The answer may lie in the origin of the simulation. Mental representations of hunting and fatness are used when the simulation is initially formed. … There is a sense in which proxytypes for HUNTING and FAT are contained in the simulation, but they meld with other proxytypes. (2002, 150–151)

Some important points emerge from these neo-empiricist descriptions of occurrent propositional thoughts. First, occurrent thoughts are produced either by *blending*, melding, or fusing together (all these metaphors are equivalent) the perceptual representations that result from people's knowledge in long-term memory (their simulators or their networks of perceptual representations) or by blending such representations with the perceptual representations produced by our perceptual systems (see Barsalou's quotation above). Unfortunately, in contrast to Smolensky's (1988) tensor-product brand of connectionism, which involves similar ideas (discussed at length in Fodor and McLaughlin's 1990 companion paper to Fodor and Pylyshyn 1988), neo-empiricists have not described this blending process at great length, using metaphors instead. For instance, it is opaque what rule, if any, governs this process. But it is clear that, while the symbols that express propositions (e.g., the proposition that the dog is barking at the cat) *may* be made out of subsymbols corresponding to the constituents of the propositions expressed (e.g., a symbol representing a dog), in a neo-empiricist framework they *need not*: as Prinz puts it in the quotation above, "There is no separate representation for hunting or for fatness" (2002, 151).

As a result, the concepts that are involved in entertaining a thought need not be cotokened as distinct symbols together with the symbol expressing this thought. In Prinz's example, while the concept of fatness is involved in thinking about hunting a fat gnu, no distinct symbol for fatness occurs together with the thought that one is hunting a fat gnu.

Furthermore, the subsymbols that are genuine parts of the symbols expressing thoughts (e.g., the symbol of a gnu in Prinz's example) do not recur identically across thoughts; rather, they are context-sensitive. This context-sensitivity takes two forms. First and foremost, unlike indexicals and relative predicates, the *vehicles* of concepts vary across concepts (Barsalou 1999, 598–599; Prinz 2002, 149). A cup is represented differently in different contexts; that is, perceptual simulations of cups vary depending on the context. In effect, then, according to neo-empiricists, concepts are not portable representations—they are not representations that can be transported identically from thought to thought. Second, just like indexicals[2] or relative predicates in natural languages, the semantic properties of concepts are determined by, and thus vary across, contexts. This second form of context-dependence results from the first: whereas, according to neo-empiricists, the extension of a concept is probably not meant to vary across contexts (tokens of CUP all have the same extension, viz., the set of cups), the way the members of its extension are thought about—what Prinz calls "the cognitive content" of the concept—varies across contexts, since it depends on how one simulates perceiving them (one thinks differently about cups in different contexts). This characteristic of perceptual symbols is reminiscent of the representations in connectionist networks, since in a connectionist network tokens of the same concept, say, CUP, correspond to different patterns of activation in different contexts—a point emphasized by Fodor and Pylyshyn (Fodor and Pylyshyn 1988; Smolensky 1988, 1991; Fodor and McLaughlin 1990).

Not only are the parts of symbols expressing propositions not portable, their delineation can also be vague: there may not be any fact of the matter about where one subsymbol starts and another ends.

Finally, current thoughts (beliefs, desires, judgments, etc.) have parts (e.g., a symbol of a gnu is a part of the thought that I am hunting a gnu), but no constituents. Fodor and Pylyshyn (1988) were instrumental in bringing to the fore the distinction between the parts and the constituents of representations. A constituent is either simple or complex. Complex constituents are recursively formed out of other constituents by putting them together according to some rule. These rules form the syntax of the representation, or symbol, system. Syntactic rules are semantically relevant and determine the semantic properties of complex symbols, that is, the interpretation of a complex symbol is a function of its constituents and of its syntax. Naturally, although every constituent is a part of a representation, not all parts are constituents. This point can be easily grasped by considering sentences in natural languages. Although "dog

barks" is a part of "the dog barks at the cat," it is not one of its constituents; in contrast, "the dog" and "barks at the cat" are constituents of this
sentence because each of them was formed out of primitive symbols
(which are by definition constituents) by applying some syntactic rule.
Just like representations in connectionist networks (patterns of activation
of nodes within a layer; see Fodor and Pylyshyn 1988), simulations have
parts, but these are not constituents, for their parts are not put together,
or concatenated, according to some syntactic rule. Whatever rule is meant
to govern the blending of perceptual symbols according to neo-empiricists, the subsymbols for a cat and for a dog in the thought that the dog
is barking at the cat are not put together according to a syntactic rule:
the symbol for the dog is not meant to be the argument of a definite
description operator, and the resulting symbol is not meant to be identified as the first argument of the predicate for barking or as standing in
a subject relation to this predicate.[3]

To summarize, according to neo-empiricists, occurrent propositional
thoughts can, but need not, be made out of subsymbols for the concepts
that were involved in producing the thoughts. The subsymbols that are
parts of the propositional symbols are not portable across thoughts; their
delineation can be vague; and they do not stand in syntactic relations.
Thus, they are not constituents.

2.2 Thoughts in Long-Term Memory

Before critically examining neo-empiricists' theory of occurrent propositional thoughts, we should say a few words about non-occurrent thoughts,
that is, about the thoughts in long-term memory. Even when one is not
currently entertaining the thought that dogs bark, that thought is somehow
stored in long-term memory. How do neo-empiricists characterize this type
of thought?

Prinz (2002, chap. 6) deals with this question at some length. He proposes that in a given network of representations in long-term memory,
perceptual representations are connected by various links, of which he
distinguishes five types: hierarchy (a link that connects two perceptual
representations when a perceptual representation results from zooming
in on another representation), transformation (a link that connects
two perceptual representations when one can result from a possible transformation of the other), binding (a link that connects two perceptual
representations when the first represents a property, part, or behavior of
the instance of the other), situational (a link that connects two perceptual
representations when these typically co-occur), and predicative (a link that

connects a singular and a predicative perceptual representation). For instance, to represent in long-term memory the proposition that dogs bark, perceptual representations of dogs are connected to auditory representations of barking by a binding link.

3   Thoughts in Long-Term Memory Require Amodal Symbols

Neo-empiricists eschew amodal symbols altogether: According to them, thoughts are perceptual through and through. We should thus be on the lookout for amodal symbols accidentally smuggled in to account for the nature of thought and cognitive processing.

Neo-empiricists often treat multimodal representations as perceptual (Prinz 2002), an assumption amodal theorists are bound to object to; they do the same for words (Prinz 2002), which is barely less controversial. In this section, however, I want to focus on a less obvious place where amodal symbols may be smuggled in: the links hypothesized to connect perceptual representations in a long-term memory network. These links are content-ful: they represent specific relations between the referents of the linked perceptual relations. For instance, the binding link can represent parthood (among several other possible relations). Consistent with the claim that the hypothesized links are contentful, they are labeled in Prinz's pictorial representation of a long-term memory network (fig. 6.4 in Prinz 2002). But, because these links are not perceptual representations, it would seem that they are amodal symbols in disguise.

It would not do to respond by allowing for some amodal symbols in addition to perceptual representations (in the spirit, if not the letter, of dual-code theories; see, e.g., Dove 2010), for, if some representations were amodal, why would they be limited to representing the relations represented by Prinz's hypothesized links? Why not amodal representations of abstract entities, such as democracy and causation, and of unobservable entities, such as atoms? And if the latter are allowed, why not representations of dogs, cats, going to the dentist, and so on?

A better response would be to deny that the hypothesized links are symbols, and thus, a fortiori, that they are amodal symbols. In this spirit, Prinz denies that entertaining the negation of a proposition involves any representation in addition to the perceptual representation involved in entertaining the proposition itself. Rather, entertaining a negated proposition (e.g., that dogs do not purr) consists in a particular way of matching the perceptual representation expressing the nonnegated proposition (e.g., that dogs purr) to perceptual experiences (Prinz 2002, 181–183). Negation is thus not a representation, it is a particular operation completed on representations. Prinz proposes to treat disjunctive thoughts and quantified thoughts similarly (183–184).[4] This strategy could perhaps be extended to links.

However, it is not easy to specify operations corresponding to the different links that need to be hypothesized, and one can doubt that all of them can be reduced to operations. Without any concrete proposal, it is difficult to assess the prospects of this strategy. More important, the strategy used by Prinz to account for negation in a neo-empiricist framework does not directly apply to the links hypothesized to make up long-term memory networks since it concerns occurrent thoughts, and not the thoughts in long-term memory. Furthermore, it is not at all clear that we can we extend Prinz's strategy to these latter thoughts. To deal with negation, disjunction, and quantification, Prinz focuses on different ways of matching occurrent simulations that express propositions to perceptions, but this strategy is not applicable to the representation of relations such as parthood, for the issue there is how to represent propositions in the first place. Perhaps Prinz could suggest that the links are priming relations: when a perceptual representation is retrieved from long-term memory, the perceptual representations it is linked to are likely to be retrieved too. But this suggestion is a nonstarter, since priming relations are not contentful; furthermore, supposing that they were, priming relations would still be insufficiently fine-grained to represent all the relations that need to be represented.

There is a general lesson to be learned from this discussion. Perceptual representations in long-term memory could stand for individuals or classes, but to represent a proposition, it seems that we need amodal symbols that put these perceptual representations together. Without them, perceptual representations remain unstructured.

However, a neo-empiricist has another card up her sleeve: she can respond that thoughts in long-term memory need not be explicitly represented, and thus that amodal symbols are not needed to structure perceptual symbols into thoughts in long-term memory. Rather, to have non-occurrent thoughts just is to have a capacity to entertain occurrent thoughts. For instance, to believe that dogs bark when that thought is not occurring just is to be able to entertain the occurrent thought that dogs bark. We now turn to assessing neo-empiricists' views about occurrent thoughts.

4   Neo-Empiricists Mischaracterize Thinking

4.1   The Individuation of Thoughts

In the neo-empiricist view of thoughts described in section 2, it is not obvious what makes a given occurrent thought the thought it is since the symbols corresponding to the constituents of the proposition expressed need not be cotokened with the symbol expressing the proposition itself. Indeed, one wonders what distinguishes two thoughts that would only differ by the missing symbol. To reuse Prinz's example, what distinguishes the perceptual representation of myself hunting a gnu from the perceptual representation of myself hunting a fat gnu if no symbol for fatness is cotokened with the perceptual representation in the second case?

Here is what Prinz has to say in response to this issue: "While one cannot separate out the representations of hunting or of fatness, one can identify the contributions that Boris's knowledge make to this simulation" (2002, 150). This response is ambiguous, and can be read in two different ways. First, Prinz seems to hold that the two thoughts under consideration differ—different perceptual representations are entertained—because our long-term knowledge about fatness is involved in producing the latter, but not the former. On this view, then, cotokening is not needed to individuate thoughts because perceptual representations are exactly as fine-grained as the propositions they express.

So understood, Prinz's response suffers from two problems. It seems to rest on a confusion between the vehicle and the content of perceptual representations. No doubt, a perceptual representation of a fat gnu represents the gnu differently from a mere representation of a gnu—namely, it represents it as fat—but this claim concerns the content of the two perceptual representations, not their vehicles, whereas the individuation issue raised here concerns their vehicles. Second, bracketing this first problem, the context-dependence of the tokens of a given concept, which was also discussed in section 2, raises another problem for Prinz's response: because the occurrences of a concept such as CUP differ across contexts, a mere difference in perceptual simulations is not sufficient to individuate occurrent thoughts. So, perceptual representations are more fine-grained than the propositions they express (assuming that not every difference in the vehicle of a concept makes a difference in the thought expressed).

The second reading of Prinz's solution deals with this problem: what distinguishes the thought that I am hunting a gnu from the thought that I am hunting a fat gnu is that the concept for fatness in long-term memory was involved in producing the second thought, but not the first one, and what makes all the thoughts expressing propositions that have fatness as a constituent thoughts about fatness despite the variation in how fatness is represented is that all of them come from the same concept in long-term memory. As Prinz puts it: "The answer may lie in the origin of the simulation. Mental representations of hunting and fatness are used when the simulation is initially formed" (2002, 151). Barsalou too embraces this solution:

If the same individual conceptualizes *bird* differently across occasions, how can stability be achieved for this concept? One solution is to assume that a common simulator for *bird* underlies these different conceptualizations both between and within individuals. ... Consider how a simulator produces stability within an individual. If a person's different simulations of a category arise from the same simulator, then they can all be viewed as instantiating the same concept. Because the same simulator produced all of these simulations, it unifies them. (1999, 588)

That is, what individuates thoughts—what makes the perceptual representations expressing propositions the thoughts they are—is a historical property, namely, the origins of these thoughts.

4.2 Recognizing the Identity of Thoughts

Prinz's and Barsalou's solution to the individuation problem is only partially satisfactory since there are in fact *two* individuation problems and their solution addresses only one of them. The first problem, the one we started with, concerns the individuation of thoughts proper—what makes a thought the thought it is—and, as we have just seen, neo-empiricists appeal to the origins of the thought to deal with this issue. The second problem concerns how the identity of thoughts or of thought components is recognized: how does one know what thought one is thinking, how does one recognize that one is thinking the same thought on two different occasions, how does one know what one is thinking about, and how does one recognize that one is thinking about the same thing on two different occasions?

Let's examine this second problem in more detail. As we have seen, constituents of the proposition expressed by an occurrent perceptual simulation are not necessarily expressed by distinct subsymbols cotokened with the occurrent perceptual simulation (as was the case for the fatness of the gnu in Prinz's example), and the tokens of a concept, such as CUP, differ across occasions. In addition, historical properties, in virtue of which, according to neo-empiricists, thoughts are the thoughts they are, are not presently instantiated: the origin of an entity—where it comes from—is not one of its presently instantiated properties. Of course, two entities often differ because they have different origins. For instance, an authentic Vermeer painting and a modern copy (by, say, van Meegeren) would be made with different pigments. But that's not the point; rather, the point is that the origin of an entity is a temporal relation between a current entity and a past event, not one of its currently instantiated properties. For this reason, an occurrent thought does not wear its origin on its sleeve, so to speak, although some of its present properties may be evidence about where it comes from. But then, if thoughts are individuated by their origin and if the origin of a thought is not presently instantiated, how do we know what thought we are currently entertaining or what we are currently thinking about? For instance, how do I know that I want to hunt a *fat* gnu? Do I need to infer what thought I am now thinking from some of its presently instantiated properties?

This problem can be formulated at the personal or at the impersonal level. At the personal level, the question is: how do *I* know what I am now thinking about if the origin of my thought determines its identity and if its origin is not a presently instantiated property? At the impersonal level, the question becomes: when a thought is entertained, how do cognitive processes manage to process it appropriately, that is, as the thought it is rather than as if it were some other thought? How is it that the proper inferences are drawn from the occurrence of a particular thought if its identity depends on its origin and if its origin is not a presently instantiated property? These two formulations are plausibly interconnected.

So, the problem is the following: if no symbol for fatness is cotokened when I am simulating hunting a fat gnu, how do I infer from the thought that I will be hunting a fat gnu that any gnu that will be killed will be heavy and difficult to carry? And if the simulation of a human being varies across contexts, how can I infer that Socrates is mortal from simulating the thought that Socrates is a human, and that human beings are mortal (however these thoughts are simulated)? If neo-empiricists do not have a satisfying answer to these questions, then neo-empiricism fails to explain the phenomenon of thinking: it fails to explain what it is to move from thought to thought, even if it has a satisfactory account of the individuation of thoughts (section 4.1). Naturally, neo-empiricists are not allowed to smuggle in amodal symbols to answer these questions. Amodal theorists who follow Fodor and Pylyshyn (1988) do not face a similar challenge: since the symbol for fatness is necessarily cotokened when I am thinking about fatness and since this symbol is portable, inferences involving thoughts about fatness (or about human beings) are simply due to the sensitivity of cognitive processes to this symbol. (The usual familiar story.)

Neo-empiricists could respond, as Prinz does, that the simulation of hunting a fat gnu and that of hunting a gnu differ and that cognitive processes are sensitive to such differences. They could also hold that cognitive processes treat the varying tokens of a given concept (e.g., the varying simulations of a cup) identically. However, first, the nature of the difference between the simulation of hunting a fat gnu and the simulation of hunting a gnu has not been properly specified, since Prinz seems to confuse the content and the vehicle of simulations. Second, before being entitled to this response, neo-empiricists need to explain how cognitive processes perform this feat. To meet this challenge, neo-empiricists could appeal to learning: Exactly as connectionist networks can treat different patterns of activation similarly when they have been trained to do so, and exactly as they do not need vehicles to be composed of distinct constituents to draw appropriate inferences, cognitive processes could have learned to distinguish simulations that are not distinguished by their constituents and to treat varying occurrences of a concept identically.

However, appealing to learning is unsatisfying. First, because, in contrast to connectionism, where learning can be formally studied as well as simulated, neo-empiricists theories have not characterized in any detail the properties of the processes defined over simulations, it is not clear what these processes can and cannot do, including what they can and cannot learn. Bracketing this first, minor issue, it would also seem that people are able to draw proper inferences from their occurrent thoughts or that they are able to infer from premises because of the nature of their cognitive architecture (the way their mind is organized), not because of a contingent history of learning, a point made against neural networks by Fodor and Pylyshyn (1988). If we are able to infer a claim from several premises only because processes have learned to treat different tokens of a concept similarly, then in principle we could think that Socrates is a human being and that human beings are mortal without being able to infer that Socrates is mortal if we had not learned to treat the two occurrences of HUMAN identically. But our mind just doesn't seem to work that way: By virtue of the way our mind is built, when we are thinking that Socrates is a human being and that human beings are mortal, we are in a position to infer than Socrates is mortal, although various performance factors (such as distraction, limited working memory, etc.) may actually prevent us from drawing this inference. What's more, assuming the productivity of thought with Fodor and Pylyshyn (1988), there must be thoughts that we have not

learned to distinguish or varying tokens of the same concept that we have
not learned to treat identically. Finally, appealing to learning is inconsistent with the systematicity of our inferential capacities (Fodor and Pylyshyn 1988). If inferential capacities are systematic, then anybody who can
infer p from p & q can also infer p from p & $(q$ & $r)$. But if recognizing a
thought for what it is requires learning, then there is no guarantee that
whoever can make the first inference can also make the second inference,
since she may not have learned to recognize the latter thought. So, if
thought is really systematic, then neo-empiricism is unacceptable.

4.3 Wrapping Up

Neo-empiricism suffers from some of the problems identified by Fodor and
Pylyshyn (1988) in their discussion of connectionism. If the neo-empiricist
view about occurrent thoughts is right, then occurrent thoughts are individuated by their origins, but, if they are, then it is unclear how we are
able to draw the inferences that our thoughts justify. If neo-empiricists
were to allude to learning to solve this problem, then they would probably
have to acknowledge that the capacity to draw a conclusion from occurrent
thoughts is a matter of a contingent history of learning rather than a
characteristic of our cognitive architecture. Some thoughts would have to
be such that we are unable to draw the appropriate conclusions from them
if thoughts are productive, while our inferential capacities would turn out
to be nonsystematic.

Conclusion

Neo-empiricists' views about thought are unacceptable. They cannot
explain what it is to have thoughts in long-term memory without positing
amodal symbols, and their theory of occurrent thoughts (occurrent simulations) renders thinking—our ability to move from thought to thought—
either mysterious or a matter of a contingent history of learning.

Notes

1. For additional discussion, see Rice 2011.

2. Of course, unlike indexicals, the reference of a perceptual symbol does not change
across contexts. Rather, its sense or cognitive content is context-sensitive.

3. In addition, constituents are often taken to be portable representations that
necessarily co-occur with the thoughts they constitute.

4. This strategy does not seem to be able to deal with the thought that it is not the case that it is not the case that *p*. If negation is a particular way of matching a represented proposition to perceptual experiences, then it is unclear what the negation of a negation is. If neo-empiricists respond that the negation of a negation is merely the assertion of a proposition, then they can't distinguish the thought that it is not the case that it is not the case that *p* from the thought that *p*.

References

Barsalou, L. W. 1999. Perceptual symbol systems. *Behavioral and Brain Sciences* 22:577–609.

Barsalou, L. W. 2010. Grounded cognition: Past, present, and future. *Topics in Cognitive Science* 2:716–724.

Dove, G. O. 2009. Beyond perceptual symbols: A call for representational pluralism. *Cognition* 110:412–431.

Dove, G. O. 2010. Another heterogeneity hypothesis. *Behavioral and Brain Sciences* 33:209–210.

Dove, G. O. 2011. On the need for embodied and dis-embodied cognition. *Frontiers in Psychology* 1:1–13.

Fodor, J. A. 1975. *The Language of Thought*. Cambridge, MA: Harvard University Press.

Fodor, J. A. 2008. *LOT 2: The Language of Thought Revisited*. Oxford: Oxford University Press.

Fodor, J. A., and B. McLaughlin. 1990. Connectionism and the problem of systematicity: Why Smolensky's solution doesn't work. *Cognition* 35:183–204.

Fodor, J. A., and Z. W. Pylyshyn. 1988. Connectionism and cognitive architecture: A critical analysis. *Cognition* 28:3–71.

Gallese, V., and G. Lakoff. 2005. The brain's concepts: The role of the sensory-motor system in conceptual knowledge. *Cognitive Neuropsychology* 21:455–479.

McCaffrey, J., and E. Machery. 2012. Philosophical issues about concepts. *Wiley Interdisciplinary Reviews: Cognitive Science* 3:265–279.

Machery, E. 2006. Two dogmas of neo-empiricism. *Philosophy Compass* 1:398–412.

Machery, E. 2007. Concept empiricism: A methodological critique. *Cognition* 104:19–46.

Mahon, B. Z., and A. Caramazza. 2008. A critical look at the embodied cognition hypothesis and a new proposal for grounding conceptual content. *Journal of Physiology, Paris* 102:59–70.

Markman, A., and H. Stilwell. 2004. Concepts a la modal: An extended review of Prinz's *Furnishing the mind. Philosophical Psychology* 17:391–401.

Prinz, J. J. 2002. *Furnishing the Mind: Concepts and Their Perceptual Basis.* Cambridge, MA: MIT Press.

Pylyshyn, Z. W. 1984. *Computation and Cognition.* Cambridge, MA: MIT Press.

Rice, C. 2011. Concept empiricism, content, and compositionality. *Philosophical Studies* 155:1–17.

Smolensky, P. 1988. On the proper treatment of connectionism. *Behavioral and Brain Sciences* 11:1–74.

Smolensky, P. 1991. Connectionism, constituency, and the language of thought. In *Meaning in Mind: Fodor and His Critics*, ed. B. M. Loewer and G. Rey, 201–227. Oxford: Blackwell.

IV

14 Systematicity and Interaction Dominance

Anthony Chemero

1 Introduction

In their massive 1988 article in *Cognition*, Jerry Fodor and Zenon Pylyshyn introduced the world to something they called "systematicity" and used it over the course of sixty-eight pages to bludgeon the then nascent reintroduction of artificial neural network research into the cognitive sciences. This was not their first attempt to squash a new research program in the cognitive sciences. They published, in the same journal, a similarly lengthy attempt at taking down Gibsonian ecological psychology (Fodor and Pylyshyn 1981). Luckily, both these research programs survived the attempted infanticide and are cheerful adults with grant money, conferences, and journals of their own.

In this not at all massive essay, I will look back at this historical episode and its aftermath. Even though the 1988 article did not succeed in ending the artificial neural network research program, it has been enormously influential, having been cited more than 2,600 times (by August 2012, according to Google Scholar). This outsized influence can only be a testament to the esteem in which cognitive scientists hold Fodor and Pylyshyn. It certainly cannot be attributed to the scientific argument that Fodor and Pylyshyn make in the article. This is the case because they invented the concept of systematicity whole cloth. In 1988, there was no laboratory evidence demonstrating the systematicity of human thought. After defending this contentious claim, I will argue for something much less controversial. I will discuss a recent line of research in cognitive science that is inconsistent with systematicity. This line of research uses nonlinear dynamical systems models to demonstrate that certain cognitive abilities are interaction dominant. Interaction-dominant systems are temporary, task-specific collections of physiological components, in which the intrinsic behavioral tendencies of the components are overwhelmed by the effect of the interactions among the components. I will argue that interaction-dominant systems cannot exhibit systematicity.

2 The Original Systematicity Argument

Fodor and Pylyshyn's original argument against the possibility of connectionist networks as a model of the cognitive architecture goes roughly as follows.

1. Human thought is systematic.
2. Systematicity requires representations with compositional structure.
3. Connectionist networks do not have representations with compositional structure.
4. *Therefore*, connectionist networks are not good models of human thought.

This argument is one of the most important and influential in the recent history of cognitive science. It drew stark battle lines soon after Rumelhart, McClelland, and the PDP Research Group (1986) drew attention to connectionist networks. It is also an argument that has been convincing to many people.

The key empirical premise in the argument is premise 1, the claim that human thought is systematic. Fodor and Pylyshyn describe what they mean by the claim many times in the course of their article, by saying that human cognitive abilities come in clusters such that if a human has one ability in a cluster, he or she also has the others. Fodor and Pylyshyn put it as follows:

What we mean when we say that linguistic capacities are *systematic* is that the ability to produce/understand some sentences is *intrinsically* connected to the ability to produce/understand certain others. ...

You don't, for example, find native speakers of English who know how to say in English that John loves the girl but don't know how to say in English that the girl loves John. (Fodor and Pylyshyn, 1988, 37)

Later, Fodor and McLaughlin (1990) claim that systematicity is a matter of psychological law:

As a matter of psychological law, an organism is able to be in one of the states belonging to the family only if it is able to be in many of the others. ... You don't find organisms that can think the thought that the girl loves John but can't think the thought that John loves the girl. (Fodor and McLaughlin 1990, 184)

Fodor and Pylyshyn take the second premise of the argument to follow from the first. Systematicity and compositionality are "best viewed as two aspects of a single phenomenon" (1988, 41). In particular, systematicity is assumed to require compositionality, where compositionality is the claim that "a lexical item must make approximately the same contribution to each expression in which it occurs" (42). So for thought to be systematic, its representations must be made up of syntactic parts that have semantics, and those parts must be able to be rearranged to form other representations with the semantics of the parts maintained. That is, the representation "Joanie loves Chachi" is composed of the parts "Joanie," "loves," and "Chachi," and those parts can be rearranged to form the representation "Chachi loves Joanie" with "Chachi," "loves," and "Joanie" still having the same referents. "Systematicity depends on compositionality, so to the extent that natural language is systematic it must be compositional too" (43).

Premise 3 is a claim about connectionist networks, and a contentious one. Many defenders of connectionism, such as Smolensky (1991), van Gelder (1990), and Chalmers (1990), have argued that connectionist networks can have representations with compositional structure. In any event, if premise 3 is granted, the conclusion follows directly. Although I think it's obvious that Smolensky, van Gelder, and Chalmers are right that connectionist networks can exhibit systematicity, I will not pursue that here. Instead, I will attempt to call premise 1 into question. I will follow Dennett (1991) and Clark (1997) in arguing that, although some portions of human language (especially written language) are systematic, human thought in general is not. I will do this by presenting empirical evidence that many human abilities are not accomplished by compositional representational structures. Since systematicity requires compositionality, this will be empirical evidence that many human cognitive abilities are not systematic. I will present this evidence in sections 4 and 5. The next section, section 3, is a rant.

3 "The Great Systematicity Hoax"

Not long after the publication of "Connectionism and the Cognitive Architecture," Tim van Gelder regularly gave a talk called "The Great Systematicity Hoax." Unfortunately, that talk was never turned into a published paper. The main point of the talk was that the term "systematictity" is too poorly defined to be of any use in scientific and philosophical reasoning.

Moreover, van Gelder argued, there is good reason to think that human cognition is not, in fact, systematic. The arguments in Ken Aizawa's (2003) book *The Systematicity Arguments*, I believe, go a long way toward solidifying the concept of systematicity. This suggests that van Gelder's first criticism, that systematicity is ill defined, no longer holds up. His second one fares much better, and it holds up because the title of van Gelder's talk is accurate—the idea that human thought is systematic was simply made up by Fodor and Pylyshyn, without any specific evidence in its favor.

Presumably, Fodor and Pylyshyn wrote "Connectionism and the Cognitive Architecture" sometime in 1986 or 1987, since it was published in the first issue of *Cognition*'s 1988 volume. The question one might ask at this point is whether there was any actual evidence, gathered by laboratory scientists, demonstrating that human thought was in fact systematic. To find out, I conducted a keyword search for the term "systematicity" in Web of Science (see http://thomsonreuters.com/web-of-science/). As of summer 2011, there were 323 articles published with the keyword "systematicity," but only eight of those were before 1988. The eight articles, as they were output by the Web of Science search engine, appear in table 14.1.

Of these articles, three are in philosophy (Rockmore 1987; Sasso 1981; Bazhenov 1979), but none of these is related to philosophy of mind or cognitive science. Rockmore 1987 and Sasso 1981 are in the history of philosophy; Bazhenov 1979 is about scientific theory change. Three of the articles (Blevy-Roman 1983; Tarone 1982; Keller-Cohen 1979) appear in the journal *Language Learning*, and are about second language learning. Two of the articles look like promising sources of evidence for Fodor and Pylyshyn's claims about systematicity. Rubenstein, Lewis, and Rubenstein 1971 is actually about human language. But its findings do not support Fodor and Pylyshyn's views of systematicity. Rubenstein et al. report a behaviorist study of reaction times to homophones, and their paper includes no claims related to the systematicity of human thought. Gentner and Toupin 1987 looks promising, at least in that it is the work of prominent cognitive scientists and is about human cognition. However, the article is actually about the relationships among sources and targets in analogical reasoning, and not at all about the relationship among lexical items. There was no prior published work that used the term "systematicity" in the way that Fodor and Pylyshyn (1988) did. Of course, it could be the case that only the term "systematicity" was made up by Fodor and Pylyshyn, and they were referring to a well-studied psychological phenomenon. We can check this, in the standard scholarly fashion, by looking at the references in Fodor and Pylyshyn's article. They cite sixty-five articles,

Table 14.1

1. **Author(s):** Rockmore, T.
Title: Hegel and the Science of Logic—The Formation of Principles of Systematicity and Historicism—Russian—Motroshilova, N. V.
Source: *Archives de Philosophie* 50 (3): 495–496, Jul.–Sep. 1987
ISSN: 0003-9632

2. **Author(s):** Gentner, D.; Toupin, C.
Title: Systematicity and Surface Similarity in the Development of Analogy
Source: *Cognitive Science* 10 (3): 277–300, Jul.–Sep. 1986
ISSN: 0364-0213

3. **Author(s):** Bleyvroman, R.
Title: The Comparative Fallacy in Interlanguage Studies—The Case of Systematicity
Source: *Language Learning* 33 (1): 1–17, 1983
ISSN: 0023-8333

4. **Author(s):** Tarone, E. E.
Title: Systematicity and Attention in Interlanguage
Source: *Language Learning* 32 (1): 69–84, 1982
ISSN: 0023-8333

5. **Author(s):** Sasso, R.
Title: Scope of "De Deo" of Spinoza ("Ethics", I) + An Example of the Workings of Systematicity in Philosophical Discourse
Source: *Archives de Philosophie* 44 (4): 579–610, 1981
ISSN: 0003-9632

6. **Author(s):** Kellercohen, D.
Title: Systematicity and Variation in the Non-Native Child's Acquisition of Conversational Skills
Source: *Language Learning* 29 (1): 27–44, 1979
ISSN: 0023-8333

7. **Author(s):** Bazhenov, L. B.
Title: Systematicity as a Methodological Regulative Principle of Scientific Theory
Source: *Voprosy Filosofii* 6: 81–89, 1979
ISSN: 0042-8744

8. **Author(s):** Rubenste, H.; Lewis, S. S. ; Rubenste, M. A.
Title: Homographic Entries in Internal Lexicon—Effects of Systematicity and Relative Frequency of Meanings
Source: *Journal of Verbal Learning and Verbal Behavior,* 10 (1): 57–62, 1971
ISSN: 0022-5371

but only one chapter of one work is listed as evidence that human thought is systematic. In note 24, Fodor and Pylyshyn cite chapter 4 of Pinker 1984 as showing that children always display systematicity with respect to the syntactic structures that appear in noun phrases (Fodor and Pylyshyn 1988, 39). It turns out, however, that this particular chapter actually includes evidence that children learning language do not always display systematicity. So Fodor and Pylyshyn (1988) gave absolutely no evidence that human thought is systematic.

This utter lack of empirical evidence puts Fodor and Pylyshyn's argument in the class of what I've called "Hegelian arguments" in the past (Chemero 2009). That name is derived from the legend that Hegel argued, in his *Habilitation*, that the number of planets was necessarily seven. Hegel did not actually argue for this claim. Instead, he used a number series from Plato's *Timaeus* to argue that, as a matter of logic, there could be no eighth planet between Mars and Jupiter—this despite contemporary observational evidence that there was in fact a planet between Mars and Jupiter. Despite observational evidence to the contrary, Hegel argued that astronomers should just stop looking for a planet between Mars and Jupiter, because there could not be one. Fodor and Pylyshyn argue that everyone should stop doing connectionist network research because there could never be a connectionist model of real cognition. Like Hegel, Fodor and Pylyshyn base their argument on what has to be the case, and not on empirical evidence. Hegelian arguments are arguments made without systematically gathered empirical evidence that no theory of a particular kind could ever account for some phenomenon of interest. They are aimed at shutting down research programs. This sort of argument is surprisingly common in the cognitive sciences, in a way that it is not in other sciences. (See, e.g., Chinese rooms, qualia, symbol grounding, bat phenomenology, etc.) It is, to put it mildly, surprising that so many cognitive scientists were embroiled in a debate that had so little basis in scientific reality.

As much as one might deplore the argumentative tactics, there is one way in which the time and effort spent discussing systematicity will have been worthwhile. If, despite Fodor and Pylyshyn's lack of evidence for it, human cognition is in fact systematic in the way that Fodor and Pylyshyn describe, we should be grateful to them for pointing it out for us. Especially, if one begins from a classical cognitivist point of view, it seems highly plausible that (some? most? all?) human cognition is systematic. Indeed, Ken Aizawa (in personal communication) has suggested that the systematicity of human cognition is so obvious as to not require experimental evidence demonstrating it. Yet, many things that seemed obvious to casual observation turn out to be untrue after scientific investigation: the Earth moves, after all.

4   Interaction-Dominant Systems

A recent line of research in the cognitive sciences explains many cognitive phenomena in a way that is incompatible with their exhibiting compositionality and, therefore, systematicity. There is now significant evidence suggesting that many cognitive abilities arise in *interaction-dominant systems*. This new research program promises to shed light on questions about cognition and the mind that previously seemed "merely philosophical." In the remainder of this section I will briefly (and inadequately) set out the basic concepts and methodologies of this new research program. In the next section, I show how those concepts and methodologies have been put into use in the cognitive sciences, focusing on the ways in which interaction dominance is inconsistent with systematicity.

First, some background: an *interaction-dominant system* is a highly interconnected system, each of whose components alters the dynamics of many of the others to such an extent that the effects of the interactions are more powerful than the intrinsic dynamics of the components (van Orden, Holden, and Turvey 2003). In an interaction-dominant system, inherent variability (i.e., fluctuations or noise) of any individual component A propagates through the system as a whole, altering the dynamics of components B, C, and D. Because of the dense connections among the components of the system, the alterations of the dynamics of B, C, and D will lead to alterations to the dynamics of component A. The initial random fluctuation of component A, in other words, will reverberate through the system for some time. So too would nonrandom changes in the dynamics of component A. This tendency for reverberations gives interaction-dominant systems what is referred to as "long memory" (Ding, Chen, and Kelso 2002). In contrast, a system without this dense dynamical feedback, a *component-dominant system*, would not show this long memory. For example, imagine a computer program that controls a robotic arm. Although noise that creeps in to the commands sent to from the computer to the arm might lead to a weld that misses its intended mark by a few millimeters, that missed weld will not alter the behavior of the program when it is time for the next weld. Component-dominant systems do not have long memory. Moreover, in interaction-dominant systems, one cannot treat the components of the system in isolation: because of the widespread feedback in interaction-dominant systems, one cannot isolate components to determine exactly what their contribution is to particular behaviors. And because the effects of interactions are more powerful than the intrinsic dynamics of the components, the behavior of the components in any particular interaction-dominant system is not predictable from their behavior in isolation or from their behavior in some other interaction-dominant system. Interaction-dominant systems, in other words, are not *modular*. Indeed, they are in a deep way *unified* in that the responsibility for system behavior is distributed across all of the components. As will be discussed below, this unity of interaction-dominant systems has been used to make strong claims about the nature of human cognition (van Orden, Holden, and Turvey 2003, 2005; Holden, van Orden, and Turvey 2009).

In order to use the unity of interaction-dominant systems to make strong claims about cognition, it has to be argued that cognitive systems are, at least sometimes, interaction dominant. Interaction-dominant systems self-organize, and some will exhibit *self-organized criticality*. Self-organized criticality is the tendency of an interaction-dominant system to maintain itself near critical boundaries, so that small changes in the system or its environment can lead to large changes in overall system behavior. It is easy to see why self-organized criticality would be a useful feature for behavioral control: a system near a critical boundary has built-in behavioral flexibility because it can easily cross the boundary to switch behaviors (e.g., going from out-of-phase to in-phase coordination patterns). (See Holden et al. 2009 for discussion.) It has long been known that self-organized critical systems exhibit a special variety of fluctuation called *1/f scaling* or *pink noise* (Bak, Tang, and Wieseneld 1987). 1/f scaling or pink noise is a kind of not-quite-random, correlated noise, halfway between genuine randomness (white noise) and a drunkard's walk, in which each fluctuation is constrained by the prior one (brown noise). 1/f scaling is often described as a fractal structure in a time series, in which the variability at a short timescale is correlated with variability at a longer timescale. In fact, interaction-dominant dynamics predicts that this 1/f scaling would be present. As discussed above, the fluctuations in an interaction-dominant system percolate through the system over time, leading to the kind of correlated structure to variability that is 1/f scaling. Here, then, is the way to infer that cognitive and neural systems are interaction dominant: exhibiting 1/f noise is evidence that a system is interaction dominant.

This suggests that the mounting evidence that 1/f scaling is ubiquitous in human physiological systems, behavior, and neural activity is also evidence that human physiological, cognitive, and neural systems are interaction dominant. To be clear, there are ways other than interaction-dominant dynamics to generate 1/f scaling. Simulations show that carefully gerrymandered component-dominant systems can exhibit 1/f scaling (Wagenmakers, Farrell, and Radcliff 2005). But such gerrymandered systems are not developed from any physiological, cognitive, or neurological principle, and so are not taken to be plausible mechanisms for the widespread 1/f scaling in human physiology, brains, and behavior (van Orden, Holden, and Turvey 2005). So the inference from 1/f scaling to interaction dominance is not foolproof, but there currently is no plausible explanation of the prevalence of 1/f scaling other than interaction dominance. Better than an inference to the best explanation, that 1/f scaling indicates interaction-dominant dynamics is an inference to the only explanation.

Applying this to the main point of this essay, interaction-dominant systems cannot exhibit compositionality. Only when dynamics are component dominant is it possible to determine the contributions of the individual working parts to the overall operation of the system. When dynamics are interaction dominant, it is impossible to localize the aspects of particular operations in particular parts of the system. A system exhibits compositionality when it has parts that make the same contribution to every representation in which they appear. Interaction-dominant systems have not been defined in terms of representation specifically, but in terms of the contribution of parts to any system completing a task. The key is that, with interaction-dominant dynamics, it is impossible to say what the contributions of the individual parts are. Moreover, individual parts will make different contributions to different interaction-dominant systems— whatever intrinsic behavioral dynamics the components bring to the system are overrun by the effects of their interactions with other components. So, even if we could figure out what part of the interaction-dominant system that realizes a particular instance of the thought that "Joanie loves Chachi" plays the role of "Chachi," we have no reason to believe any of the following about that part: that it would play the same role in the realization of any subsequent thought; that it would necessarily play any role in subsequent realizations of thoughts about Chachi; even that it would play the role of "Chachi" should it be involved in subsequent realizations of the thought that "Joanie loves Chachi." The contribution of any participant in an interaction-dominant system is thoroughly context dependent. The part that plays the role of "Chachi" in one context might become "Potsie" or "spoon" or "fly-or-BB" in another. Interaction-dominant systems cannot have compositional structure. And, since compositionality is required for systematicity, interaction-dominant systems cannot exhibit systematicity.

This matters for discussions of systematicity in cognition only to the extent that cognition seems to be accomplished via interaction-dominant dynamics. In the next section I will argue that at least some of it is. Maybe lots.

5   Interaction Dominance in Cognition

The evidence that cognitive tasks are accomplished by interaction-dominant systems has been gathered primarily, but not entirely, by those who argue that cognitive systems are necessarily embodied and sometimes extend beyond the biological body.[1] 1/f scaling has been observed in the brain, and in a wide variety of cognitive and behavioral tasks, from tapping, to key pressing, to word naming, and many others (see van Orden, Kloos, and Wallot 2009 for a review). This indicates that task-specific, temporary coalitions of components encompassing portions of the participants' brain and body were responsible for the performance of the experimental task. That the portions of the cognitive system that engages in tasks such as these is not fully encapsulated in the brain is perhaps not that surprising, since each has a strong motor component. But we also see 1/f scaling in "purely cognitive" phenomena. In one example, Stephen, Dixon, and Isenhower (2009; Stephen and Dixon 2009) have shown that problem-solving inference is accomplished by an inter-action-dominant system. They found that learning a new strategy for solving a problem coincides with the appearance of 1/f scaling, as measured in eye movements. This indicates that even leaps of insight do not occur in the brain alone—the eye movements are part of the interaction-dominant system that realizes the cognitive act. Findings such as this affect not only the extent of the biological resources required for cognitive faculties, but also the separation of cognitive faculties from one another.

There is reason to think that this expansion of the cognitive system does not stop at the boundaries of the biological body. For example, Dotov, Nie, and Chemero (2010, in press; Nie, Dotov, and Chemero 2011) describe experiments designed to induce and then temporarily disrupt an extended cognitive system. Participants in these experiments play a simple video game, controlling an object on a monitor using a mouse. At some point during the one-minute trial, the connection between the mouse and the object it controls is disrupted temporarily before returning to normal.

Dotov et al. found 1/f scaling at the hand-mouse interface while the mouse was operating normally, but not during the disruption. As discussed above, this indicates that, during normal operation, the computer mouse is part of the smoothly functioning interaction-dominant system engaged in the task; during the mouse perturbation, however, the 1/f scaling at the hand-mouse interface disappears temporarily, indicating that the mouse is no longer part of the extended interaction-dominant system. These experiments were designed to detect, and did in fact detect, the presence of an extended cognitive system, an interaction-dominant system that included both biological and nonbiological parts. The fact that such a mundane experimental setup (using a computer mouse to control an object on a monitor) generated an extended cognitive system suggests that extended cognitive systems are quite common.

These, of course, are not the only examples of interaction dominance in cognition. Dixon, Holden, Mirman, and Stephen (2012) constructed a fractal theory of *development*, explicitly connecting their theory to complex systems work in developmental biology. In addition, interaction-dominant dynamics have been found in *word naming* (van Orden et al. 2003), *word recall* (Rhodes and Turvey 2007), *visual search* (Stephen and Mirman 2010), *memory* (Maylor et al. 2001; Brown et al. 2007), and *social cognition* (Richardson et al. 2007; Harrison and Richardson 2009). (See van Orden, Kloos, and Wallot 2009 and Kello et al. 2010 for reviews.)

1/f scaling is ubiquitous in the brain as well. Heterogeneous coupling and multiscale dynamics are widespread features of the brain. Brain connectivity is organized on a hierarchy of scales ranging from local circuits of neurons to functional topological networks. At each scale the relevant neural dynamics are determined not just by processes at that scale, but by processes at other smaller and larger scales as well. Such multilevel clustered architectures promote varied and stable dynamic patterns via criticality and other dynamical and topological features. There is therefore also growing evidence that neural circuits are interaction dominant. Several recent studies have found evidence of 1/f scaling in human neural activity (e.g., Freeman, Rogers, Holmes, and Silbergeld 2000; Bressler and Kelso 2001; Bullmore et al. 2001; Linkenaker-Hansen et al. 2001; Buzsaki 2006; Freeman 2006, 2009; Freeman and Zhai 2009; see Bogen 2010 for a philosophical discussion). He, Zempel, Snyder, and Raichle (2010) have extended these latter findings by demonstrating that human arrhythmic brain activity is *multifractal*, in that it contains mutually nested and coupled frequency scales—lower frequencies of brain activity modulate the amplitude of higher frequencies. This is a dynamic property not only characteristic of interaction-dominant systems, but exhibited only by interaction-dominant systems (Ihlen and Vereijken 2010).

The recognition of interaction-dominant dynamics in the brain, in behavior, and in brain-body-environment systems has exploded in the twenty-first century. There is no reason not to expect this to continue. Finding that many or most of our cognitive abilities are realized by interaction-dominant systems would be bad news for Fodor and Pylyshyn's claim that exhibiting systematicity is a requirement on any theory of the cognitive architecture.

6   Implementation: A Last Refuge for Systematicity?

Fans of systematicity might seem to have a ready objection at this point. Fodor and Pylyshyn argue that systematicity is a requirement on any theory of the cognitive architecture. Connectionist networks do not exhibit systematicity, so the cognitive architecture cannot be an architecture of connectionist networks. Still, they continue, it is possible that connectionist networks are a reasonable theory of how the cognitive architecture is implemented in the brain. Something similar might be going on with interaction dominance. It is possible, the objection would go, that interaction dominance might characterize the implementation of the cognitive architecture, without thereby characterizing the cognitive architecture itself, which might be component dominant and hence potentially supportive of compositionality.[2] In order to claim that interaction dominance is a feature of the low-level implementation of a high-level component-dominant system, one would have to encapsulate the interactions *within* components. That interaction dominance was so encapsulated was the early view of David Gilden, who first found 1/f scaling in cognition (Gilden, Thornton, and Mallon 1995; Gilden 2001). Gilden (2001) used subtractive methods to isolate 1/f scaling to memory processes, and claimed that 1/f scaling is part of the *formation* of representations. This is exactly what would be required to claim that interaction-dominant systems are part of the implementation of the component-dominant systems that make up the cognitive architecture, allowing the cognitive architecture to potentially exhibit compositionality and systematicity.

This attempted save of systematicity is implausible. It must be noted that, as with the simulations described in section 4, the hypothesis that certain memory processes involved in forming representations alone are interaction dominant has essentially no basis in known physiology or behavior. This does not make it impossible that representation formation alone is interaction dominant, but it does suggest the need for some sort of strong evidence. The evidence that Gilden (2001; van Orden et al. 2003) provides involves the subtraction of background noise to isolate the $1/f$ scaling of the supposedly encapsulated memory processes. First, although I will not discuss it here, there is good reason to be skeptical of any use of subtractive methods to isolate components of cognitive systems (see, e.g., Uttal 2001; van Orden, Pennington and Stone 2001; among many others). In contrast with this meager evidence that interaction dominance is isolated to one or a few encapsulated systems, there is voluminous evidence that interaction dominance is ubiquitous in physiology, the brain, and behavior at all scales. (See section 5.) There is every reason to doubt that interaction-dominant dynamics is a feature only of the implementation of the cognitive architecture, but not of cognition itself. To claim the contrary, it would need to be shown that the interaction-dominant dynamics seen at each scale is in fact explicable in terms of $1/f$ scaling encapsulated within one or more of the components of the component-dominant system at that scale. Of course, it is logically possible that such a case could be made, but no one should be betting that it will happen.

7   Systematicity Revisited

At this point, it isn't clear how much of cognition is accomplished by interaction-dominant systems, but it is clear that it is more than a little. To whatever extent cognition is interaction dominant, it won't exhibit systematicity. This makes it seem as if the time spent arguing over whether or not connectionist networks can exhibit systematicity was wasted. That said, it is worthwhile to realize, as it seems to me we have by now, that theories of what is going on inside the brain during cognition do not have to imply inherent systematicity. It won't be breaking any new ground for me to say that what is required is the ability to explain the limited ways in which human intelligent behavior exhibits systematicity in those limited cases in which it does, and that this won't necessarily require that the brain-internal portions of the cognitive system exhibit systematicity (see, e.g., Rumelhart et al. 1986; Dennett 1991; Clark 1997; Wilson 2004). This is one of the key claims of embodied cognitive science. Those cases in which human intelligent behavior exhibits systematicity seem to be primarily those in which humans work with an external formal system, such as logic, arithmetic, and some portions of (especially written) natural language. We humans can do these things, but they do not make up most of what we do. In fact, working with external formal systems is something that requires years of instruction—reading, writing, and arithmetic. And even then, we're not very good at them. Even PhDs in philosophy are much more competent at tying their shoes than they are at proving theorems. It is, perhaps, a historical quirk that cognitive scientists missed this point for so long. The legacy of Fodor and Pylyshyn's 1988 paper is to have stalled by a few years the widespread acceptance of the fact that human cognitive abilities are not housed in brains alone.

Acknowledgments

Thanks to Paco Calvo, Maurice Lamb, and Luis Favela for helpful comments on this chapter.

Notes

1. This section draws on Anderson, Richardson, and Chemero 2012.

2. Thanks to Paco Calvo for pointing this out to me.

References

Aizawa, K. 2003. *The Systematicity Arguments*. New York: Springer.

Anderson, M., M. Richardson, and A. Chemero. 2012. Eroding the boundaries of cognition: Implications of embodiment. *Topics in Cognitive Science* 4:717–730.

Bak, P., C. Tang, and K. Wiesenfeld. 1988. Self-organized criticality. *Physical Review A.* 38:364–374.

Bogen, J. 2010. Noise in the world. *Philosophy of Science* 77:778–791.

Bressler, S., and J. A. S. Kelso. 2001. Cortical coordination dynamics and cognition. *Trends in Cognitive Sciences* 5:26–36.

Brown, G. D. A., I. Neath, and N. Chater. 2007. A temporal ratio model of memory. *Psychological Review* 114:539–576.

Bullmore, E., C. Long, J. Suckling, J. Fadili, G. Calvert, F. Zelaya, T. A. Carpenter, and M. Brammer. 2001. Colored noise and computational inference in neurophysiological (fMRI) time series analysis: Resampling methods in time and wavelet domains. *Human Brain Mapping* 12:61–78.

Buzsaki, G. 2006. *Rhythms of the Brain*, 119–135. New York: Oxford University Press.

Chalmers, D. 1990. Syntactic transformations on distributed representations. *Connection Science* 2:53–62.

Chemero, A. 2009. *Radical Embodied Cognitive Science*. Cambridge, MA: MIT Press.

Clark, A. 1997. *Being There*. Cambridge, MA: MIT Press.

Dennett, D. 1991. *Consciousness Explained*. Boston: Little, Brown.

Ding, M., Y. Chen, and J. A. S. Kelso. 2002. Statistical analysis of timing errors. *Brain and Cognition* 48:98–106.

Dixon, J., J. Holden, D. Mirman, and D. Stephen. 2012. Multifractal development of cognitive structure. *Topics in Cognitive Science* 4(1):51–62.

Dotov, D., L. Nie, and A. Chemero. 2010. A demonstration of the transition from readiness-to-hand to unreadiness-to-hand. *PLoS ONE* 5:e9433.

Dotov, D., L. Nie, and A. Chemero. In press. Readiness-to-hand, unreadiness-to-hand, and multifractality. *Journal of Mind and Behavior*.

Fodor, J., and B. McLaughlin. 1990. Connectionism and the problem of systematicity: Why Smolensky's solution doesn't work. *Cognition* 35:183–205.

Fodor, J., and Z. Pylyshyn. 1981. How direct is visual perception? *Cognition* 8:139–196.

Fodor, J., and Z. Pylyshyn. 1988. Connectionism and cognitive architecture: A critical analysis. *Cognition* 28:3–71.

Freeman, W. J. 2006. Origin, structure, and role of background EEG activity. Part 4. Neural frame simulation. *Clinical Neurophysiology* 117:572–589.

Freeman, W. J. 2009. Deep analysis of perception through dynamic structures that emerge in cortical activity from self-regulated noise. *Cognitive Neurodynamics* 3 (1):105–116.

Freeman, W. J., L. J. Rogers, M. D. Holmes, and D. L. Silbergeld. 2000. Spatial spectral analysis of human electrocorticograms including the alpha and gamma bands. *Journal of Neuroscience Methods* 9:111–121.

Freeman, W. J., and J. Zhai. 2009. Simulated power spectral density (PSD) of background electrocorticogram (ECoG). *Cognitive Neurodynamics* 3 (1):97–103.

Gilden, D. L. 2001. Cognitive emissions of $1/f$ noise. *Psychological Review* 108: 33–56.

Gilden, D. L., T. Thornton, and M. W. Mallon. 1995. 1/f noise in human cognition. *Science* 267:1837–1839.

Harrison, S., and M. Richardson. 2009. Horsing around: Spontaneous four-legged coordination. *Journal of Motor Behavior* 41:519–524.

He, B., J. Zempel, A. Snyder, and M. Raichle. 2010. The temporal structures and functional significance of scale-free brain activity. *Neuron* 66:353–369.

Holden, J., G. van Orden, and M. T. Turvey. 2009. Dispersion of response times reveals cognitive dynamics. *Psychological Review* 116:318–342.

Ihlen, E., and B. Vereijken. 2010. Interaction-dominant dynamics in human cognition: Beyond 1/fα fluctuation. *Journal of Experimental Psychology: General* 139: 436–463.

Kello, C., G. Brown, R. Ferro-i-Cancho, J. Holden, K. Linkenakker-Hansen, T. Rhodes, and G. van Orden. 2010. Scaling laws in cognitive sciences. *Trends in Cognitive Sciences* 14:223–232.

Linkenkaer-Hansen, K., V. V. Nikouline, J. M. Palva, and R. J. Ilmoniemi. 2001. Long-range temporal correlations and scaling behavior in human brain oscillations. *Journal of Neuroscience* 21:1370–1377.

Maylor, E. A., N. Chater, and G. D. A. Brown. 2001. Scale invariance in the retrieval of retrospective and prospective memories. *Psychonomic Bulletin and Review* 8:162–167.

Nie, L., D. Dotov, and A. Chemero. 2011. Readiness-to-hand, extended cognition, and multifractality. In *Proceedings of the 33rd Annual Meeting of the Cognitive Science Society*, ed. L. Carlson, C. Hoelscher, and T. Shipley, 1835–1840. Austin, TX: Cognitive Science Society.

Rhodes, T., and M. Turvey. 2007. Human memory retrieval as Levy foraging. *Physica A* 385:255–260.

Richardson, M. J., K. L. Marsh, R. Isenhower, J. Goodman, and R. C. Schmidt. 2007. Rocking together: Dynamics of intentional and unintentional interpersonal coordination. *Human Movement Science* 26:867–891.

Rumelhart, D. E., McClelland, J. L., and the PDP Research Group. 1986. *Parallel Distributed Processing: Explorations in the Microstructure of Cognition*, vol. 1: *Foundations*. Cambridge, MA: MIT Press.

Smolensky, P. 1991. Connectionism, constituency, and the language of thought. In *Meaning in Mind: Fodor and His Critics*, ed. B. M. Loewer and G. Rey. Oxford: Blackwell.

Stephen, D. G., and J. A. Dixon. 2009. The self-organization of insight: Entropy and power laws in problem solving. *Journal of Problem Solving* 2:72–102.

Stephen, D., J. Dixon, and R. Isenhower. 2009. Dynamics of representational change: Entropy, action, cognition. *Journal of Experimental Psychology: Human Perception and Performance* 35:1811–1822.

Stephen, D. G., and D. Mirman. 2010. Interactions dominate the dynamics of visual cognition. *Cognition* 115:154–165.

Uttal, W. 2001. *The New Phrenology*. Cambridge, MA: MIT Press.

van Gelder, T. 1990. Compositionality: A connectionist variation on a classical theme. *Cognitive Science* 14:355–384.

van Orden, G., J. Holden, and M. T. Turvey. 2003. Self-organization of cognitive performance. *Journal of Experimental Psychology: General* 132:331–351.

van Orden, G., J. Holden, and M. Turvey. 2005. Human cognition and 1/f scaling. *Journal of Experimental Psychology: General* 134:117–123.

van Orden, G., H. Kloos, and S. Wallot. 2009. Living in the pink: Intentionality, well-being, complexity. In *Handbook of the Philosophy of Science*, vol. 10: *Philosophy of Complex Systems*, ed. Cliff Hooker, 639–683. Amsterdam: Elsevier.

van Orden, G., B. Pennington, and G. Stone. 2001. What do double dissociations prove? *Cognitive Science* 25:111–172.

Wagenmakers, E.-J., S. Farrell, and R. Ratcliff. 2005. Human cognition and a pile of sand: A discussion on serial correlations and self-organized criticality. *Journal of Experimental Psychology: General* 134:108–116.

Wilson, R. 2004. *Boundaries of the Mind*. New York: Cambridge University Press.

15 From Systematicity to Interactive Regularities: Grounding Cognition at the Sensorimotor Level

David Travieso, Antoni Gomila, and Lorena Lobo

1 Introduction

The systematicity debate initially turned on the issue of the best explanation for the systematicity of cognition—then a property taken for granted, so that it barely required anything more than cursory exemplification. Connectionists challenged the idea that a "language of thought" of atomic constituents, plus formal combinatorial rules, was the only (best) approach to account for that claimed property of cognition. In these post-cognitivist times, we rather think that the proper reaction to the Fodor and Pylyshyn's (1988) challenge is to deny that cognition is systematic in general. Systematicity rather seems a property intrinsically dependent upon language rather than cognition in general, since the typical examples of systematicity are in fact syntax-bound; in addition, when we examine nonverbal cognition, we don't find the kind of systematicity required by the argument. Current post-cognitivist approaches to cognition, which emphasize embodiment and dynamic interaction, in its turn, also challenge the cognitivist assumption that the explanandum that a theory of cognition has to account for includes systematicity as a basic property of cognitive processes.

While the general strategy to claim that cognition is systematic was based on structural parallels between language and thinking, inferring for thinking what was supposed to be obvious for language, it has also been proposed that the systematicity of cognition can be found in perception— on the understanding that perception is a cognitive process, and one that is clearly independent of language. In *The Language of Thought*, Fodor (1975) exemplified his general approach to cognition by appealing to three areas of research: language processing, decision making, and perception. However, at that time the systematicity as explanandum and the combinatorial structure of representational elements as explanans were not clearly distinguished, and in fact, there was no explicit discussion of whether perception is systematic on the same model—the linguistic one— that was used to claim that cognition is systematic. But the basic idea is reasonably clear: perception is to count as systematic if it can be shown that perceptual capacities come in groups, if having one perceptual capacity involves having some others—just as understanding one sentence is connected to understanding many others. Marr's (1982) theory, which tried to account for perception in terms of basic representations and formal rules to combine and transform them, displayed the same kind of explanatory strategy as that of a combinatorial syntax of representational units, and was taken by Fodor as support for his language of thought (LOT) approach. So, it could be said that one way to support the systematicity of perception is to show that it can be better explained by a version of this classical cognitive architecture.

In this chapter, after briefly reviewing our previous arguments against the systematicity of cognition in general, and for a change in the order of dependence between thought and language—so that the systematicity of thought, when it is found, is parasitic on the systematicity of language—we will discuss the attempt to argue for the systematicity of perception and the contention that it is better explained by a combinatorial syntax of primitive representational units. We will first discuss the example offered to claim that perception is systematic, quite apart from a commitment to Marr's approach to perception: the phenomenon of amodal completion, as presented by Aizawa (this vol.). We will argue that it falls short of proving that perception is systematic, because what is claimed to be a group of interconnected perceptual abilities is better viewed as just one; furthermore, the best explanation of amodal completion is not in terms of a compositional structure plus inferential processes, because it is a global, emergent pattern that arises from context-dependent interactions. The way it is discussed by Aizawa, moreover, can't sustain the claim that perception is systematic in the required sense, because it is just a version of the standard illustration of cognitive systematicity by propositional interdependencies—those that we claim are language dependent.

Next, we will discuss whether there is indeed a way to sustain the claim that perception is in fact systematic. Again, our response will be negative. Our strategy will consist in showing that, in general, it is wrong to try to explain perception in terms of basic representational, language-like units, plus formal rules of inference over those units. For the classicists, the very existence of an account of this kind was taken as indirect proof that the explanandum was indeed systematic. We agree that if indeed the best explanation available for a given cognitive process were in terms of a classical architecture, then it would be right to expect it to be systematic. However, we will try to show that classical accounts, such as Marr's, are not the best explanations available. To this end, we will discuss one of the areas where clear apparent systematicity can be found, such as spatial perception (other similar phenomena include viewpoint invariance, or shape generalization). We will challenge a basic assumption of Marr's approach in this regard: that percepts involve an integration within a common spatial representational framework, which has the properties of Euclidean space. And we will discuss the phenomenon of spatial distortion, to claim that the strategy to account for it in terms of basic representations plus formal inferences is flawed. This discussion will also allow us to show that, although systematic dependencies are not found in perception in general, some robust regularities are central to perception. To account for them, though, we will claim that the best explanation is a nonclassical one—one along the lines of the ecological approach to perception: the approach that looks for the higher-order informational invariants, found in the sensorimotor loop, that cognitive systems exploit to guide their behavior. The fruitfulness of this approach will be exemplified by briefly considering two phenomena: sensory substitution and direct learning.

We can sum up our argument by confronting the basic structure of Fodor and Pylyshyn's argument for a LOT. The argument was: (1) Cognition is systematic; (2) the best (only) explanation of systematicity is compositional structure; therefore, (3) the best (only) explanation of cognition requires compositional structure. We reject premise 1 and modify premise 2: to the extent that regularities are found in cognition, its best explanation requires taking into account the dynamics of the interaction and the context-dependencies that constrain the way the elements may interact. Therefore, the conclusion has to be modified too; the best explanation is one that captures, and accounts for, this kind of informational pattern captured in the interaction dynamics. It is true that we still lack a common unifying framework for all the post-cognitivist approaches that may accommodate this idea, but it has to be conceded at least that we now have a promising way to avoid the roadblocks that have turned the classical approach into a stagnant research program.

2   Cognition Is Not Systematic in General; It Is Language Dependent

In previous work (Gomila 2011, 2012; Gomila, Travieso, and Lobo 2012), we have developed a variation of the response strategy to Fodor and

Pylyshyn's challenge initially offered by Dennett (1991, 1993) and later by Clark (1997): cognition is not systematic in general; rather, cognition exhibits systematicity when it is intrinsically connected to language. Thus, the LOT approach gets the wrong dependencies. It views the systematicity of language as derived from the systematicity of thought, when it is the other way around: thinking is systematic when, and because, it is linguistically structured.

We have argued for this view by means of two main arguments. On the one hand, we have shown that systematicity is syntax bound, which suggests that it is language dependent. Thus, for instance, the typical illustrations of systematicity involve sentences like "John loves Mary" and "Mary loves John," while it is not realized that possible combinations like "loves John Mary" should also be possible if systematicity were not syntax bound. Conversely, "John loves Mary" is not constitutively connected to "Mary is loved by John": the latter takes two more years to develop, despite the fact that the concepts involved are the same. This sort of example calls into question the basic notion of systematicity: if a mind can think X and can think Y, then it can think any combination of X and Y. They suggest that what can be thought is syntax bound, because syntax is required for the articulation of thoughts (Hinzen 2012). On the other hand, we have argued that nonverbal cognitive creatures do not exhibit the kind of systematicity at stake, because while they are good in perception, they are not so good at imagination and inference (Tomasello and Call 1997). The literature on human conceptual development, in addition, suggests that nonverbal creatures are capable of conceptual abilities much earlier than thought by Piaget, but they do not combine these elements systematically in their behavioral interactions, even if some of them may prove productive, until much later (Spelke 2003; Carruthers 2005)—when they have acquired their lexical labels (Lupyan 2008) and the ability to combine them, according to their syntactical categories, within sentential structures, as in the case of the critical role of the acquisition of the sentential complement to make false belief understanding possible (de Villiers and de Villiers 2009). Similarly, comparative psychology also supports this restructuring effect of language: studies of ape cognition indicate that nonhuman primates are good at perception, but not so good at other cognitive processes such as reasoning (Premack 2004). Only the symbol-trained individuals show some degree of systematic understanding. Nonverbal minds, in general, exhibit highly specialized cognitive abilities, which resemble the kind of listlike way of learning a language that is the opposite of our systematic way of learning it.

In contrast to Johnson (2004) and Chemero (2009), though, we acknowledge that systematic connections play an important role in language, and language-dependent thinking, precisely because of the structuring role of the syntax: the knowledge of a language is not made up of a list of elements, or a set of independent components, but of intrinsically interconnected ones. This organization is the key to the openness of such knowledge, which is manifested in our ability to understand or say previously unheard preferences, and for linguistically coded concepts, to think and understand new thoughts. In summary, some cognitive processes, those that are propositionally articulated, do exhibit systematicity, but only because they rely on the systematicity of language.

3 Systematicity in Perception: Is Amodal Completion Enough?

To resist our argument, then, one requires evidence of systematic processes having nothing to do with language. In addition, this evidence cannot be piecemeal and argued case by case, if we are to avoid the risk of inviting the response that the alleged examples lack generality. For these reasons, the question of whether perception exhibits the kind of systematicity in question becomes pressing, as a last-resort place to look for systematicity. Not surprisingly, Fodor (1975) claimed that evidence of systematicity came from three areas: language, decision making, and perception, on the grounds of the classical theories developed at that time in those different areas—again, seeing the explanandum as closely related to the explanans. Given the connection of conscious decision making to a linguistically structured process, and its limited role within animal cognition in general, as well as current evidence that nonverbal cognition does not exhibit the sort of systematicity of verbal cognition, perception becomes the critical place to look for systematicity.

However, to demonstrate that perception exhibits systematicity is not an easy task, because the very plausibility of the description of a set of abilities as systematically related depends, to an important degree, on the sort of explanation that can in fact unify those abilities. To put it differently, the very individuation of the abilities claimed to be systematically connected can't be question begging or ad hoc but has to be principled— that is, grounded in the best account available. Thus, to illustrate this point, it could be claimed that there is a systematic relation between the ability to use a red pen and the ability to use a black pen—that if you can use one, you can also use the other. But, regardless of whether the claim is true, such a proposal invites the reply that these are not basic abilities, and therefore, they do not constitute a proper explanandum that calls for a principled account in the first place. Using a red pen and using a black pen are better seen not as two systematically related abilities but rather as exemplifications of a single one.

This kind of consideration undermines, in our view, such "easy" ways to contend that perception is systematic as McLaughlin's (1993) approach; he lists several such "connections," some connected with beliefs, some with perception. The latter are these:

(3) the capacity to see a visual stimulus as a square above a triangle and the capacity to see a visual stimulus as a triangle above a square, and

(4) the capacity to prefer a green triangular object to a red square object and the capacity to prefer a red triangular object to a green square object. (McLaughlin 1993, 219)

Similarly, Cummins (1996) claims:

Consider, for example, the perception of objects in space:

(SP)   Anyone who can see (imagine) a scene involving objects 1 and 2 can see (imagine) a scene in which their locations are switched.

Again, any system employing a complete scheme will satisfy (SP), but it is surely the case that spatial representation underlies a number of substantive systematicities as well. (Cummins 1996, 604)

The problem with this way of identifying systematic dependencies is that it does not guarantee that it captures different though intrinsically connected perceptual capacities or abilities. It rather seems a way to parcel out a capacity—the perception of objects in space—in terms of the multiple ways in which it can be exemplified. This is made clearer when the distinction between the transparent and the opaque reading of the verb "see" is taken into account. On the transparent reading, "an agent sees a scene involving objects 1 and 2" means that the agent gets visual information from the scene, without any commitment to the content of his visual experience. On the opaque reading, the description aims at how the scene is grasped by the subject. This makes clear that it is not two different visual capacities that are involved in these examples (capacities claimed to be intrinsically interdependent), but just one: an ability of visual perception that is exercised on different occasions, as the transparent reading makes it clear. It is at the level of the contents of the visual experiences—the opaque reading—that the dependencies are found, but this has to do with the concepts available to the subject, that is, with thinking, not with perception. It follows that what does the work of supporting a systematic connection in

McLaughlin's examples is not the visual process itself (in fact, just one capacity, not several), but the conceptual contents. Were McLaughlin and Cummins to concede the point, then they would be conceding that their argument is for the systematicity of cognition, not perception.

A more stringent approach is put forth by Aizawa (this vol.). His strategy is to focus on a perceptual process, the phenomenon of amodal completion, as a more principled way to individuate capacities, supported by scientific research, and to allege their systematic connections. Amodal completion occurs when the subject perceives a complete object despite its being partially occluded in the subject's visual field. Thus, a famous example of amodal completion is that we see the dog behind the fence, not just stripes of it. In addition, amodal completion also applies to object perception, since each object partially occludes itself. We perceive the whole object, thus completing that part which in fact does not stimulate our retinas (we see our car, not just the surface that is facing us). This is the well-studied phenomenon of "filling in," uncovered by such eminent names of the history of psychology as Michotte and Kanisza (for a review of current views, see Shipley 2001).

The way Aizawa chooses to deploy this phenomenon, though, is not from its scientific characterization, but in a way parallel to that of McLaughlin and Cummins: Aizawa finds evidence of systematicity in the propositional contents that characterize the perceptive experience. Taking as an example a "Pac-Man" shaped figure, which is perceived as a black square occluding a gray circle (see Aizawa, this vol., fig. 3.1), he states that it proves the existence of four systematically interrelated capacities:

Consider the following fourfold combination of capacities:

i. The capacity to see a black square occluding a gray circle
ii. The capacity to see a gray square occluding a black circle
iii. The capacity to see a black circle occluding a gray square
iv. The capacity to see a gray circle occluding a black square. (Aizawa, this vol.)

From this, he concludes:

Amodal completion is, then, systematic, because there is a grammar to mental representations that enables the formation of a collection of mental representations, such as:

BLACK SQUARE OCCLUDING GRAY CIRCLE
GRAY SQUARE OCCLUDING BLACK CIRCLE
BLACK CIRCLE OCCLUDING GRAY SQUARE
GRAY CIRCLE OCCLUDING BLACK SQUARE. (Aizawa, this vol.)

Figure 15.1
Amodal completion fails for a heart figure, due to the Gestalt law of good continuation that constrains the interpolation process.

In this way, despite the interest of considering the phenomenon of amodal completion from a scientific point of view, to find out whether it can be best accounted for in terms of a set of primitive components, systematically combined, as one would expect from a defense of classicist systematicity, Aizawa chooses in fact to follow the same strategy as McLaughlin: it is the propositional contents of the visual experiences that might exhibit systematicity in the desired sense, not the perceptual capacity itself, which again, seems to be one and the same. To appreciate this, observe what happens when the occluded shape is not that of a well-known geometrical figure, but, for example, that of a heart (see figure 15.1).

In this case, it is not true that seeing a black heart occluding a gray square, say, is systematically connected to seeing the square occluding the heart, because no heart is completed in this case: amodal completion is not systematic in the required sense. Notice that, while the perceptual experience does not support the contention that the set of claimed interdependent capacities are systematic at the perceptual level, it is still possible to formulate them at the propositional level, which means again that these systematic patterns are in fact to be found at the conceptual level, connected to language understanding:

Black square occluding a gray heart
Gray square occluding a black heart
Black heart occluding a gray square
Gray heart occluding a black square

The fact that amodal completion does not work in this case clearly indicates that the perceptual process does not proceed by the systematic combination of perceptual primitives (corresponding to the basic conceptual units of the language of thought), plus some inferential process that transforms the corresponding mental formulas in that language (which is what Aizawa should demonstrate in order to sustain his claim that perception is systematic in the proper way), but by some kind of global organization of a Gestalt effect. Briefly, the perceptual process of amodal completion depends on the possibility of establishing an edge (in particular, a closed edge delimiting an object), and in the case of a partial occlusion by another object (another closed edge), the visual system interpolates the edge so as to close it and make the partially occluded object perceptible as such. This mechanism has limitations, of course, such as the incapacity to interpolate acute angles, because of what the Gestaltists called the good continuation law. Therefore, the heart is not completed, as there is no "good continuation" to join the two edges that touch the square. In conclusion, amodal completion is not a case of a systematic process, thought to depend upon a set of primitive elements that get combined, but the emerging result of the interaction and context-dependency of forms, curvatures, points of view, and dynamic information. Of course, a "classical" account of amodal completion can still be defended, but it cannot be considered the best available because it lacks the resources to explain these and other interactive effects.

Thus, for instance, the interpolation process is dependent on the strength of the edge detected, that is, sensitive to the contrast of background and figure—something unexpected from a classical account. When background and figure are the same color or lightness (gray), the object is no more perceivable and the completion process does not work, as there is no partial occlusion of the edge. However, even subtle differences in contrast between object and background modulate the salience of the completion process. In figure 15.2 we have selected a few examples of the many variations that can take place, to demonstrate the sensitivity of the process to these contextual factors. The point is that it is not true in general that if one is able to see X occluding Y, then one is also able to see Y occluding X, for any X and Y—as required by the notion of systematicity.

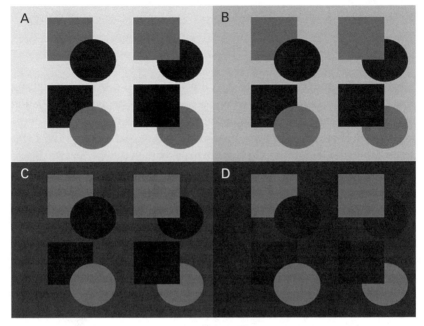

Figure 15.2
Amodal completion of a square and a circle with gray (125) and black (0) under four different gray backgrounds (a = 235, b = 200, c = 50, d = 20). Changing the contrast between figure and background changes the salience of the completed object accordingly. It is part of the explanation of the phenomenon that the interpolation of the edge depends on the strength of the contrast of the edge.

This is not specific of hearts or amodal completion per se. Much the same happens in the complementary phenomenon of subjective or illusory contours, such as the famous Kanizsa triangle (see figure 15.3), which also illustrates amodal completion for the "Pac-Men" at the angles.

It is not that our visual system works with basic representations of circles and triangles, which make their way into the visual content of the experience despite the partiality of the visual information available (as required by the notion of systematicity in perception at stake), but that some organizational pattern emerges out of a specific, and context-dependent, specification of values. Again, the illusion disappears if just one of the three angular "Pac-Men" is visible. Or consider figure 15.4. We do not perceive a triangle occluded by a cross, but a triangle appears if the cross is covered. Again, this demonstrates that seeing a figure is not the outcome of a set

Figure 15.3
The Kanisza triangle.

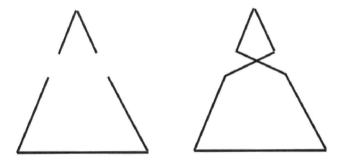

Figure 15.4
(a) The disappearing triangle. (b) The cross blocks the completion of the triangle, as in the first figure.

of abilities of seeing its elementary parts, compositionally integrated through an inferential process. As a consequence, seeing a certain relationship between A and B does not guarantee that A and B hold as algebraic variables for any such elementary part, nor does it guarantee that the converse relationship between B and A can also be seen—as required by systematicity. The conceptual articulation of the visual content of the experience, on the other hand, which obviously allows the expression of all such combinatorial possibilities, has to do with linguistic abilities, as argued in the previous section.

4   From Systematicity to Regularities in Spatial Perception

We have discussed the case of amodal completion, to reject Aizawa's allegation that it demonstrates that perception is systematic in the required sense. At several points we have underlined the fact that Aizawa's way to describe the relevance of the phenomenon of amodal completion to the question of whether perception is systematic fails to address the right kind of requirement: he remains concerned in the end with the "rough and ready" apparent systematic relations among the propositional contents of one's visual experience pointed out by McLaughlin, instead of trying to find the relevant scientific evidence that allows a principled individuation of capacities, or—given the close connection between the description of the explanandum and the explanans—trying to put forward an explanation of amodal completion that appeals to representational primitives plus inferential rules of transformation of such representations, as the supposedly "best possible" explanation of the phenomenon. For one indirect argument in favor of the systematicity of perception consists in the claim that the best theory of perception is committed to a classical architecture. As a matter of fact, this was Fodor's strategy, when he appealed to the influential approach of Marr (1982). So, in this section we want to raise the question of whether a compositional-inferential program, such as Marr's, is the best explanation available in the case of spatial perception.

Marr's program shared the basic notion that spatial perception is a process that starts from sensory inputs and ends in a representation of a spatial configuration. What's new to Marr's approach is the way he articulated that process as an inferential one, as a formal transformation of very specific kinds of representations according to formal rules. Marr focused on how a 3D representation is built up from the 2D raw data of the retinal image; in particular, he dealt with the transformation of the egocentric frame of reference of the image from a particular point of view to an allocentric representation of the object perceived. The allocentric representation of the object has to be located in a spatial representation of the environment, where geometric transformations may be performed and the different objects are integrated in a single representation. This spatial representation is thought to be systematically used to solve spatial behavior in a way that resembles the use of lexical units in language processes. Spatial representations can be rotated, symmetrically transformed, and permuted, and can undergo the rest of classical transformations of Euclidean geometry.

However, several geometrical, haptic, and visual tasks, such as spatial matching tasks (Cuijpers, Kappers, and Koenderink 2003; Kappers 1999; Fernández-Díaz and Travieso 2011), reveal powerful perceptual distortions in how such spatial integration takes place. In what follows, we will show that the classical strategy to account for such distortions as anomalous inferences is flawed, and that a superior explanation can be provided within the approach of ecological psychology.

A relevant task to consider in this respect is the parallelity test (Kappers 1999), a task where it is checked whether this crucial axiom of Euclidean geometry holds for spatial perception. The task consists in rotating a test rod so as to place it parallel to a reference rod. The result of this test is that we make strong errors in grasping parallelity, both in psychophysical estimations and when performing parallelity estimations in different modalities (see figure 15.5). In the haptic domain, when performing the task while blindfolded, the deviations from parallelity appear in different planes (i.e., horizontal, frontal, and sagittal), and both in bimanual (i.e., one hand touching the reference rod and the other the test rod) and unimanual (i.e., one hand touches the reference rod and then goes to the test rod and rotates it). These distortions happen both for vision and for touch (Cuijpers, Kappers, and Koenderink 2003) and probably for hearing, although in this case estimations are made using pointing tasks where participants are asked to point/orient the rods to sound sources (Arthur, Philbeck, Sargent, and Dopkins 2008).

The way to address these results from a classical, Marr-inspired approach is to attribute these deviations to an erroneous application of inferences. Such errors, according to this approach, produce deformations of the spatial representation, resulting in a perceptual illusion. These erroneous inferences are thought to be due to the fact that the allocentric representation inherits a bias produced by the initially egocentric frame of reference. More specifically, in a sophisticated version of this explanatory strategy,

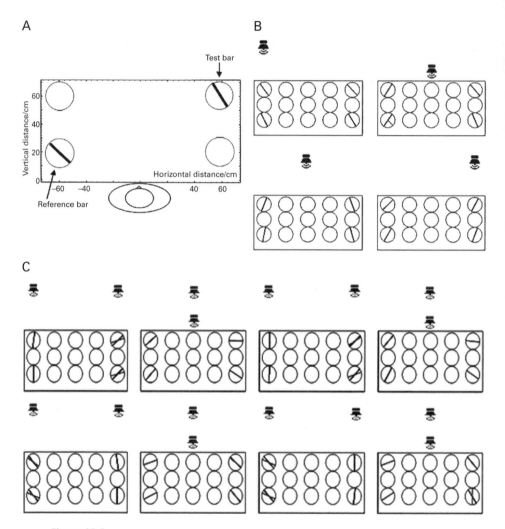

Figure 15.5

Deviations in the haptic parallelity test (adaptation from Fernández-Díaz, and Travieso 2011). In this example, subjects are asked to put the test bar parallel to the reference bar (figure 15.5a). Figure 15.5b shows mean results on an acoustic pointing task. Finally, figure 15.5c shows differences in a parallelity crossmodal haptic-acoustic task where the interactions vary depending whether the sound source is aligned with the reference or the test rods.

Cuijpers et al. (2003) suggest that the resulting spatial representation, the cognitive map where spatial behavior is planned, fits better with another geometry that does not include the parallelity axiom of Euclidean geometry. In particular, spatial representation is thought to be better described through a Riemannian space with a constant negative curvature, so that the deviations are intrinsic to the allocentric representation itself. For this explanation to go through, though, it should hold in all cases. However, our results show that this is not the case for the haptic system. The strong and systematic deviations in parallelity matching tasks disappear when the task is a mirror one (i.e., when participants are asked to mirror the orientation of the reference rod to a test rod to the mid-sagittal axis, that is, the mid-body). And when the haptic parallelity task is performed on the back of the subject, the pattern of deviations is inverted (Fernández-Díaz and Travieso 2011).

On the other hand, the assumption that all modalities are integrated in an amodal spatial representation (Nanay 2011) does not hold either. If the curved space descriptions were consistent (with different curvatures) for visual, acoustic, and haptic modalities, they would interact systematically, that is, crossmodal interaction should be consistent with the spatial map they produce. In fact, this result has been found in visual-haptic crossmodal interaction, where vision improves haptic performance reducing the deviations from parallelity (Newport, Rabb, and Jackson 2002). But contrary to the assumption, it has been shown that, depending on the task, crossmodal information can be used to increase or decrease perceptual errors in parallelity matching tasks (Fernández-Díaz and Travieso 2011). Even temporal delays affect the geometrical matching results (Zuidhoek, Kappers, van der Lubbe, and Postma 2003). What this suggests is that different tasks rely on different spatial information, which is locally specified, rather than globally integrated into a unique spatial map. Similar results have been found in many other sensorimotor tasks, to the point that relevant authors in the field try to explain "why we don't mind to be inconsistent" (Smeets and Brenner 2008).

In conclusion, although the use of a Riemannian space allows a better description of results than that of a Euclidean one, it is not true that the negative curvature is stable or constant, because it may change depending on sensory modality, perceptual task, body configuration, or whether the information available is dynamic or static, or depending on the temporal constraints of the task. Moreover, no evidence justifies holding that this mathematical description reflects a spatial integrated representation that is fed by the different modalities. Geometrical tasks are context and task dependent to the point that different solutions are used to solve the different tasks, sometimes relying on the proprioception of the movements, others on the visual control of the hand or on dynamic information.

However, it is still evident that our behavior in space is stable and well adapted. Although, as we have already argued, classical cognitivism's account of this stability is inadequate, the multiple regularities that characterize our patterns of sensorimotor interaction with our environment still require an explanation. In our view, such regularities are grasped through our forms of action in the environment, in particular in the patterns of information made available in the sensorimotor loop that sustains such interactions, as the ecological approach in perception holds (Gibson 1979; Turvey and Carello 1986). The idea of a sensorimotor loop, to start with, refers to the necessity of a dynamic interaction of the perceiver and the environment in order to generate, and be able to pick up, those regularities—the "invariants" in neo-Gibsonian terms. When the perceiver moves, there are sensory patterns that change lawfully depending on the context. Through those changes, robust high-order regularities are revealed: those regularities that remain constant across the changes. Once the cognitive system has grasped them, it can access the relevant environmental properties for the task at hand, without the need for further representational inferences or explicit representations.

Let us consider an example. Dynamic touch is the perceptual ability to estimate different properties of objects, like length, through invariants in the rotational mechanics that are accessible via the proprioceptive system. Typically, this ability is exhibited in estimating properties such as the length of a rod by grasping it from one end (when it can be wielded but not seen). Estimating the length of an object that is only grasped at a certain position has to rely on the resistance of the object to rotation at the point of wielding. This resistance is described in rotational mechanics by the inertia tensor, a numerical quantity (a 3×3 matrix for 3D movements) for the object's resistance to rotation, which is related to its point of rotation and its mass distribution. It is the equivalent to mass in the expression $F = m \cdot a$, but in the rotational form:

$$I = \int \rho(s) \; \delta(s)^2 \; dV, \tag{1}$$

where "$\rho(s)$" is the mass-density function and "$\delta(s)$" is the distance or radius to the rotation axis.

If mass can be estimated by applying a force and perceiving the resulting acceleration in an object, the inertia tensor can be estimated by applying a torque and perceiving the resulting angular acceleration. Different torques will produce different angular acceleration values, whereas the inertia tensor will be constant over those transformations. But the most important point of this example is that in order for anybody to access those invariants it is necessary to wield the object: to generate a sensorimotor loop. This way, the sensory changes contingent upon the perceiver's movements give access to those invariants across time. At the same time, the invariants guarantee that the objects' "responses" to the perceiver actions are regular (i.e., the mass-density function and the length of the radius do not change unless the object is grasped at another point or immersed in another medium). For example, the stronger the force applied in the torque during the wielding movement, the more angular acceleration the object reaches, according to the lawful relations that the inertia tensor specifies. As a matter of fact, it has been demonstrated that our estimations of length, weight, and other properties of bodies is adjusted to the predictions of the inertia tensor (Turvey 1996), and that different features of the wielding affect what is perceived, such as applied force (Debats, van de Langenberg, Kingma, Smeets, and Beek 2010), amplitude, and speed of the wielding (Lobo and Travieso 2012), or the orientation of those movements (Armazarski, Isenhower, Kay, Turvey, and Michaels 2010; Michaels and Isenhower 2011ab).

These findings, and many others along the same lines, suggest a superior way to account for the regularities in our perceptual experience, one not derived from some reservoir of conceptual primitives that get recombined through inferential processes. Quite the opposite: the regularities appear at the interaction of the physical object with the actions of the perceiver on it. These invariants are the *high-order informational patterns in the sensorimotor loop*, which are thus grasped by the perceiver. Ecological psychology (Gibson 1979; Turvey and Carello 1986) tries to uncover such invariants that humans are attuned to by engaging actively with the world. This amounts to a radical departure from the traditional view of perception as the recovery of the distal stimulus out of an impoverished proximal stimulus that reaches the retina. From this standpoint, the sensory information available in perception is not the amount of light at a certain location of the retina, but movement and intensity gradients, stable relative proportions, or changes in these gradients. The fact that through interaction we become sensitive to these higher-order parameters runs against the traditional view that they are calculated, or derived, from lower-level ones (Michaels and Carello 1981). In the next section, we will provide some further illustration of this approach to perception by considering how it can be applied in two areas: sensory substitution and direct learning. In so doing, we expect to make even clearer why the ecological approach is superior to the classical one.

5 Sensory Substitution

When the field of sensory substitution was developed in the late 1960s and early '70s, the implicit idea that drove the project, in congruence with classical cognitivism, was to produce an alternative input to the visual system through the haptic system. Thus, the first devices developed (i.e., the Optacon and TVSS), were designed to get spatial information (letters in the case of the Optacon and images in the case of the TVSS) from cameras whose light detection was transformed in vibration delivered to the body in a "tactile retina." Participants thus would receive an alternative sensory stimulation, upon which they could apply the systematic inferences of perceptual processing. In other words, sensory substitution was originally conceived as just the substitution of the sensory input, leaving the rest of the perceptual process intact. Some comments by Bach-y-Rita (1972), probably the most influential pioneer in sensory substitution, are revealing in this respect: "You see with your brain, not with your eyes"; "The nerve impulses coming from the eye are no different than those from the big toe"; or "Just give the brain the information and it will figure it out" (all from Bach-y-Rita 1972).

These first devices were designed with an atomistic conception of the stimulus: the amount of light at each position, so that the device had to transduce that magnitude into a certain amount of vibration. All other visual properties were supposed to arise from inferential processes. Therefore, research on these devices was focused on detection thresholds for intensity and frequency of vibration, the minimum distance between vibrators, and other variables in the skin psychophysics (see Jones and Sarter 2008 and Gallace, Tan, and Spence 2007 for a review).

However, many users were not able to experience the externality of perception, that is, to interpret the sensory signals produced by the device as of external objects. A cursory comment by Bach-y-Rita (1972) indicates that he came close to realizing the cause: he noticed that, with a portable version of the TVSS, blind participants had difficulties in establishing the contingencies between the camera movements and the vibration pattern. In addition, the generalization in the use of these devices still is an unresolved question, and this is probably related to the fact that most of these devices give symbolic information to the user, like a signal meaning that someone is calling, or discrete signals like beeps indicating the direction to move.

These problems, in our opinion, can be overcome by adopting the eco-logical psychology approach (Travieso and Jacobs 2009). From this perspec-tive, to be successful, the design of a haptic-to-vision sensory substitution device has to offer the possibility of establishing sensorimotor contingen-cies so that high-order informational patterns become available to the perceiver. Thus, we have built up a simple haptic-to-visual sensory substitu-tion prototype. The guiding idea is that some of the higher-order invariants that can be extracted through sensorimotor interaction in the visual modality can also be found through sensorimotor interaction in the haptic modality. In other words, the sensory substitution device can work for these intermodal magnitudes. In one of its possible configurations, the device has twenty-four stimulators located vertically on the chest of the user; and they vibrate as a function of the distance to the first obstacle (see figure 15.6 for a schema of the apparatus). The user is free to move wearing the apparatus, and we test the detection of obstacles and objects by means of psychophysical methods.

Our experimental results show that participants with the sensory sub-stitution device are able to learn how to use the device and to improve their behavior with it (Lobo, Barrientos, Jacobs, and Travieso 2012). They manage to do so when they can use dynamic vibratory information

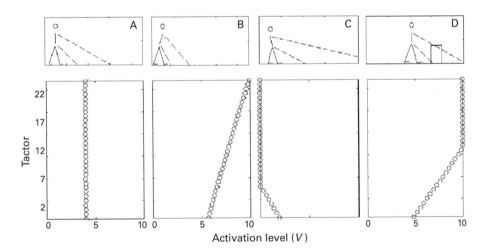

Figure 15.6

Illustration of the sensory substitution device (adaptation from Díaz, Barrientos, Jacobs, and Travieso 2011). The upper row shows the alignment of the virtual sensors detecting the distance of the first obstacle encountered. The lower row shows the corresponding pattern of stimulation. As can be seen, different patterns appear when the subject leans back and forth, or when he or she encounters obstacles.

contingent upon their own movements; in this way, they can detect movement gradients and the presence of nearby surfaces and their size (Díaz, Barrientos, Jacobs, and Travieso 2011, 2012). For example, our results show that using contingent and continuous informational patterns, participants are able to better detect steps than with static and discrete information, where they are limited to vibration differences between adjacent vibrators. Their discrimination was better than the predictions in the psychophysical literature, which establish three or four centimeters as the minimum distance to distinguish between vibrators, and five to seven distinguishable levels of vibrations at the chest (Verrillo 1973; Van Erp 2007). We also found that the use of dynamic vibratory information allows a better detection of steps when it is contingent on the user's movement, compared to the very same pattern of information given to another perceiver non-contingent on his movements (Díaz et al. 2011, 2012), a test that was not possible in previous devices.

Again, the use of our sensory substitution prototype shows that spatial perception is improved when dynamic patterns of information are available and that these dynamic patterns should be contingent on the perceiver's movements, reaffirming the interaction-dominant character of sensorimotor processes. In this respect, the ecological approach to sensory substitution can overcome one of the main handicaps of former devices, where the necessity of understanding sensorimotor contingencies was revealed to be problematic. On the contrary, establishing those contingencies is now the base of the new functioning.

However, it could be conceded that the ecological approach works well for sensorimotor processes, while disputing that it can contribute to our understanding of higher cognition. This is a very big question, of course, but at the moment it can at least be said that learning is within the reach of our approach. In the next section, we present how we have begun to study learning.

6 Direct Learning

Work with the sensory-substitution device has made it clear that participants learn to use it: they need time to discover the relevant invariants in the sensory input across their own movements. This idea has been generalized beyond the area of sensory substitution, and we have begun to study this process of learning, of discovering and using the relevant magnitude for a particular task. By analogy with the "direct perception" notion of ecological psychology, it has been called "direct learning" (Jacobs and

Michaels 2007; Jacobs, Ibáñez-Gijón, Díaz, and Travieso 2011; Michaels, Arzamarski, Isenhower, and Jacobs 2008). The main hypothesis guiding this theory is that, just as a crucial aspect of understanding perception is knowing the information on which it is based, as the ecological theory states, in order to develop a theory of learning we should also know the information on which it is based and how the subject manages to discover it. We have already seen that the perception-action loop can be described as an interrelated process. Perception is the process of accessing informational variables that emerge during interaction with the environment, and action is the behavior guided by the information variables given by perception. Thus, a learning process is a change in behavior in order to improve adaptation.

The theory of direct learning has initially focused on two types of perceptual learning: calibration and education of attention. Calibration is the process of changing the adjustment of a kind of action to an informational variable. That is, if an informational variable is used to guide a certain behavior, as, for example, the trajectory of a moving object in interception, we can become more accurate and efficient in following that trajectory as a way to improve our interception abilities. Formally expressed:

$$a= f(I). \tag{2}$$

This means that the action is a function of the perceptual information, and calibration is conceived as the modulation of f, to improve the adjustment of the action to the informational variable. A classic example of calibration is that of wearing prismatic goggles, by which the visual field is displaced several degrees. Whereas these goggles initially produce errors in aiming, pointing, and so on, subjects quickly adapt through a process of calibration, which allows them to compensate for the deviations in the visual field.

The second type of learning is the education of attention. Far from the classical definition of attention as a centrally guided cognitive process, here attention is operationally defined as a change in the variable used to guide the action system. That is, the education of attention is a change in the informational basis of action. In reference to equation (2), the education of attention is a change of the variable I. Methodologically, this line of research consists in selecting, given a certain perception-action situation, an informational space of candidate variables to guide the action in that situation. These candidates can be a position in a fixed image in an experimental task, or dynamic information from an image produced by a moving subject, or those rotational variables in the above-mentioned example of dynamic touch. This set of variables constitutes an informational space (Jacobs and Michaels 2007; Jacobs and Travieso 2010). An informational space is a continuous space in which each point represents an informational variable, and the trajectories represent the change produced by learning. The study of the learning process consists, therefore, in the study of changes in the use of the variables in the informational space. The method to estimate variables used by the subject is regression or correlation between the informational variables and the action variables (or psychophysical estimations) during the different phases of the study. Representing the different variables used by the subject in the different experimental sessions, the evolution of the perception-action loop can be observed (Jacobs et al. 2011). This evolution is due to the utility of the variables, which is assessed through the feedback received: if the action is not successful when based on an informational variable (e.g., if the tennis player misses the shot when he fixates on the net), next time a different informational variable may be selected. A measure of the precision of actions is required in order to compare and switch to the informational variable that best guides action.

This approach has proved useful for describing perceptual learning processes such as emergency breaking (Fajen and Devaney 2006), haptic estimations (Michaels, Arzamarski, Isenhower, and Jacobs 2008), landing in-flight simulators (Huet, Jacobs, Camachon, Goulon, and Montagne 2009; Huet, Jacobs, Camachon, Missenard, Gray, and Montagne 2011), and dynamic tasks like catching balls (Morice, François, Jacobs, and Montagne 2010). It has also been proposed for the design of training regimes in sports (Ibáñez-Gijón, Travieso, and Jacobs 2011).

7 Conclusion

In this chapter, we have argued that perception is not systematic, in the way required for a formal combinatorial explanation to claim some superiority. The regularities that do exist in spatial perception, which are not properly systematic, are better explained as the result of interactions with a rich and regular environment, along the lines of ecological psychology, and the search for higher-order informational patterns in the sensorimotor loop. From this point of view, the patterns need not be explicitly represented for the cognitive system to be sensitive to them, nor is it required that the cognitive system builds an integrated explicit representation of all the sensorimotor information that can influence the behavior of the system.

Is this approach another form of eliminating cognition from the picture, or confusing it with behavior, as suggested by Aizawa? We don't think so. It is rather committed to a different understanding of cognition, as a dynamical, interactive process, instead of the logicism of the classical cognitivist view, or the simple associationism of the connectionist view. It is true that the different alternative approaches to cognitivism are not in complete agreement about how to conceive of cognition, and therefore, have not converged on a unique new paradigm. But it should be recognized that classical cognitivism has made only meager progress and faces insurmountable problems.

Acknowledgments

This research was supported by the Spanish Ministry of Economy and Competitiveness through project FFI2009–13416-C02, and by Fundación Séneca-Agencia de Ciencia y Tecnología de la Región de Murcia, through project 11944/PHCS/09.

References

Armazarski, R., R. W. Isenhower, B. A. Kay, M. T. Turvey, and C. F. Michaels. 2010. Effects of intention and learning on attention to information in dynamic touch. *Attention, Perception & Psychophysics* 72 (3):721–735.

Arthur, J. C., J. W. Philbeck, J. Sargent, and S. Dopkins. 2008. Misperception of exocentric directions in auditory space. *Acta Psychologica* 129:71–82.

Bach-y-Rita, P. 1972. *Brain Mechanisms in Sensory Substitution*. New York: Academic Press.

Carruthers, P. 2005. *The Architecture of the Mind*. Oxford: Oxford University Press.

Chemero, A. 2009. *Radical Embodied Cognitive Science*. Cambridge, MA: MIT Press.

Clark, A. 1997. *Being There: Putting Brain, Body, and World Together Again*. Cambridge, MA: MIT Press.

Cuijpers, R. H., A. M. L. Kappers, and J. J. Koenderink. 2003. The metrics of visual and haptic space based on parallelity judgments. *Journal of Mathematical Psychology* 47:278–291.

Cummins, R. 1996. Systematicity. *Journal of Philosophy* 93 (12):591–614.

Debats, N. B., R. W. van de Langenberg, I. Kingma, J. B. J. Smeets, and P. J. Beek. 2010. Exploratory movements determine cue weighting in haptic length perception of handheld rods. *Journal of Neurophysiology* 104:2821–2830.

Dennett, D. 1991. *Consciousness Explained*. Boston: Little, Brown.

Dennett, D. 1993. Learning and labeling. *Mind and Language* 8:540–548.

Díaz, A., A. Barrientos, D. M. Jacobs, and D. Travieso. 2011. Vibrotactile flow and the detection of step-on-places with a sensory substitution device. In *Studies in Perception and Action XI*, ed. E. Charles and L. Smart. Mahwah, NJ: Erlbaum.

Díaz, A., A. Barrientos, D. M. Jacobs, and D. Travieso. 2012. Action-contingent vibrotactile flow facilitates the detection of step-on-places with a partly virtual sensory substitution device. *Human Movement Science* 31 (6):1571–1584.

de Villiers & de Villiers. 2009. Complements enable representation of the contents of false beliefs: the evolution of a theory of a theory of mind. In *Language Acquisition*, ed. S. Foster-Cohen. London: Palgrave Macmillan.

Fajen, B. R., and M. C. Devaney. 2006. Learning to control collisions: the role of perceptual attunement and action boundaries. *Journal of Experimental Psychology. Human Perception and Performance* 32:300–313.

Fernández-Díaz, M., and D. Travieso. 2011. Performance in haptic geometrical matching tasks depends on movement and position of the arms. *Acta Psychologica* 136:382–389.

Fodor, J. 1975. *The Language of Thought*. Cambridge, MA: Harvard University Press.

Fodor, J., and Z. Pylyshyn. 1988. Connectionism and cognitive architecture. *Cognition* 28:3–71.

Gallace, A., H. Z. Tan, and C. Spence. 2007. The body surface as a communication system: The state of the art after 50 years. *Presence* 16:655–676.

Gibson, J. J. 1979. *The Ecological Approach to Visual Perception*. Boston: Houghton Mifflin.

Gomila, A. 2011. The language of thought: Still a "game in town"? Critical Review of Fodor's LOT2. *Teorema* 30:145–155.

Gomila, A. 2012. *Verbal Minds: Language and the Architecture of Cognition*. Amsterdam: Elsevier.

Gomila, A., D. Travieso, and L. Lobo. 2012. Wherein is human cognition systematic? *Minds and Machines* 22 (2):101–115.

Hinzen, W. 2013. Narrow syntax and the language of thought. *Philosophical Psychology* 26 (1):1–23. doi:10.1080/09515089.2011.627537.

Huet, M., D. M. Jacobs, C. Camachon, O. Missenard, R. Gray, and G. Montagne. 2011. The education of attention as explanation of variability of practice effects: Learning the final approach phase in a flight simulator. *Journal of Experimental Psychology. Human Perception and Performance* 37 (6):1841–1854.

Huet, M., D. M. Jacobs, C. Camachon, C. Goulon, and G. Montagne. 2009. Self-controlled concurrent feedback facilitates the learning of the final approach phase in a fixed-base simulator. *Human Factors* 51:858–871.

Ibáñez-Gijón, J., D. Travieso, and D. M. Jacobs. 2011. El enfoque neogibsoniano como marco conceptual y metodológico para el diseño de programas de entrenamiento deportivo. *Revista de Psicología del Deporte* 20:667–668.

Jacobs, D. M., and C. F. Michaels. 2007. Direct learning. *Ecological Psychology* 19:321–349.

Jacobs, D. M., and D. Travieso. 2010. Training regimens, manifolds, and vector fields. *International Journal of Sport Psychology* 41:79–80.

Jacobs, D. M., J. Ibáñez-Gijón, A. Díaz, and D. Travieso. 2011. On potential-based and direct movements in information spaces. *Ecological Psychology* 23 (2):123–145.

Johnson, K. 2004. On the systematicity of language and thought. *Journal of Philosophy* 101 (3):111–139.

Jones, L. A., and N. B. Sarter. 2008. Tactile displays: Guidance for their design and application. *Human Factors* 50:90–111.

Kappers, A. M. L. 1999. Large systematic deviations in the haptic perception of parallelity. *Perception* 28:1001–1012.

Lobo, L., and D. Travieso. 2012. El patrón de exploración modula la percepción de longitudes a través del tacto dinámico. *Psicothema* 24 (1):55–61.

Lobo, L., A. Barrientos, D. M. Jacobs, and D. Travieso. 2012. Learning to step on a target object with a sensory substitution device on the leg. Poster at the XII European Workshop on Ecological Psychology, Madrid, Spain, June.

Lupyan, G. 2008. The conceptual grouping effect: Categories matter (and named categories matter more). *Cognition* 108:566–577.

Marr, D. 1982. *Vision*. San Francisco: Freeman.

McLaughlin, B. P. 1993. Systematicity, conceptual truth, and evolution. *Royal Institute of Philosophy* 34 (Supplement):217–234.

Michaels, C. F., R. Arzamarski, R. W. Isenhower, and D. M. Jacobs. 2008. Direct learning in dynamic touch. *Journal of Experimental Psychology: Human Perception and Performance* 34 (4):944–957.

Michaels, C. F., and C. Carello. 1981. *Direct Perception*. Englewood Cliffs, NJ: Prentice Hall.

Michaels, C. F., and R. W. Isenhower. 2011a. Information space is action space: Perceiving the partial lengths of rods rotated on an axle. *Attention, Perception & Psychophysics* 73 (1):160–171.

Michaels, C. F., and R. W. Isenhower. 2011b. An information space for partial length perception in dynamic touch. *Ecological Psychology* 23:37–57.

Morice, A. H. P., M. François, D. M. Jacobs, and G. Montagne. 2010. Environmental constraints modify the way an interceptive action is controlled. *Experimental Brain Research* 202:397–411.

Nanay, B. 2011. Systematicity and multimodality: some ground-level problems for anti-representationalism. Presentation at the Systematicity and the Post-Connectionist Era: Taking Stock of the Architecture of Cognition, San José, Spain, May.

Newport, R., B. Rabb, and S. R. Jackson. 2002. Noninformative vision improves haptic spatial perception. *Current Biology* 12:1661–1664.

Premack, D. 2004. Is language the key to human intelligence? *Science* 303: 318–320.

Shipley, Th. F., ed. 2001. *From Fragments to Objects: Segmentation and Grouping in Vision.* Amsterdam: Elsevier.

Smeets, J., and E. Brenner. 2008. Why we don't mind to be inconsistent. In *Handbook of Cognitive Science: An Embodied Approach*, ed. P. Calvo and A. Gomila, 207–217. Amsterdam: Elsevier.

Spelke, E. 2003. What makes us smart? Core knowledge and natural language. In *Language in Mind: Advances in the Study of Language and Thought*, ed. D. Gentner and S. Goldin-Meadow, 277–312. Cambridge, MA: MIT Press.

Tomasello, M., and J. Call. 1997. *Primate Cognition.* Oxford: Oxford University Press.

Travieso, D., and D. M. Jacobs. 2009. The ecological level of analysis: Can neogibsonian principles be applied beyond perception and action? *Integrative Psychological & Behavioral Science* 43:393–405.

Turvey, M. T. 1996. Dynamic touch. *American Psychologist* 51 (11):1134–1152.

Turvey, M. T., and C. Carello. 1986. The ecological approach to perceiving-acting: A pictorial essay. *Acta Psychologica* 63:133–155.

Van Erp, J. B. F. 2007. *Tactile Displays for Navigation and Orientation: Perception and Behaviour.* Leiden, the Netherlands: Mostert & van Onderen.

Verrillo, R. T. 1973. Vibrotactile intensity scaling at several body sites. In *Cutaneous Communication Systems and Devices*, ed. F. Gildard, 9–14. Austin, TX: Psychonomic Society.

Zuidhoek, S., A. M. L. Kappers, R. H. J. van der Lubbe, and A. Postma. 2003. Delay improves performance on a haptic spatial matching task. *Experimental Brain Research* 149:320–330.

16   The Emergence of Systematicity in Minimally Cognitive Agents

Paco Calvo, Emma Martín, and John Symons

1   Introduction

The honeybee bounces against the pane of glass, the moth circles the lightbulb, and the dog chases its tail. Honeybees, moths, and dogs are each capable of a complex and interesting set of behaviors. But sometimes we notice animals failing to accomplish their goals and being unable to adapt their behavior successfully in light of their failures. At moments like these it is natural to think less of the family dog, the honeybee, or the moth. This is not one of our dog's more impressive moments, and while the dog is not a stupid creature, chasing its tail certainly appears to be a stupid behavior.

When a behavior is obviously automatic, repetitive, or arbitrary, we tend to downgrade the level of agency we ascribe to the animal or system in question. By contrast, when a system adapts to changing environmental conditions, contributes to the pursuit of some identifiable goal, can be combined in flexible ways with other behaviors, and has a variety of other systematic features, we are inclined to judge that the behavior is the result of some underlying intelligence or agency.

This chapter suggests that our intuitive judgments about the underlying intelligence or agency of nonlinguistic cognitive agents are prompted by a set of systematic features that mark what we will call *intelligent behavior*. These systematic features of intelligent behaviors do not necessarily license the claim that there is any single coordinating or governing intelligence in the agent. However, we will argue that intelligent behavior is indicative of meaningful engagement with the environment. This meaningful engagement is phylogenetically and ontogenetically prior to the kinds of intellectual and cognitive capacities that we expect from adult humans.

In the pages that follow, we will explain what it means to locate systematicity in the behavior of infralinguistic and minimally cognitive agents.

Along the way, we will unpack the idea of meaningful engagement with the environment and will offer some ideas as to how such engagement might serve as the basis for the emergence of more sophisticated forms of cognition and agency.

2 Systematicity

Jerry Fodor, Zenon Pylyshyn, Brian McLaughlin, and others argued that the productive and systematic features of thought should be explained in terms of a combinatorial system of mental representations with a syntactically and semantically classical character. They argue that since human language exhibits some essentially systematic features and since language is an expression of human thought, human thought must also have the systematic features we find in language. We can understand novel sentences when we hear them and can create new meaningful sentences by combining the parts of old meaningful sentences according to those systematic transformations we associate with competent use of a natural language. If one can genuinely understand the sentence "Carnap likes Quine," then one can understand the sentence "Quine likes Carnap." The fact that such transformations pose no challenge to the intelligence of adult humans is taken as evidence that thought itself has the same structure as language.

Competing approaches to the Fodorian model of mind included the view that thought works as a series of images, or that it has the same structure as action, or that thought is brainlike in the sense of being organized in a network. From the Fodorian perspective, all of the major alternatives failed to provide the explanatory power of the assumption that there exists a syntactically and semantically combinatorial system of mental representations.

As many cognitive scientists and philosophers argued in the 1990s, alternative cognitive architectures may also be able to produce behaviors that mimic the systematic features of human language.[1] However, for proponents of a Fodorian view of systematicity, merely being able to reproduce systematic properties is not the purpose of a science of cognition. The real goal is explanation rather than simulation, and from the Fodorian perspective, a classical computational architecture provides the best path toward an *explanation* of mind insofar as the classical framework, like cognition and language, is systematic to the core.

Fodorians argue correctly that explaining the nature of human thought surely involves giving some account of its systematic features. However, the exclusive focus on linguistic systematicity has meant that other forms of systematicity have been neglected. So, for example, Robert Cummins (1996) pointed to the systematic features of visual perception as an example of a nonlinguistic form of systematicity that also seems important to a full understanding of perception and associated aspects of human thought.

Our view is that we ought to extend Cummins's insight beyond perception, and we argue that systematicity can also be found in the behavioral repertoire of agents. The kind of systematic properties that we target are those that distinguish intelligent, adaptive behaviors from automatic, non-adaptive behaviors.

Very roughly, at this stage, we can intuitively recognize the difference between the way a wolf stalks its prey and the way that the wolf's stomach digests its prey: in a Dennettian spirit, one might say that understanding and explaining the hunting behavior of a wolf involves adopting something like an intentional stance toward the behavior, whereas understanding processes in the digestive system does not (Symons 2002). In the pages that follow, we focus on cases that fall somewhere in between fully automatic physiological processes and full-fledged intentional action. The kinds of simple behaviors that we will discuss in this chapter are those exhibited by plants and other minimally cognitive agents.

Some biological processes, say, the excretion of bile or the rhythm of a beating heart, adapt to changing environments in a manner that reliably comports with the goals of an agent and yet do not warrant the honorific "intelligent." As mentioned above, automatic processes of this kind differ intuitively from the actions of intelligent agents insofar as they can be explained without reference to intentional content. So, how should we understand the transition from automatic processes to full-fledged intelligent behaviors? The behavior of a plant shares some features in common with digestion while also bearing some resemblance to the kind of intentional cognitive lives of animals like wolves and human beings. Examples of sophisticated plant behavior straddle the line between automatic physiological processes and systematic cognitive (albeit minimally cognitive) phenomena. These strike us as obvious opportunities to investigate the emergence of meaningful engagement with the environment.

The greatest strength of the Fodorian approach to systematicity was its careful attention to explanation. Rivals should also have an account of what is required for an account to qualify as genuinely explanatory in the cognitive or psychological domain. In this chapter, we propose taking a neo-Gibsonian approach to the explanation of behavioral systematicity. However, the target of explanation is different for us than it is for the

Fodorian. Our goal is to provide an explanation of the *emergence* of systematic intelligence per se rather than a defense of a particular cognitive architecture. On our view, arguments concerning the virtues of cognitive architectures can be distinguished from arguments concerning explanation. Recognizing the distinction is likely to benefit progress on both topics.

On our view, explaining the emergence of intelligent behavior requires attention to marginal cases of behavioral systematicity in minimally cognitive agents like plants and insects rather than beginning with the linguistically mediated cognition of adult human beings. To this end, we critically review some recent work in the field of "plant neurobiology" for the purpose of determining whether the ecological perspective can account for the behavioral systematicity that interests us in plants and other minimally cognitive agents. The approach we present here offers a framework for a naturalistic account of the emergence of intelligent behavior over the course of natural history, and we hope that explaining the systematic features of the behavior of plants and insects can provide the basis for understanding systematicity in more familiar kinds of cognitive systems.

3   Minimal Forms of Cognitive Agency

Common sense tells us that plants are unlikely to qualify as cognitive in any meaningful sense. One reason for this is the impression that plants do not really do much. Since plants generate their own food from light or other energy sources they move on a timescale that is normally imperceptible to animals.[2] Cognitive scientists and philosophers have assumed that there must be a strong connection between movement and cognition. Thus, Patricia Churchland represents the traditional view of plant intelligence as follows:

If you root yourself in the ground, you can afford to be stupid. But if you move, you must have mechanisms for moving, and mechanisms to ensure that the movement is not utterly arbitrary and independent of what is going on outside. (1986, 13)

Elsewhere, she writes:

First and foremost, animals are in the *moving* business; they feed, flee, fight, and reproduce by moving their body parts in accord with bodily needs. This *modus vivendi* is strikingly different from that of plants, which take life as it comes. (2002, 70)

Plants are usually slow moving, but are they stupid? Time-lapse photography has permitted plant researchers to notice nonprogrammed forms of movement triggered by differential changes in volume or in rates of growth. For example, consider the light-foraging behavior of the stilt palm (Allen 1977), a plant that grows new roots in the direction of sunlight, letting the older ones die. In relation to light-foraging behavior of this sort, Anthony Trewavas (2003) writes:

> The filiform stem explores, locates and recognizes a new trunk and reverses the growth pattern. As it climbs, the internode becomes progressively thicker and leaves progressively redevelop to full size. ... This behaviour is analogous to animals that climb trees to forage, intelligently descend when food is exhausted or competition severe, and then climb the next tree. (15)

Apparently the stilt palm is not taking life as it comes. In addition, *pace* Michael Tye and others, we now know that the behavior of plants is often flexible.[3] Strikingly, plants appear to learn from experience, not by modifying their "dendrites," but rather by developing plasmodesmatal connections (Trewavas 2003). Goal-oriented overcompensatory growth, oscillations in gravitropic behavior, and acclimatization under different forms of stress are well-studied illustrations of error-correction and learning in plants.

Our ignorance of the capabilities of plants has given way to the view that many plants do not simply sit passively photosynthesizing. Instead, plants can adapt in ways that may lead to an advantage in the future based on an assessment of current conditions. Plants can respond to soil structure, volume, and neighbor competition in ways that are advantageous to them. We now know of several examples of plant behavior that can be interpreted as territorial and that plants can discriminate their own roots from alien roots.[4] A number of examples of interplant communication have also been documented. Some plants communicate aerially with conspecifics and members of different species via a number of released volatile organic compounds that cause changes in the behavior of their conspecifics.[5] Most famously, the physiological processes underlying the collective response of acacia trees to being eaten by giraffes are well understood.[6]

Insights into the adaptive behavior of some plants have encouraged the development of a controversial field known as "plant neurobiology." Plant neurobiologists induce considerable discomfort in many of their colleagues when they use terms like *communication, planning, navigating, discriminating, perceiving,* and *remembering* to describe plant behavior. According to plant neurobiologists, higher plants have physiological processes that are analogous to animal nervous systems. They argue that these systems allow plants to act in intelligent ways.[7] But concerns immediately arise. Can the adaptive behavior of plants be considered cognitive in any meaningful sense? It is easy to find examples of plant neurobiologists providing inappropriately high-level cognitive explanations for the behavior of plants. Furthermore, many mainstream plant biologists have criticized plant neurobiology, describing it as "founded on superficial analogies and questionable extrapolations" (Alpi et al. 2007).

On our view, the reasons that careful observers of plant behavior are tempted to use anthropomorphic language stems from the systematic features of plant behavior. In the behavior of plants, we see the first infralinguistic glimmerings of the kind of intelligence that we see in higher animals. We now know that the behavior of plants, specifically their movement and the changes we detect in their morphology are often flexible and non-automatic. For the most part, this behavior takes place very slowly, but with the benefit of time-lapse photography, we can clearly see how some plants respond to change in ways that solve problems in dynamic and competitive environments.

An objection to our approach would be to deny that there is anything *cognitive* associated with plant behavior. Admittedly, we set the bar low for the purposes of this chapter insofar as we consider *motility* and the possession of a dedicated *sensorimotor organization system* as sufficient conditions for minimal cognition. Minimal cognition initially consists of exploiting the spatiotemporally dispersed characteristics of metabolically relevant environmental features. This is achieved courtesy of free and reversible bodily movement and is enabled by organized sensorimotor activity.[8] Plants may be taken to exemplify minimal cognition insofar as they manipulate their respective environments in meaningful ways.[9]

Our goal for the remainder of this chapter will be to show that these systematic features of behavior can be fruitfully understood in Gibsonian terms in a way that illuminates the meaningful interaction of the plant and its environment.

The behavior of plants is intelligent insofar as it is engaged with the environment in ecologically meaningful ways. But unpacking precisely what it means to call something "ecologically meaningful" is a nontrivial challenge. Why would we even feel tempted to invoke notions like meaning in this context? Let's consider an example: when we observe a vine in time-lapse photography exploring its environment we notice that its behavior has a systematic structure. Consider its ability to reach for a surface, test its suitability, withdraw if unsuitable, adjust its position slightly, and then repeat the behavior as necessary. The intuitive sense that the plant is striving, or has a plan, is obviously anthropomorphic. However, this intuitive reading of the plant's behavior is our response to the visual evidence that plants must cope with a dynamic landscape of threats and opportunities in their environment. While time-lapse photography allows us to notice plant behavior, and plant physiology can reveal the mechanisms at work in the plant, explaining the apparent meaningfulness and intelligence that is exhibited in plant behavior requires a different approach. We suggest that we can understand the intuitive meaningfulness of plant behavior by attending to ecological principles without thereby falling prey to anthropomorphism.

4 Fodorian Objections to the Ecological Approach

Systematicity is a characteristic feature of intelligent behavior. The source of this systematicity is the manner in which the organism engages with its environment. When an organism acts, it will do so for metabolically relevant reasons. We assume that if a given organism has the ability to acquire a metabolically relevant piece of information from its environment, it will thereby acquire a variety of systematically related patterns. On our view, even at the lowest levels, metabolically relevant patterns have systematic features. These systematic features of behavior are integral to the agent's capacity to respond selectively to a changing environment. Systematically organized patterns in behavior and the environment are precisely what support the agent's capacity to freely and reversibly navigate its local environment. Getting clear on what we mean by "metabolically relevant" information, and why a Gibsonian perspective is helpful here, is the central task of this section of the chapter.

To meet the charge of anthropomorphism, our account of systematicity in minimally cognitive systems must demonstrate that it does not rely on some prior cognitively penetrated system of relations. On our view, the emergence of higher-level cognition depends on systematic features in behavior and the environment. (See Symons 2001.) We contend that inquiry into the mechanisms that underlie the minimally cognitive capacities that are necessary for navigating environments containing metabolically relevant information will provide an account of how systematicity first appears over the course of natural history.

Following Fodor, philosophers have been very suspicious of claims like ours. During the 1980s, Fodor and Pylyshyn criticized Gibson's account of perception before moving on to criticize connectionist cognitive architectures later in the decade. Their 1981 paper "How Direct Is Visual Perception? Some Reflections on Gibson's 'Ecological Approach'" argued that the only way for ecological theories of perception to account for vision is by allowing for cognitive penetrability in the form of inferential processing. Their classic 1988 paper "Connectionism and Cognitive Architecture: A Critical Analysis" argued that the only way for connectionist theory to account for the systematicity of thought is by committing itself to a classical combinatorial structure.

Their 1988 paper addressed the possibility of a representational realist alternative to a classical model of the mind. In their 1981 paper, they criticized the possibility of a nonrepresentational alternative to a constructivist theory of perception. The ecological counterpart to representation and perceptual processing are information pickup and perceptual resonance, respectively. The connectionist counterparts to propositional, context-independent forms of representation and computation are context-dependent, vectorial representations and vector-to-vector transformations, respectively.

In what follows, we read Fodor and Pylyshyn's criticism of nonclassical cognitive architectures and their criticism of ecological theories as related defenses of a classical computational model of mind. Both lines of criticism challenge competitors to demonstrate how nonclassical alternatives could provide explanations of genuinely cognitive phenomena. As we shall see, shifting from the systematicity of thought to the systematicity of overt behavior allows an alternative explanatory framework that circumvents some of their concerns.

It is helpful to briefly introduce Gibson's theory of perception by contrast with the classical view before arguing for the applicability of Gibsonian ideas to the systematicity of behavior. Theories of perception can be divided roughly into those that are congenial to some form of Helmholtzian constructivism (Rock 1983) and those that adopt an ecological approach (Gibson 1979). Proponents of the former regard perception primarily as the outcome of a logic-like process of inference, whereby perception is hypothesized to be mediated or indirect. Following the Gibsonian lead, ecological theories of perception assume that an agent's perception is organized around its actions. Opportunities for action are perceived directly by agents as they interact with their local environment.[10]

The two core principles of Gibsonian psychology that we stress are the *specificational* account of information and the idea that *affordances* are what are perceived by agents.[11] The notions of specification and affordance play a technical and idiosyncratic role in Gibson's thought so we shall introduce them first. When properties of the world match unambiguously the patterns of ambient energy arrays available to a perceptual system, the energy arrays serve to *specify* the properties. Matches of this kind result from constraints that operate at an ecological scale between the agent and its environment. This is the most obvious contrast with constructivists, for whom there is an inherently ambiguous relation between the pattern in the energy array and the world. The assumption that this ambiguity must exist is precisely the reason that traditional constructivist theories call for an inferential treatment of perception.

How do Gibsonian theories explain visual perception without recourse to inferential processes? The optical variable tau (τ) (Lee 1976) provides a canonical illustration of the way information of optic arrays specifies properties of the environment unambiguously. Consider the distance between a car and an intersection as the driver approaches a stop sign. How does the driver judge when to apply the brakes to stop the car? What type of information can the driver rely on? One answer is provided by tau theory. David Lee defines tau as the inverse of the relative rate of expansion on the retina of an incoming object (e.g., a traffic sign).[12] In this way, the ecological psychologist treats tau as an optical invariant insofar as it specifies time-to-contact, not just between driver and stop sign, but between any animal and the object in its vicinity in terms of rate of retinal expansion in the direction of motion. Presented formally,

$$\tau = \theta/(\Delta\theta/\Delta t)$$

where θ stands for the angular size of the incoming object, and $\Delta\theta/\Delta t$ stands for the image's rate of expansion, as projected into the eye. The ecological psychologist's working hypothesis is that the optic flow field that obtains in the changing ambient optic arrays during navigation permits the agent to grasp the rate at which action gaps are closing (Lee 1998). In our example, a tau-theoretic approach does not demand articulation in terms of the agent's beliefs concerning the actual speed of the car or the size of the signal.[13]

The second major principle of ecological psychology is the idea that we perceive affordances. Gibson explains affordances as follows:

The affordances of the environment are what it offers the animal, what it provides or furnishes, either for good or ill. The verb to afford is found in the dictionary, but the noun affordance is not. I have made it up. I mean by it something that refers to both the environment and the animal in a way that no existing term does. It implies the complementarity of the animal and the environment. (1979, 127)

By contrast, from a Helmholtzian perspective, agent-independent variables, such as (absolute) distance, size, or speed, serve as the basic building blocks of perception. On this view, we perceive the distance to an object or its size and subsequently infer the object's availability for some action. By contrast, affordances are ecological properties, meaning that they are individuated by reference to the agent. Compare the ecological property of being reachable to the physical property of being one meter in length. Affordances are "opportunities for behavior"; properties of the local environment that permit agents to interact in ways that are relevant to the agent itself. The availability of an object for grasping is a property that makes no sense apart from its relation to an agent.

Ecological theories of perception take their start from the notion that what a biological agent perceives depends initially at least on the aspects of the environment that are relevant for it—more specifically, on those aspects of the environment that are available for biologically relevant interactions. We perceive by directly resonating with informational invariants that specify opportunities for behavior in the form of affordances. Different biological agents will perceive different affordances—that is, different opportunities for behavioral interaction within their respective environmental niches.[14]

Having outlined some of the central tenets of Gibson's view in very broad strokes, we are ready to turn to the objections. Ecological psychology is widely criticized as failing to do justice to the explanatory role played by systematic features of cognition. Consider speech perception, in particular the debate over "rule learning" in infants. Marcus et al. (1999) performed a series of experiments with seven-month-old infants. After having exposed them for two minutes to strings of artificial syllables conforming to a simple grammar—for example, "le le di," "ga ga li," and the like, from an AAB grammar, or "wi je je," "ga li li," and so on, from an ABB one—infants' speech perception skills were tested by analyzing their listening preferences for pairs of novel strings, one of which belonged to the same category they had been habituated to—for example, "wo wo fe" (AAB) versus "wo fe fe" (ABB). The results show that infants listened longer to those that did not conform to the pattern they had been exposed to during habituation.[15]

Marcus et al. interpret their results as showing that infants exploit abstract knowledge that allows them to induce the implicit grammar common to different sequences of syllables:

We propose that a system that could account for our results is one in which infants extract abstract-like rules that represent relationships between placeholders (variables) such as "the first item X is the same as the third item Y" or more generally that "item I is the same as item J." (Marcus et al. 1999, 79)

Connectionist responses focused on the possibility that an associative learning mechanism might induce *sameness* out of the statistical dependencies between syllable tokens in the linguistic corpus (Elman 1999). Their working hypothesis was that infants might be exploiting discrepancies based on expectations in order to make successful predictions in accordance with Marcus et al.'s data. If connectionist networks were to do so associatively, without recourse to universally open-ended rules or to devices that store particular values of variables to perform variable binding, no classical implementation would be required (Calvo and Colunga 2003).[16]

Likewise, the ecological psychologist may frame the challenge of explaining how infants perceive speech in terms of the *direct* perception of *sameness*-related properties. After all, the class of artificial syllables in Marcus et al.'s experiments corresponds to a set of objects in the infant's environment. We may thus try to make sameness available in the form of parameters that specify the property of "item I belonging to the same set as item J." As this is an empirical matter, an infant's perceptual systems could in principle resonate to any such property. Certainly, the Gibsonian may conjecture, the resonator in question might be a complex one, but so be it.

A combined reading of the two papers by Fodor and Pylyshyn mentioned above (1981, 1988) provides the arguments against the possibility that sameness is transduced in this way. According to Fodor and Pylyshyn (1988):

Connectionist theories acknowledge *only causal connectedness* as a primitive relation among nodes; when you know how activation and inhibition flow among them, you know everything there is to know about how the nodes in a network are related. By contrast, Classical theories acknowledge not only causal relations among the semantically evaluable objects that they posit, but also a range of structural relations, of which constituency is paradigmatic. (12)

On the other hand, as Fodor and Pylyshyn (1981) observe:

The reason that productive properties are *prima facie* not transduced is that, in many of the most interesting cases, membership in the associated set is inferred from a prior identification of the internal structure to the stimulus. (177)

Swap "classical" and "connectionist" for "constructivist" and "ecological psychology," and "nodes" and "network" for "transductors" and "organism," respectively; bear in mind that a combinatorial syntax and semantics for mental representations and structure-sensitivity of processes operate as a constraint on candidate mechanisms,[17] and the challenge takes the following form: given that ecological psychology acknowledges causal connectedness only as the relation of transduction in the environment-organism coupling, these systems face the challenge of explaining how properties that are relevant to systematic features of behavior are transduced. Specifically, for example, how can sameness be directly perceived, without resorting to a prior internal structure of some sort that allows for the identification of membership itself?

The ecological theorist of perception acknowledges only causal relations among the meaningful opportunities for behavior that they posit in the form of affordances—in the form of the direct relation of interaction that obtains between organism and environment when observed at the appropriate ecological scale. From a Fodorian perspective, the ecological theorist of perception turns a blind eye to any *internal* structural relations, and as such, the approach fails to deliver the kind of structural richness available to classical computational models. Were we to target ecological parameters that specify sameness non-inferentially, the Fodorian would complain that we have not provided an explanation as to how productive or systematic properties obtain.

There are a number of responses to this line of argument. One of us has explored the idea that arguments against ecological approaches that appeal to the complexity of some information-processing task sometimes proceed with an explanatory project in mind different from Gibson's (Symons 2001, 2007). Another general line of response is to see Fodorian criticisms as due to an unwarranted concern with the linguiform structuring of higher cognitive abilities. Linking cognitive phenomena to speech perception and similar cognitive tasks begs the question against nonclassical alternatives insofar as nonclassical approaches would regard language-level systematicity as an achievement for an agent rather than as a precondition for cognitive agency.

The critical issue here is clearly the question of the target of explanation. As is well known, the Fodorian perspective denies that there is an extra-linguistic source of explanation for linguistic capacities. However, if we hope to explain the emergence of systematicity per se, then we will need to shift our focus from higher-level cognition to nonhuman, infraverbal, minimal forms of cognition.[18] In the sections that follow, we will explain the kinds of systematic features of behavior in minimally cognitive agents that we believe can serve as the scaffolding for the kind of linguistic systematicity that interested Fodor et al. in the 1980s and '90s. Once again, the target of explanation is the emergence of the kinds of systematic features that are exhibited by intelligent behavior.

5   A Neo-Gibsonian Approach to Minimally Cognitive Agents

We propose what we are calling a neo-Gibsonian response to Fodor and Pylyshyn's challenge.[19] In this section we elaborate on the principles of "general tau theory" (Lee 2009) to give form to a neo-Gibsonian approach to the systematicity of behavior in minimally cognitive agents.

General tau theory (Lee 2009) is a theory of the *skilled* control of goal-directed movement. The chief concern of tau theory is the control of movement as a system interacts with objects in its local environment. This requires the control of "action gaps" between a given current state and the desired goal state. The closure of action gaps has been studied extensively, with general tau theory serving to account for such varied phenomena as the visual control of braking or steering by drivers (Lee 1980), of diving by gannets (Lee and Reddish 1981), or of docking on feeders by humming-birds (Lee et al. 1991). These are examples of the "time-to-collision" problem (Lee 1976). What unites these apparently disparate phenomena is the coordinated control of action gaps, and in these cases, tau is the informational variable that underlies goal-directed behavior.

According to general tau theory, for a movement to be *goal directed* the agent must be able to control the action gap between the current state of the system and a goal state. The ecological variable tau is the informational currency for the purpose of the controlling action gaps. The tau of an action gap can be sensed courtesy of a corresponding sensory gap (Lee 2009). As we saw earlier, direct sensing of the surrounding optic flow-field by a driver's perceptual system provides a canonical example. This is also how a gannet, as it dives into the ocean, is capable of retracting its wings at the appropriate moment (Lee and Reddish 1981).

In addition, general tau theory explains the coordinated control of two different action gaps, X and Y, via the *coupling* of their respective taus. A canonical example of tau-coupling is provided by the way we can intercept with our hand (H) a given target (T) at a certain goal (G), via the mainte-nance of a constant proportion between the taus corresponding to the two action-gaps: $\tau(HT) = k\tau(HG)$ (see Lee 1998). Or consider, for the sake of illustration, two boats moving in the sea at constant speeds along linear paths. Will the two boats collide with each other at a future time? This question may be approached ecologically by considering whether the two motion gaps, corresponding to their respective trajectories toward the hypothetical point of collision, close simultaneously or not. Were they to close simultaneously, courtesy of the maintenance of a constant ratio over a period of time, the two boats would enter into a collision trajectory. Note how drastically a tau-coupling-based gloss differs from constructivist attempts to describe the closure of action gaps. In Fauconnier and Turner's (1998) "network model of conceptual integration," for instance, cognitive integration ("blending") would underlie the inferential capacity to control action gaps, X and Y. Mark Turner vividly illustrates it with the metaphor of a "shrinking triangle."[20] We could imagine the two boats and the hypothetical point of collision forming an imaginary triangle. For the two boats to collide with each other would translate into the imaginary triangles that would form at time t and at subsequent time steps being proportional. Thus, were we to superpose the resulting proportional triangles and animate the image, we would "see" a shrinking triangle that collapses into the point of collision. According to Turner, decisions involved in collision avoidance call for a cognitive blend, as the shrinking triangle illustrates. But we need not rely on inferential treatment of any sort. General tau theory allows us to talk in terms of the coupling of the taus of different action gaps, and this is information that is specificational. Simultaneous closure delivers the goods *directly*.[21]

General tau theory serves to provide form and constraint to a neo-Gibsonian theory of minimal cognition. According to the ecological theory of perception introduced thus far, there are invariant properties of objects in the environment that, when appropriately tuned to in terms of ambient energy arrays, result in direct perception. With a touch of analyticity, this is what Fodor and Pylyshyn have in mind when they read Gibson as meaning to say that

for any object or event x, there is some property P such that the direct pickup of P is necessary and sufficient for the perception of x. (Fodor and Pylyshyn 1981, 140)

Necessity and sufficiency aside, the issue boils down to the way to interpret "ambient energy arrays" so as to bridge P and x, but there's the rub. Surely, the ecological psychologist has a clear picture in mind as to what "ambient energy array" means. Nevertheless, and with apologies to the reader familiar with the specialized literature, we shall spell out our particular take in the form of a set of principles that will serve to introduce our neo-Gibsonian approach to minimal cognition. Starting with the less-disputed principles, and moving in increasing order of controversy to the more contentious ones, the neo-Gibsonian theory of minimal cognition we propose (i) is not modality specific; (ii) includes intra-organismic properties; (iii) is substrate neutral; and (iv) portrays perception as a function of the global ambient energy array. To these four principles we now turn.

• *A neo-Gibsonian theory of minimal cognition is not modality specific.*

For the purpose of introducing the ecological worldview we have focused thus far on ambient *optic* arrays. However, there is ample evidence that information remains specificational regardless of the sensory modality involved. We may say that vision is only one facet of the question of action-gap closure. Considering a gap to be closed, such as the gap between a hummingbird and a feeder, its tau is the time needed to close it, to land on the feeder, at the current rate-of-closing. Likewise, the tau of a cell is the time needed to swim to a cathode by sensing electric fields at its current rate-of-closing (Delafield-Butt et al. 2012). Other tau-based studies include the steering bats perform courtesy of echolocation (Lee et al. 1995) and the gliding pitch between notes in expressive musical performance (Schogler et al. 2008). Illustrations abound. As Lee puts it:

To dispel a common misconception, tau is *not* the inverse of the rate of dilation of an optical image, any more than gravity is the apple falling on Newton's head. The apple falling is an example of the general principle of gravity. The image dilation is an example of the general principle of tau. (2004, 8)

It is thus changes in the sensory gaps of any modality what informs as to which opportunities for behavioral output are present in the form of the closure of the action gap in question.

• *A neo-Gibsonian theory of minimal cognition includes intra-organismic properties.*

The ecological psychologist distinguishes between ambience and environment. Ambience relates to the surrounded organism *and* the surrounding environment. In this way, the unity of interest is the reciprocity of the whole ecological scenario itself. Organism and environment are not detached, and in their interaction it is ambient (not environment) energy flows that count. However, once granted such a reciprocal relation, a rather literal reading of Gibson's well-known aphorism—*It's not what is inside the head that is important, it's what the head is inside of*—precludes us from noting that direct ecological interaction between organism and environment, on the one hand, and intra-organismic, say, neural processing properties, on the other, are not antagonistic. With the emphasis on the relation between an organism and its surroundings, it is easy to see that the ecological ambience does not need to be exclusively exogenous.

The neglect of endogenous ambience is clearly widespread among Gibsonians. A notably rare exception is Lee (2009). Affordances are neither external nor internal by necessity. Affordances are dispositional properties to close action gaps. We may then consider both endogenous and exogenous sources alike. The ecological ambience of a cell, or a population of cells, implies a reciprocity of the cell, say, a neuron or a population of cortical neurons, and its surroundings, which may be the extracellular, cortical, or subcortical environment. Thanks to this, intrinsic guidance can take place. What counts is that information remains specificational, not whether it is constrained by the scale of perception and action or by the spatiotemporal scale of endogenous cellular processes.

• *A neo-Gibsonian theory of minimal cognition is substrate neutral.*

Once the distinction between exogenous and endogenous ambiences is in place, it is important to note that intrinsic guidance need not be (exclusively) *neurally* based. This is often the case with the (neurocomputational) constructivist. Llinás and Churchland's (1996) concept of "endogenesis," for instance, lays the stress on the fact that cognitive activity is the result of endogenous neural processes. But when the ecological psychologist decides to go inward, the methodological constraint that operates is that the spatiotemporal scale of processes remains ecological, and not whether the substrate is neural or not. It may, for instance, be hormonal instead, granted that specificational information that can be detected for the purpose of appropriate resonance exists at the scale of hormonal processes. It is the fact that a property is appropriately defined at an ecological scale that counts. Bluntly, there is no reason why the neo-Gibsonian should be a neural chauvinist.

• *A neo-Gibsonian theory of minimal cognition portrays perception as a function of the global ambient energy array.*

According to our first principle, a neo-Gibsonian theory of minimal cognition is not modality specific in the sense that specification may take place in a number of energy arrays. Plausibly, then, the specification of reality may not reside in a single ambient energy array. It is possible, however, that researchers have decided to focus on the optic, instead of, say, the acoustic array simply on experimental or methodological grounds. In this way, a further twist comes with the idea that specification, unconstrained, may exist in the global energy array itself, in some higher-order format that cuts across sensory modalities (Stoffregen and Bardy 2001). If there are intrinsic forms of tau-guiding action gaps, we may be searching for transmodal integration via synchronous interactions between neurons; a process that under the hypothesis that information is picked up from the global energy array itself could be couched non-inferentially. This would not be a radically innovative way of proceeding. Neuroscience is familiar with cross-modal integration (Kujala 2001). What is at stake, rather, is the question of whether there are transducers for global variables. This, despite Fodor and Pylyshyn's skeptical appraisal, is an open empirical question. Still, it is important to point out that, one way or the other—whether specification is transmodal or unimodal—perception would remain unmediated.[22]

As we saw at the outset of this section, Fodor and Pylyshyn (1981) stressed that the direct pickup of properties was both necessary and sufficient for the perception of objects and events, but questioned that ambient energy arrays may permit the specification of information in the form of affordances. In our view, the ecological link between properties and objects or events may be licensed or not as a function of the way we read "ambient energy arrays." According to our neo-Gibsonian reading, direct perception is a function of the global ambient energy array. This appraisal, combined with the lack of specificity with respect to modality, acknowledging intra-organismic properties, and an eschewal of neural chauvinism, sets the stage for assessing the plausibility of direct perception and behavioral systematicity in the remainder of the chapter. Direct perception is the emergent result of organism–environment interplay. In the next section, we discuss the question of whether a neo-Gibsonian theory of minimal cognition may apply to plants.

6 Plants as Perceiving and Behaving Organisms

Gibson did not believe that plants were capable of perception, and he might have worried that our claim that plant behavior can be understood according to the principles of ecological theory of perception as equivalent to a reductio argument against his view.[23] However, contrary to Gibson's own view of the capacities of plants, if perception is understood in terms of resonance to specificational information then there is ample ground to argue that plants perceive. Plants are animate, move about, and have an internal system for organizing behavior that, in some important respects, is similar to the animal nervous system.[24] Circumnutation in climbing vines, a helical and rotational form of oscillatory movement already studied by Darwin (1875), provides one of the best well-known illustrations of the endogenously governed exploratory strategies of the surrounding environment performed by plants. Furthermore, the behavior of plants is often systematic in the sense of being reversible, non-automatic, and repeatable in a manner that responds to metabolically salient features of the environment.

Not all ecological theorists shared Gibson's dim view of the perceptual capacities of plants. In a reply to Fodor and Pylyshyn (1981), Turvey et al. (1981) consider from an ecological point of view *Monstera gigantea*, a climbing vine whose seeds are able to perceive an affordance ("climbability") skototropically as they grow toward darkness. As we saw earlier, it is essential to distinguish between ambience and environment, the former being inherently relational with respect to perception-in-action. Consider how climbing plants can be understood as perceivers (Isnard and Silk 2009). Vines may well perceive Gibsonian affordances, possibilities for action, such as when a support is directly perceived as affording climbing. To understand climbability it is necessary to bear in mind that a vine and its support are functionally coupled subsystems. The vine should not be seen as a kind of organism that acts, by climbing, onto a separate kind of thing that is the support. As Gibson observes:

The words *animal* and *environment* make an inseparable pair. Each term implies the other. No animal could exist without an environment surrounding it. Equally, although not so obvious, an environment implies an animal (or at least an organism) to be surrounded. (1979, 8)

Replacing "animal" for "plant," we see that the rest of Gibson's claim holds in a relatively straightforward manner. Thus, a vine could not live without an environment that furnishes it with rocks, tree trunks, and all sorts of supports that are directly perceived as affording climbing. The complementarity of the plant and its vicinity means that the plant-in-its-environment serves as the proper unit of analysis.

On the other hand, we saw that action-gap closure is not only a matter of distance, but can also cover angle, pitch, frequency, and so on. In addition to the well-studied case of vision, action-gap closure also figures in other modalities including haptics and echolocation, among others. It is thus ambient energy arrays of any form that can serve this purpose. Plants tune to a wealth of information beyond the vectors of light and gravity. In the case of plants, we cannot ignore forms of sensory input such as electrical, magnetic, chemical, acoustical, and vibrational. Consider plant bioacoustics, a field of research that informs us that plants may have benefited at an evolutionary scale from the perception of sound and vibrations (Gagliano et al. 2012). And yet, more intriguingly, there is evidence that plants even exploit bioacoustics to communicate with insects (Barlow 2010). Overall, to the best of our knowledge, plants can sense, and integrate, up to twenty-two different biotic and abiotic vectors (Trewavas 2008). At first sight, then, if plants perceive, it seems there is no reason to exclude the possibility that direct perception takes place as a function of the plant's global ambient energy array.

We may furthermore dig deeper and address intrinsic guidance by plant structures, once we recognize the role of the plant's endogenous ambience. In fact, neuroid conduction (Mackie 1970) applies to protists, plants, and animals alike insofar as they all have nonneural cells with electric signaling properties. Thus, the fact that plants lack neurons is not a handicap from the ecological perspective.[25]

Plant neurobiology (Brenner et al. 2006) has consolidated in the last few years as a discipline that studies plant behavior from the analysis of the integrated signaling and electrophysiological properties of plant networks of cells, with special attention to the involvement of action potentials, long-distance electrical signaling, and vesicle-mediated transport of auxin, among other pythohormones. As Baluška et al. (2006) point out:

Each root apex is proposed to harbor brain-like units of the nervous system of plants. The number of root apices in the plant body is high, and all "brain units" are interconnected via vascular strands (plant neurons) with their polarly-transported auxin (plant neurotransmitter), to form a serial (parallel) neuronal system of plants. (28)[26]

If plant behavior is partly the result of endogenous nonneural processes, then vascular strands and auxin correlates may serve to guide endogenously goal-directed climbing behavior toward tree trunks or other supports under the principles of tau-coupling and the intrinsic tau-guidance of action gaps. Following the analogy with animal models, it may be the case that taus of auxin action gaps underlie the type of hormonal information that directs tropistic responses in plants. Tau information may guide climbing, and plant neurobiology may well show that the type of activity that underlies sensorimotor coordination across the plant may also be tau-based.[27]

A condition for minimal cognition was that the navigational capacities and sensorimotor organization were organized globally. The "root-brain," a concept inspired in the discovery of the "transition zone" (TZ) within the root-apex, may play an important role here. Plant neurobiologists consider TZ—the only area in plant roots where electrical activity/fields are maximal, and where electrical synchronization obtains (Masi et al. 2009)—to be a "brainlike" command center (Baluška et al. 2004, 2009), with a polar auxin transport circuit underpinning patterns of root growth. TZ integrates not only hormonal endogenous input, but also sensory stimulation. Plant roots determine root growth that results in alignment or repulsion movements as a function of the global structure of its

vicinity. We may then consider the role of the root-brain in the integration of sensorimotor pathways for the purpose of adapting flexibly as a
number of global tropistic responses take place in the form of differential
growth. We may thus take into account the role of the root-brain in the
specification of plant–environment reciprocity and consider plant perception as a function of the global ambient energy array. To think of perception as a function of the global ambient energy array may mean searching
for TZ cells at the root-brain that respond selectively to the embracing
activity, with invariants spanning across the vectors of gravity, light, and
the like, on the one hand, and across endogenous hormonal stimuli, on
the other.

Our hypothesis is that root-brains resonate to high-order invariants.
This hypothesis is compatible not only with perception being a function
of the global ambient energy array, but also with the other neo-Gibsonian
principles described above. Note that, for convenience, we simplified
matters by considering tropistic behavior as if triggered by energy arrays
on a case by case basis. However, Gibson provided many different examples
of lower- as well as higher-order invariants, among them, gravity (Gibson
1966, 319) and the penumbra of a shadow (Gibson 1979, 286). A more
realistic portrayal should be amenable to multiple sources of perturbation.
A possibility then is that TZ could be sensitive to structure in the global
array directly, in the same way we have hypothesized animals are (Stoffregen & Bardy, 2001). In this way, we do not need to interpret synchronized firing at TZ cells as inferential processing. Rather, TZ cells, under the
ecological lens, resonate to information from the global array. As we saw,
there are intrinsic forms of tau-guiding action gaps. Considering that interactions in the plant brain between TZ cells offer a way of integrating
endogenous and exogenous information, affordances may be systematically perceived in terms of higher-level combinations of invariants.

Bearing this background in mind, we conclude this section with a
consideration of honeybee behavioral systematicity by way of contrast.
We then consider in the next section how to scale up from plant and
insect direct perception and behavioral systematicity to human-level
performance.

Having assumed that the globally organized exploitation of the environment by means of motility enabled by various sensorimotor organizations
marks the borderline between purely reactive and minimally cognitive
behavior, we may now assess the capabilities of insects as they navigate
their metabolically relevant environment, an intelligent form of behavior
usually couched in constructivist terms. Consider the swarming behavior

of honeybees (Visscher 2003), where an individual contributes to the creation of a "spatiotemporal order characterized by the alignment of directions and maintenance of equal speeds and distances" (Ciszak et al. 2012). Honeybees can estimate and communicate distance and direction of hive-to-nectar and nectar-to-hive through the "waggle dance" (von Frisch 1993). Abilities of this sort are not exclusive to bees. More generally, we may say that insects that are able to acquire a piece of information that is metabolically relevant have the resources to acquire related pieces of information that are also metabolically relevant.

Can we account for the behavior of honeybees ecologically? At first sight, it may seem that we cannot. Peter Carruthers, for instance, claims that

bees have a suite of information-generating systems that construct representations of the relative directions and distances between a variety of substances and properties and the hive, as well as a number of goal-generating systems taking as inputs body states and a variety of kinds of contextual information, and generating a current goal as output. (2004, 215)

This should remind us of Marcus's take on infant speech perception. Bees, like infants, appear to exploit abstract knowledge that allows them to induce underlying regularities. This process appears to call for the implementation of operations defined over abstract variables algebraically. Constructivism appears again to be the default stance. To find both the nectar and their way back to the hive honeybees exploit directional data computationally by generating a spatiotemporal representation of the sun's course (Dyer and Dickinson 1994).

The basis for an ecological explanation is on offer. By running experiments in which bees are forced to fly in narrow tunnels, Esch et al. (2001) have been able to experimentally manipulate optic flow fields. Interestingly, their results are inconsistent with absolute distance or any other bee-independent variable being computed in the waggle dance, as would need to be the case under an algebraic interpretation. Rather, the estimation of distance is consistent with reliance on self-induced optic flow in the open. Srinivasan et al. (2000) provide converging evidence in terms of bee-dependent properties, and other insects such as desert ants (Ronacher and Wehner 1995) appear to follow a strategy with similar ecological credentials. Collett and Collett (2002) also interpret honeybee navigation in Gibsonian terms. This research is compatible with ecological forms of navigation that make no cognitive use of representational-*cum*-computational maps or operations of any sort.

As we saw earlier, minimal cognition initially consisted of exploiting the environment, courtesy of free and reversible bodily movements enabled by various sensorimotor systems of organization. Of course, the analogy with plants cannot be cast straightforwardly in finding-their-way-back-home terms.[28] Circumnutation, or other forms of plant movement, are specific to plants' needs and constraints. Still, the analogy is functional, and operates when drawn in relation to the exploration of roots for nutrients, for instance. Roots literally navigate systematically through their local environment.

In effect, the navigational repertoire of plants is considerable, and a number of highly sophisticated and intriguing navigational capabilities may be considered beyond tendril-climbing, such as search and escape movements performed by roots in response to competition (see Baluška, Mancuso et al. 2010 and references therein), or the photophobic behavior of crawling maize roots, a behavioral response that cannot be interpreted as a simple form of unidirectional negative phototropism (Burbach et al. 2012). Root-swarming behaviors exhibit similar levels of behavioral complexity. Roots not only must navigate the soil structure, but must also coordinate their root system so as to optimize nutrient intake in addition to other adaptive considerations such as territoriality. In these cases, information must be shared across the plant root system. In fact, communication takes place not only between root apices of the same plant, but also with respect to the root systems of neighboring ones. In this way, the concept of adaptive swarming behavior applies to plants, as they solve by social interaction problems that outstrip the individual when viewed in isolation (Baluška, Lev-Yadun, and Mancuso 2010).

Two forms of interaction between roots that have been studied meet our criteria for minimal cognition. As Ciszak et al. (2012) report, roots may align/repulse and grow in the same/different direction in the absence of physical contact (distance-based alignment/repulsion). On the other hand, root crossing may take place when roots first attract each other, to repulse afterward. Overall, complex patterns of collective behavior have been observed, with groups of roots being able to choose the same or opposing growth direction (see Ciszak et al. 2012 for details). We may thus consider how the position and orientation of individual roots relates in an emergent manner to the tendencies of other roots to respect alignment in growth.

These studies are consistent with the idea that roots that are able to acquire metabolically relevant information have the resources to acquire related information that is also metabolically relevant. Roots can estimate and communicate distance, direction of nutrient vectors in soil patches, and potential competition through angle adjustment in navigation, and the adjustment of angle can be signaled by electric fields that roots generate, and sensed in turn by other individual roots or root systems.[29] In this way, swarming behavior results in systematic patterns of navigation that we understand are in principle subject to a methodological treatment akin to the one pursued by Collett and Collett (2002), Esch et al. (2001), Srinivasan et al. (2000), and other authors, in the case of insect ecological navigation.

Summing up, minimally cognitive systems exploit the spatiotemporally dispersed characteristics of metabolically relevant environmental features by performing free and reversible bodily movements. As a result, both insects and plants generate a flow (plausibly by root circumnutation, in the case of plants) that is informationally rich insofar as navigational paths are ecologically specified. Invariant information is generated through navigation, a capacity that is itself guided by the structure of that very information. It is this reciprocity between perception and action that tells against a cognitivist rendering of minimal cognition, and against an inferential treatment of systematicity as conceived for such minimal agents.[30]

7   The Ecological Approach to (Minimal) Cognition

Finally, is it possible to scale up from a neo-Gibsonian approach to minimal cognition to higher-level forms of systematicity? One option is to maintain that higher cognition inherits its combinatorial power from the structuring role of public language itself (Dennett 1995; Clark 1997; Symons 2001). The systematicity of thought might then be seen as the felicitous outcome of an agreed-on systematicity of language. Sympathizers with this route may come, for instance, from the connectionist corner (Bechtel and Abrahamsen 2002) or from a dynamic and interaction-dominant perspective (Gomila et al. 2012).[31]

Rather than denying systematicity at the behavioral level, one could imagine a split into an ecological lower-level system and a constructivist higher-level one. We will not consider the former option, as we have granted the systematicity of behavior under our "minimal cognition" approach.[32] The two visual systems model (Goodale and Milner 1992) provides the canonical illustration of the second option (Norman 2002), where the ventral ("what") and dorsal ("where") pathways serve different purposes. Whereas the ventral pathway, being inferential and memory based, connects vision with cognition proper and therefore fits nicely with constructivist concepts, the dorsal pathway, being in charge of the control of motor behavior, is more in line with ecological principles.

Clark's (2013) "hierarchical generative model" provides yet another type of conciliatory strategy that although, to the best of our knowledge, has not been considered for the purpose of responding to the systematicity challenge, will serve by way of contrast to pin down our methodological proposal. Clark (2013) invites us to consider a "unified theory of mind and action" by combining Bayesian (top-down) and connectionist (bottom-up) methodologies into a single architecture. According to "probabilistic models" of thought (Griffiths et al. 2010), configurations of sequences of symbols obtain as a result of probabilistically tinkering with a number of parameters that determine syntactically correct structures. Together with particular mechanisms in the hippocampus and neocortex (McClelland et al. 2010) that account for the possibility of combining rapid parsing and associative learning, a conciliatory solution might be forthcoming.[33]

By contrast with these methodological choices, we believe that ecological psychology has the resources to avoid the potential pitfalls of these strategies: neglectfulness of systematicity itself or, possibly, becoming implementational. Our claim is not that ecological psychology might be able to account for those cognitive phenomena that resist a description in terms of systematicity. Rather, we believe that a great deal of behavior (behavior that we identify with minimally cognitive flexible responses) is thoroughly systematic, and that the neo-Gibsonian may be able to explain it. This includes the perception of speech. Gibson notes:

Now consider perception at second hand, or vicarious perception: perception mediated by communications and dependent on the "medium" of communication, like speech sound, painting, writing or sculpture. The perception is indirect since the information has been presented by the speaker, painter, writer or sculptor, and has been *selected* by him from the unlimited realm of available information. This kind of apprehension is complicated by the fact that direct perception of sounds or surfaces occurs along with the indirect perception. The sign is often noticed along with what is signified. Nevertheless, however complicated, the outcome is that one man can metaphorically see through the eyes of another. (quoted in Fowler 1986, 23–24)

Gibson's comments may appear at first sight to drive us straightforwardly into constructivism. But, in Gibson's usage, being "indirect" is not tantamount to cognitive penetration. Despite his distrust of mental abstractions, in *The Senses Considered as Perceptual Systems* Gibson already hints at a way to bridge language and ecological psychology with an eye to articulating the direct perception of meaning. For example, he discusses "the pick up of symbolic speech" (1966, 90ff.) and "the effect of language on perception" (280–282). Unfortunately, Gibson did not elaborate much beyond some preliminary remarks.[34] We close our chapter by inviting the reader to consider the supracommunicative role of language as an ecological *tool*.

An ecological approach cannot be in the business of developing a generative grammar for verbal behavior; its aim is rather to foster an understanding of language use that is integral to the emergent and self-organized (linguistic and nonlinguistic) behavior of humans. In this way, the development of an ecological approach to language ought to start with the realization that language is not that *special* (at least, not in the sense implied by cognitivism).[35] Clark (1998), in this same spirit, proposes a supracommunicative role of language, although his approach is alien to ecological psychology concerns. Clark starts by clearing up potential misunderstandings in the very title of his article "Magic Words":

> Of course, words aren't magic. Neither are sextants, compasses, maps. ... In the case of these other tools and props, however, it is transparently clear that they function so as to either carry out or to facilitate computational operations important to various human projects. The slide rule transforms complex mathematical problems (ones that would baffle or tax the unaided subject) into simple tasks of perceptual recognition. ... These various tools and props thus act to generate information, or to store it, or to transform it, or some combination of the three. ... Public language, I shall argue, is just such a tool. (1998, 162)

Clark (1998) is proposing a supracommunicative role for language. He criticizes noncomputational models of cognition for emphasizing communicative aspects of language to the detriment of its exploitation as an external artifact. His is nonetheless an approach that calls for the augmentation, courtesy of "magic words," of our computational powers.[36] Being noncomputational, our interest resides in identifying the role of language in serving not to augment computations but to allow for an ecological manipulation of the environment for the purpose of perceiving a brand new set of affordances that nonlinguistic animals are unaware of.

Tools are part and parcel of the Gibsonian worldview. The environment contains artifacts, and these alter its layout, which results in a global change of the affordances the environment furnishes to the human animal. Tools offer a different set of opportunities for interaction. Language, as an ecological tool, is not intrinsically different from artifacts such as telescopes or microscopes, which permit us to go beyond the native capacities of our visual systems. Language further alters the environment's layout, providing novel ways to interact in the sociolinguistic environment. The ecological approach to language as a tool has to do with behaving adaptively in the face of linguistic information. In general, we perceive the affordances of artifacts; likewise, we contend, we perceive the affordances of words, as a more sophisticated type of artifact. We perceive interindividual emergent properties, such as public conversation, in pretty much the same way that we perceive rocks and their contextual affordances, or social affordances (Costall 1995), for that matter.[37] In this way, an external linguistic scaffolding provides yet another set of ecological properties to be taken on a par with others in the environment.[38]

8 Conclusion

Our approach appears to be open to a very basic challenge. This is posed clearly by Ken Aizawa (chapter 3, this vol.), who argues that post-connectionist cognitive science has drifted away from a focus on cognition and has lapsed into a kind of uncritical behaviorism. He contends that an increasing emphasis on behavior has led to confusion concerning the importance of systematicity as a distinctive mark of the cognitive. We agree with Aizawa that denying the difference between the behavior we find in plants and the systematic features of higher cognition more generally would be foolish. However, we do not accept the view that we must mark the distinction between the cognitive and noncognitive by the presence or absence of structured systematicity.

Aizawa and others have long regarded cognition as essentially representational. While this view comports with our understanding of normal, adult human intelligence and provides a useful explanatory framework for psychological research with such subjects, it fails to shed any significant light on the emergence of cognition over the course of natural history. We contend that one cannot explain the emergence of the kind of sophisticated representational capacities that are assumed to play a role in adult humans if one's account of systematicity presupposes preexisting representations.

The neo-Gibsonian strategy involves recognizing alternative paths to the emergence of cognition over the course of natural history. Rather than assuming that we arrive at cognition via representation, an explanation of the emergence of cognition supposes that we can provide explanations in terms of increasingly sophisticated causal patterns of relations (García Rodríguez and Calvo Garzón 2010). If minimally cognitive systems need not involve representations (our neo-Gibsonian take) then objections like

Aizawa's can be addressed. What we have assumed is that cognition is a form of adaptative behavior and that becoming a cognitive system involves an organism managing to succeed as an adaptive system. Of course, not all forms of adaptive behavior must be regarded as cognitive. On our view, minimal cognition involves adaptive behavior that is systematic.

Traditionally, the objection to projects like ours is that we risk conflating cognition with behavior, or ignoring cognition entirely. However, this is not the case. Instead, we have attempted to demonstrate how systematicity at the cognitive level can emerge from the kind of meaningful engagement with the environment that is phylogenetically and ontogenetically prior to the kinds of intellectual and cognitive capacities that we expect from adult humans. We assume that there are ways of meaningfully engaging with one's environment that are nonrepresentational. We also assume that the concept of meaningful engagement can come apart from the concept of representation.

Clearly, our work departs from traditional debates over the systematicity of thought insofar as it is directed toward a different explanandum. If we see the goal of cognitive science as accounting for the emergence of intelligence and cognition, then it will be natural to attend to minimally cognitive agents and the emergence of simple forms of systematicity in behavior.

Acknowledgments

This material draws from preliminary work presented at the following workshops: Varieties of Representation: Kazimierz Naturalist Workshop (Kazimierz Dolny, Poland); Smart Solutions from the Plant Kingdom: Beyond the Animal Models (Florence, Italy); and 12th European Workshop on Ecological Psychology (Miraflores de la Sierra, Spain). This research was also presented in seminars held at the Universitat Autònoma de Barcelona (Spain) and the Laboratoire de Psychologie Cognitive (CNRS and Aix Marseille Université, France). We would like to thank all these audiences, and Frantisek Baluška, Stefano Mancuso, and Dave Lee, for their inspiring work in plant neurobiology and ecological psychology, respectively. We are grateful to Anastasia Seals for comments. Very special thanks go to Ken Aizawa for his detailed criticisms and helpful suggestions on an earlier draft of this manuscript. This research was supported by DGICYT Project FFI2009-13416-C02-01 (Spanish Ministry of Economy and Competitiveness) to PC and EM, and by Fundación Séneca-Agencia de Ciencia y Tecnología de la Región de Murcia through project 11944/PHCS/09 to PC and JS.

Notes

1. Notable early proposals are those of Smolensky (1987, 1990), who took issue with the challenge by exploiting microfeatural descriptions and tensor product variable binding for the purpose of modeling weaker and stronger forms of compositionality, respectively; Chalmers (1990), who modeled structure-sensitive operations on recursive auto-associative memories (RAAM—Pollack 1990); and van Gelder (1990), who distinguished between functional and concatenative compositionality.

2. The obvious exceptions being the familiar behavior of the Venus flytrap and the mimosa.

3. Michael Tye dismisses any talk of plant intelligence by claiming that it is entirely genetically determined and inflexible: "The behavior of plants is inflexible. It is genetically determined and, therefore, not modifiable by learning. Plants do not learn from experience. They do not acquire beliefs and change them in light of things that happen to them. Nor do they have any desires" (1997, 302).

4. Chemical and electric signaling below ground (Schenk, Callaway, and Mahall 1999) underlies root segregation, something that involves a form of self-recognition (roots must make decisions as to how to segregate) and amounts to a competitive form of territoriality.

5. This has been popularized as "talking trees," or more aptly as "eavesdropping" (Baldwin et al. 2006).

6. For a review of plant biochemical warfare against the herbivore, see Mithöfer and Boland 2012.

7. Trewavas observes that a number of forms of plant memory "can be recognized by the ability to interact with, and modify, the transduction pathways to new signals. ... A more complex form of memory requires information storage of previous signalling, with the ability to retrieve the information at a much later time. Both forms occur in plants" (2003, 7). As to plant learning, Trewavas (2003) notes that "wild plants need trial-and-error learning because the environmental circumstances in which signals arrive can be so variable. ... Indications of trial-and-error learning can be deduced from the presence of damped or even robust oscillations in behaviour as the organism continually assesses and makes further corrections to behaviour" (ibid., 4).

8. Traditional antibehaviorist considerations (Chomsky 1959) do not pose a challenge to our account of minimal cognition insofar as dedicated sensorimotor organization and navigational capacities are globally organized.

9. Arguably, plants possess a far broader repertoire of cognitive capacities than those we discuss here. For elaboration of this proposal see Calvo Garzón and Keijzer 2011, and references therein.

10. For an introduction to Helmholtz's theory of perception in the context of the early history of experimental psychology, and to some of the theoretical problems in the field of perception, see Kim 2009. A good entry point to the direct perception approach is Michaels and Carello 1981.

11. For further discussion of specification and affordances in Gibson, see Richardson et al. 2008.

12. For general criticism of tau theory that we ignore for present purposes, see Tresilian 1999.

13. Most traditional work in cognitive science has assumed that information-processing tasks in visual perception must be articulated in terms of an inferential process. David Marr, for example, argued that the principal failure of the ecological theory of perception was its inability to grasp the actual complexity of visual perception. Tau theory provides reason to believe that some perceptual tasks might be simpler than Marr believed. For more on the issue of the complexity of information processing, see Symons 2007.

14. It is important to note, then, that with (unambiguous) direct perception requirements in terms of memory storage drop dramatically. Information does not need to be stored temporarily for the purpose of inferential information processing. The direct pickup of informational invariants in ambient light serves to explain visual perception. According to ecological psychology, organisms pick up invariants and "resonate" to (i.e., transduce) the ambient properties that they specify. The Gibsonian task then is to discover the type of information that is specificational for the non-inferential resolution of the perceptual problem in question.

15. See Marcus et al. 1999 for the details, and Gerken and Aslin 2005 for a review of the language development literature.

16. Although for a skeptical appraisal of the alleged success of connectionism, see Marcus 2001.

17. To wit: "In particular, the symbol structures in a Classical model are assumed to correspond to real physical structures in the brain and the combinatorial structure of a representation is supposed to have a counterpart in structural relations among physical properties of the brain. For example, the relation 'part of,' which holds between a relatively simple symbol and a more complex one, is assumed to correspond to some physical relation among brain states" (Fodor and Pylyshyn 1988, 13).

18. The shift is strategic, as will become apparent below. Other than linguistic systematicity being a canonical illustration, there is no intrinsic connection to be found between natural languages and the systematicity of thought. McLaughlin (1993), for instance, explores systematicity in non-human animals. See also Aizawa's chapter in this volume.

19. "Neo-Gibsonian" insofar as Gibson himself would not have accepted an application of ecological theories of perception to minimally cognitive agents like plants. We briefly touch on Gibson's objections below.

20. See http://vrnewsscape.ucla.edu/mind/2012-05-03_Turner_Nutshell.html.

21. Tau-coupling also permits the intrinsic tau-guidance of action gaps. Lee (2009) considers different guiding gaps, as the gap is closed with constant acceleration or with constant deceleration, as you speed up to hop onto the train, or as birds dock on perches, respectively.

22. The principle that information is specificational can be given a strong and a weak reading. The strong reading says that when a given pattern in the energy array bears a one-to-one correspondence to properties of the world, information is uniquely specified. On the weak reading, the relation between ambient energy arrays and properties of the world may be many-to-one. That is, patterns of the ambient energy array may allow for the transduction of environmental properties in a manner that, although non-unique, is unambiguous with respect to properties of the world. Note that this weak reading is all we need for perception, as a function of the global ambient energy array, to remain unmediated.

23. Gibson did not regard plants as capable of perception:

In this book, *environment* will refer to the surroundings of those organisms that perceive and behave, that is to say, animals. The environment of plants, organisms that lack sense organs and muscles, is not relevant in the study of perception and behavior. We shall treat the vegetation of the world as animals do, as if it were lumped together with the inorganic minerals of the world, with the physical, chemical, and geological environment. Plants in general are not animate; they do not move about, they do not behave, they lack a nervous system, and they do not have sensations. In these respects they are like the objects of physics, chemistry, and geology. (1979, 7)

24. This should appear obvious to many since the pioneering research on plants by Charles Darwin and his son. Nevertheless, despite their groundbreaking work (*The Movements and Habits of Climbing Plants* and *The Power of Movement in Plants*), conventional cognitive science has continued to ignore the perceptual and behavioral capacities of plants over a century later.

25. For a recent reinterpretation of the early evolution of nervous systems and what they can do that is congenial to our treatment of plants, see Keijzer et al. 2013.

26. Alpi et al. (2007) complain that "there is no evidence for structures such as neurons, synapses or a brain in plants" (136). For clarification, see Brenner et al. 2007 and Trewavas 2007.

27. Here we have in mind directional responses, but the same ecological principles may hold in the guidance of nondirectional (nastic) responses, such as the thigmonastic response of the Venus flytrap (*Dionaea muscipula*) and other carnivorous plants when they close their traps in response to touch.

28. Interestingly, Frantisek Baluska (personal communication) observes that growing roots perform circumnutation movements that in a sense resemble the waggle dance of bees, although of course not in the information-processing sense that Carruthers, for instance, would endorse.

29. For other possible forms of communication between root tips, see Baluška, Lev-Yadun, and Mancuso 2010.

30. Heft's work on ecological navigation is illustrative: "A commonly held view is that knowledge of environmental configuration must be based on cognitive operations that construct a mental representation, or a 'cognitive map,' from discontinuous perceptual encounters. Such a constructivist account seems to be required because the overall layout cannot be perceived from any single location in the environment. ... In contrast to this position, the most radical aspect of Gibson's treatment of navigation is his claim that by following paths through the environment, eventually one does come to perceive the overall layout of the environment" (Heft 1996, 124).

31. On interaction-dominance, see Chemero, this volume.

32. Gomila et al. (2012) defend this view, although see Martínez-Manrique (this volume) for an alternative.

33. We do not have the space to elaborate further on this (see Calvo et al. 2012), although we suspect that by allowing top-down processing to inform the output of the emergentist bottom-up part of the model, the price to pay will be subsumption under Fodor and Pylyshyn's charge of implementation.

34. Since then, a number of authors have made further efforts to conciliate the principles of ecological psychology with language. Verbrugge (1985) approaches language in terms of the direct perception of speech as a type of event subject to ecological laws (acoustical laws, in the case of verbal speech). Reed (1987) tries a quasi-grammatical approach to language that is congenial with Chomskian principles (Noble 1993). We shall not review this literature here (see Hodges and Fowler 2010; Fowler and Hodges 2011, for further insights).

35. Noble (1993), who elaborates on the evolutionary emergence of language from a neo-Gibsonian stance, dubs this the "language is special" doctrine.

36. Of course, such a supracommunicative role is already present in Vygotskyan approaches (Vygotsky 1978).

37. See also Travieso and Jacobs 2009.

38. Symons (2001) presents an early version of this position,. arguing that systematic patterns of intelligent behavior do not necessarily license the view that internal representations play a role in the cognitive system analogous to that played by syntactic structures in a computer program. Linguistic structures are instead regarded as external targets for the development of individual brains.

References

Allen, P. H. 1977. *The Rain Forests of Golfo Dulce*. Stanford: Stanford University Press.

Alpi, A., N. Amrhein, A. Bertl, et al. 2007. Plant neurobiology: No brain, no gain? *Trends in Plant Science* 12 (4):135–136.

Baldwin, I. T., R. Halitschke, A. Paschold, C. C. von Dahl, and C. A. Preston. 2006. Volatile signaling in plant-plant interactions: "Talking trees" in the genomics era. *Science* 311 (5762):812–815.

Baluška, F., S. Mancuso, and D. Volkmann, eds. 2006. *Communication in Plants: Neuronal Aspects of Plant Life*. Berlin: Springer-Verlag.

Baluška, F., S. Mancuso, D. Volkmann, and P. W. Barlow. 2004. Root apices as plant command centres: The unique "brain-like" status of the root apex transition zone. *Biologia* 59:9–17.

Baluška, F., S. Mancuso, D. Volkmann, and P. W. Barlow. 2009. The "root-brain" hypothesis of Charles and Francis Darwin: Revival after more than 125 years. *Plant Signaling & Behavior* 4 (12):1121–1127.

Baluška, F., Si. Lev-Yadun, and S. Mancuso. 2010. Swarm intelligence in plant roots. *Trends in Ecology & Evolution* 25 (12):682–683.

Baluška, F., S. Mancuso, D. Volkmann, and P. W. Barlow. 2010. Root apex transition zone: A signalling–response nexus in the root. *Trends in Plant Science* 15:402–408.

Barlow, P. W. 2010. Plastic, inquisitive roots, and intelligent plants in the light of some new vistas in plant biology. *Plant Biosystems* 144:396–407.

Bechtel, W., and A. Abrahamsen. 2002. *Connectionism and the Mind: Parallel, Processing, Dynamics, and Evolution in Networks*, 2nd ed. Oxford: Blackwell.

Brenner, E. D., R. Stahlberg, S. Mancuso, J. M. Vivanco, F. Baluška, and E. van Volkenburgh. 2006. Plant neurobiology: An integrated view of plant signaling. *Trends in Plant Science* 11 (8):413–419.

Brenner, E. D., R. Stahlberg, S. Mancuso, F. Baluška, and E. van Volkenburgh. 2007. Response to Alpi et al.: Plant neurobiology: The gain is more than the name. *Trends in Plant Science* 12 (7):285–286.

Burbach, C., K. Markus, Z. Yin, M. Schlicht, and F. Baluška. 2012. Photophobic behavior of maize roots. *Plant Signaling and Behavior* 7 (7):1–5.

Calvo, F., and E. Colunga. 2003. The statistical brain: Reply to Marcus' *The Algebraic Mind*. In *Proceedings of the Twenty-Fifth Annual Conference of the Cognitive Science Society*, ed. Richard Alterman and David Kirsh, 210–215. Mahwah, NJ: Erlbaum.

Calvo Garzón, P., and F. Keijzer. 2011. Plants: Adaptive behavior, root brains, and minimal cognition. *Adaptive Behavior* 19:155–171.

Calvo, P., J. Symons, and E. Martín. 2012. Beyond "error-correction." *Frontiers in Psychology* 3:423.

Carruthers, P. 2004. On being simple minded. *American Philosophical Quarterly* 41:205–220.

Chalmers, D. J. 1990. Syntactic transformations on distributed representations. *Connection Science* 2:53–62.

Chomsky, N. 1959. A review of B. F. Skinner's *Verbal Behavior. Language* 35 (1): 26–58.

Churchland, P. S. 1986. *Neurophilosophy: Toward a Unified Science of the Mind Brain.* Cambridge, MA: MIT Press.

Churchland, P. S. 2002. *Brain-wise: Studies in Neurophilosophy.* Cambridge, MA: MIT Press.

Ciszak, M., D. Comparini, B. Mazzolai, F. Baluška, F. T. Arecchi, et al. 2012. Swarming behavior in plant roots. *PLoS ONE* 7 (1): e29759. doi:10.1371/journal. pone.0029759.

Clark, Andy. 1997. *Being There.* Cambridge, Mass.: MIT Press.

Clark, Andy. 1998. Magic words: How language augments human computation. In *Language and Thought: Interdisciplinary Themes,* ed. Peter Carruthers and Jill Boucher. Cambridge, Mass: MIT Press.

Clark, A. 2013. Whatever next? Predictive brains, situated agents, and the future of cognitive science. *Behavioral and Brain Sciences* 36 (3):1–73.

Collett, T. S., and M. Collett. 2002. Memory use in insect visual navigation. *Nature Reviews: Neuroscience* 3:542–552.

Costall, A. 1995. Socializing affordances. *Theory and Psychology* 5:467–482.

Cummins, R. 1996. Systematicity. *Journal of Philosophy* 12 (93):591–614.

Darwin, C. 1875. *Insectivorous Plants.* New York: Appleton.

Delafield-Butt, J. T., G.-J. Pepping, C. D. McCaig, and D. N. Lee. 2012. Prescriptive guidance in a free-swimming cell. *Biological Cybernetics* 106 (4–5):283–293.

Dennett, D. C. 1995. *Darwin's Dangerous Idea: Evolution and the Meanings of Life.* New York: Simon & Schuster.

Dyer, F. C., and J. A. Dickinson. 1994. Development of sun compensation by honeybees: How partially experienced bees estimate the sun's course. *Proceedings of the National Academy of Sciences of the United States of America* 91:4471–4474.

Elman, J. L. 1999. *Generalization, Rules, and Neural Networks: A Simulation of Marcus et al.* University of California, San Diego. http://www.crl.ucsd.edu/~elman/Papers/ MVRVsim.html.

Esch, H. E., S. Zhang, M. V. Srinivasan, and J. Tautz. 2001. Honeybee dances communicate distances measured by optic flow. *Nature* 411:581–583.

Fauconnier, G., and M. Turner. 1998. Conceptual integration networks. *Cognitive Science* 22 (2):133–187.

Fodor, J. A., and Z. W. Pylyshyn. 1981. How direct is visual perception? Some reflections on Gibson's "Ecological Approach." *Cognition* 9:139–196.

Fodor, J. A., and Z. W. Pylyshyn. 1988. Connectionism and cognitive architecture: A critical analysis. *Cognition* 28:3–71.

Fowler, C. 1986. An event approach to the study of speech perception from a direct-realist perspective. *Journal of Phonetics* 14:3–28.

Fowler, C., and B. Hodges. 2011. Dynamics and languaging: Toward an ecology of language. *Ecological Psychology* 23:147–156.

Gagliano, M., S. Mancuso, and D. Robert. 2012. Towards understanding plant bioacoustics. *Trends in Plant Science* 17 (6):323–325.

García, R. A., and P. Calvo Garzón. 2010. Is cognition a matter of representations? Emulation, teleology, and time-keeping in biological systems. *Adaptive Behavior* 18:400–415.

Gerken, L. A., and R. N. Aslin. 2005. Thirty years of research on infant speech perception: The legacy of Peter W. Jusczyk. *Language Learning and Development* 1 (1):5–21.

Gibson, J. J. 1966. *The Senses Considered as Perceptual Systems*. Boston: Houghton Mifflin.

Gibson, J. J. 1979. *The Ecological Approach to Visual Perception*. Boston: Houghton Mifflin.

Gomila, A., D. Travieso, and L. Lobo. 2012. Wherein is human cognition systematic? *Minds and Machines* 22 (2):101–115.

Goodale, M. A., and D. Milner. 1992. Separate visual pathways for perception and action. *Trends in Neurosciences* 15:20–25.

Griffiths, T. L., N. Chater, C. Kemp, A. Perfors, and J. B. Tenenbaum. 2010. Probabilistic models of cognition: Exploring the laws of thought. *Trends in Cognitive Sciences* 14:357–364.

Heft, H. 1996. The ecological approach to navigation: A Gibsonian perspective. In *The Construction of Cognitive Maps*, ed. J. Portugali, 105–132. Dordrecht: Kluwer Academic.

Hodges, B., and C. Fowler. 2010. New affordances for language: Distributed, dynamical, and dialogical resources. *Ecological Psychology* 22:239–253.

Isnard, S., and W. K. Silk. 2009. Moving with climbing plants from Charles Darwin's time into the 21st century. *American Journal of Botany* 96 (7):1205–1221.

Keijzer, F., M. van Duijn, and P. Lyon. 2013. What nervous systems do: Early evolution, input–output, and the skin brain thesis. *Adaptive Behavior* 21 (2):67–85.

Kim, A. 2009. Early experimental psychology. In *Routledge Companion to Philosophy of Psychology*, ed. J. Symons and P. Calvo, 41–59. London: Routledge.

Kujala, T. 2001. Brain science: A more direct way of understanding our senses. *Behavioral and Brain Sciences* 24 (2):224.

Lee, D. N. 1976. A theory of visual control of braking based on information about rime-to-collision. *Perception* 5:437–459.

Lee, D. N. 1980. The optic flow field: The foundation of vision. *Philosophical Transactions of the Royal Society of London, Series B: Biological Sciences* 290:169–179.

Lee, D. N. 1998. Guiding movement by coupling raus. *Ecological Psychology* 10:221–250.

Lee, D. N. 2004. Tau in action in development. In *Action, Perception, and Cognition in Learning and Development*, ed. J. J. Rieser, J. J. Lockman, and C. A. Nelson, 3–49. Hillsdale, NJ: Erlbaum.

Lee, D. N. 2009. General Tau Theory: Evolution to date. *Perception (Special Issue: Landmarks in Perception)* 38:837–858.

Lee, D. N., and P. E. Reddish. 1981. Plummeting gannets: A paradigm of ecological optics. *Nature* 293:293–294.

Lee, D. N., P. E. Reddish, and D. T. Rand. 1991. Aerial docking by hummingbirds. *Naturwissenschaften* 78:526–527.

Lee, D. N., J. A. Simmons, P. A. Saillant, and F. H. Bouffard. 1995. Steering by echolocation: A paradigm of ecological acoustics. *Journal of Comparative Physiology, Series A: Neuroethology, Sensory, Neural, and Behavioral Physiology* 176:347–354.

Llinás, R., and P. S. Churchland, eds. 1996. *The Mind-Brain Continuum*. Cambridge, MA: MIT Press.

Mackie, G. O. 1970. Neuroid conduction and the evolution of conducting tissues. *Quarterly Review of Biology* 45:319–332.

Marcus, G. F., S. Vijayan, S. B. Rao, and P. M. Vishton. 1999. Rule learning by seven-month-old Infants. *Science* 283:77–80.

Marcus, G. F. 2001. *The Algebraic Mind*. Cambridge, MA: MIT Press.

Masi, E., M. Ciszak, G. Stefano, L. Renna, E. Azzarello, C. Pandolfi, et al. 2009. Spatiotemporal dynamics of the electrical network activity in the root apex. *Proceedings of the National Academy of Sciences USA* 106: 4048–4053.

McClelland, J. L., M. M. Botvinick, D. C. Noelle, D. C. Plaut, T. T. Rogers, M. S. Seidenberg, and L. B. Smith. 2010. Letting structure emerge: Connectionist and dynamical systems approaches to cognition. *Trends in Cognitive Sciences* 14:348–356.

McLaughlin, B. 1993. The connectionism/classicism battle to win souls. *Philosophical Studies* 71:163–190.

McLaughlin, B. 2009. Systematicity redux. *Synthese* 170 (2):251–274.

Michaels, C. F., and C. Carello. 1981. *Direct Perception*. Englewood Cliffs, NJ: Prentice-Hall.

Mithöfer, A., and W. Boland. 2012. Plant defense against herbivores: Chemical aspects. *Annual Review of Plant Biology* 63:431–450.

Noble, W. 1993. What kind of approach to language fits Gibson's approach to perception? *Theory & Psychology* 3 (1):57–78.

Norman, J. 2002. Two visual systems and two theories of perception: An attempt to reconcile the constructivist and ecological approaches. *Behavioral and Brain Sciences* 25:73–144.

Pollack, J. B. 1990. Recursive distributed representations. *Artificial Intelligence* 46: 77–105.

Reed, E. 1987. James Gibson's ecological approach to cognition. In *Cognitive Psychology in Question*, ed. A. Costall and A. Still, 142–173. Sussex: Harvester Press.

Richardson, M. J., K. Shockley, B. R. Fajen, M. A. Riley, and M. Turvey. 2008. Ecological psychology: Six principles for an embodied-embedded approach to behavior. In *Handbook of Cognitive Science: An Embodied Approach*, ed. P. Calvo and A. Gomila, 161–190. Oxford: Elsevier.

Rock, I. 1983. *The Logic of Perception*. Cambridge, MA: MIT Press.

Ronacher, B., and R. Wehner. 1995. Desert ants *Cataglyphis Fortis* use self-induced optic flow to measure distances travelled. *Journal of Comparative Physiology, Series A: Neuroethology, Sensory, Neural, and Behavioral Physiology* 177:21–27.

Schenk, H. J., R. Callaway, and B. Mahall. 1999. Spatial root segregation: Are plants territorial? *Advances in Ecological Research* 28:145–180.

Schogler, B., G.-J. Pepping, and D. N. Lee. 2008. TauG-guidance of transients in expressive musical performance. *Experimental Brain Research* 198:361–372.

Smolensky, P. 1987. The constituent structure of connectionist mental states. *Southern Journal of Philosophy* 26:37–60.

Smolensky, P. 1990. Tensor product variable binding and the representation of symbolic structures in connectionist systems. *Artificial Intelligence* 46:159–216.

Srinivasan, M. V., S. Zhang, M. Altwein, and J. Tautz. 2000. Honeybee navigation: Nature and calibration of the "odometer." *Science* 287:851–853.

Stoffregen, T. A., and B. G. Bardy. 2001. On specification and the senses. *Behavioral and Brain Sciences* 24:195–261.

Symons, J. 2001. Explanation, representation, and the dynamical hypothesis. *Minds and Machines* 11 (4):521–541.

Symons, J. 2002. *On Dennett.* Belmont: Wadsworth.

Symons, J. 2007. Understanding the complexity of information processing tasks in vision. In *Philosophy and Complexity: Essays on Epistemology, Evolution, and Emergence,* ed. C. Gershenson, D. Aerts, and B. Edmonds, 300–314. Singapore: World Scientific.

Travieso, D., and D. Jacobs. 2009. The ecological level of analysis: Can neogibsonian principles be applied beyond perception and action? *Integrative Psychological and Behavioral Science* 43:393–405.

Tresilian, J. R. 1999. Visually timed action: Time-out for "tau"? *Trends in Cognitive Sciences* 3 (8):301–310.

Trewavas, A. J. 2003. Aspects of plant intelligence. *Annals of Botany* 92:1–20.

Trewavas, A. J. 2007. Plant neurobiology: All metaphors have value. *Trends in Plant Science* 12:231–233.

Trewavas, A. J. 2008. Aspects of plant intelligence: Convergence and evolution. In *The Deep Structure of Biology: Is Convergence Sufficiently Ubiquitous to Give a Directional Signal?,* ed. S. C. Morris, 68–110. West Conshohocken, PA: Templeton Press.

Turvey, M., R. Shaw, E. S. Reed, and W. Mace. 1981. Ecological laws for perceiving and acting: A reply to Fodor and Pylyshyn. *Cognition* 10:237–304.

Tye, M. 1997. The problem of simple minds: Is there anything it is like to be a honey bee? *Philosophical Studies: An International Journal for Philosophy in the Analytic Tradition* 88:289–317.

van Gelder, T. 1990. Compositionality: A connectionist variation on a classical theme. *Cognitive Science* 14:355–384.

Verbrugge, R. 1985. Language and event perception: Steps towards a synthesis. In *Event Perception,* ed. W. Warren and R. Shaw, 157–193. Hillsdale, NJ: Erlbaum.

Visscher, P. K. 2003. Animal behaviour: How self-organization evolves. *Nature* 421:799–800.

von Frisch, K. 1993. *The Dance Language and Orientation of Bees.* Cambridge, MA: Harvard University Press.

Vygotsky, L. 1978. Interaction between learning and development. In *Mind and Society,* 79–91. Cambridge, MA: Harvard University Press.

17 Order and Disorders in the Form of Thought: The Dynamics of Systematicity

Michael Silberstein

Introduction

Fodor, for all his recent changes of mind about how the mind works, is still convinced that systematicity is real and cannot be explained in connectionist terms (Fodor 2000, 50). On the other hand, there are those who doubt the very existence of linguistic systematicity à la Fodor and Pylyshyn (e.g., Chemero, this vol.). It would be nice if we could make some empirical headway on this debate. The idea behind this chapter, then, is to exploit the fact that various related sorts of systematicity appear to be very often damaged in those with schizophrenia. Thus we have a real live case of something like systematicity going awry and can study the effects on cognition and behavior. So this is another instance of the now popular strategy to use psychopathology to shed light on theories of standard cognitive functioning. Furthermore, evidence is mounting that one important explanation for said failure of systematicity is precisely in terms of subsymbolic network properties or their absence. This suggests that, contra Fodor, although something like systematicity may be real, it need not be explained exclusively in terms of the symbolic and rules thereof. This leaves open the possibility that systematicity can be characterized and explained, at least partially, in terms of dynamical and network properties. As Abrahamsen and Bechtel put it: "If one is seeking maximal contrast to symbolic rules and representations, it is to be found not in the pastiche of connectionism but rather within the tighter confines of one of its tributaries, mathematical modeling ... the dynamical approach to perception, action and cognition" (2006, 169). Of course, there is no reason why such cognitive dynamical systems cannot for some explanatory purposes have both symbolic and subsymbolic aspects, broadly conceived.

Aside from the kind of linguistic systematicity that vexed Fodor and Pylyshyn (1988) with their specific concerns about the combinatorics of semantic content (i.e., compositionality) and the language of thought (LOT), there is possibly a much more general kind of systematicity that philosophers of cognitive science sometimes call transformational systematicity of thought, that is, the logical, causal, temporal and narrative coherence of thought as a process. Failures of transformational systematicity sometimes manifest as failures of so-called linguistic systematicity; more will be said about the possible relationships in what comes. However, unlike linguistic systematicity, it's clear that transformational systematicity really does exist and really does require explanation, because it's precisely what some schizophrenics and perhaps some other psychotics with formal thought disorders lack (Murphy 2006; Kitcher and David 2003; Radden 2004; Graham 2010).

The focus here will be on formal thought disorders in schizophrenics because such disorders of the form (as opposed to content) of thought are considered symptomatic of the disorder (Lieberman, Stroup, and Perkins 2006). Thought disorders are consistently manifested and diagnosed in both speech and writing. It is therefore customary in the literature to discuss language explicitly when explicating thought disorders and to discuss thought when explicating abnormalities of language. Therefore, our discussion will often toggle back and forth without warning between thought and language. It must be acknowledged that because schizophrenia is so heterogeneous in symptomology, not obviously a unitary disease, perhaps not even a natural kind and so poorly understood in general, that it is not an ideal basis for doing the philosophical psychopathology of systematicity. Unfortunately, however, schizophrenia is the best case we have for the study of systematicity, so we will have to make do.

In section 1, I give a characterization of formal thought disorders with enough detail to appreciate why they entail a failure of inferential or transformational systematicity of thought. In section 2, I canvas recent attempts by systems neuroscience to explain formal thought disorders using dynamical systems theory and graph theory. Finally, section 3 will make clear that what is doing the explanatory work with regard to transformational systematicity or its failure in these systems neuroscience models has nothing to do with representations with a combinatorial structure or anything LOT-like. Rather, the explanation has to do with the dynamical and graphical properties of the neural networks in question, and perhaps even more spatiotemporally extended networks. This account of transformational systematicity will involve the machinery of dynamical systems theory such as stable attractors, order-parameters, and the like, and graph theory such as small-world networks (Sporns 2011; Silberstein and Chemero forthcoming). Fodor and Pylyshyn (1988) claim that connectionism and neural network models fail precisely because they cannot account for linguistic systematicity. Their argument is that systematicity requires representations with compositional structure and connectionist networks do not have representations with compositional structure. This is quite ironic, if the systems neuroscience explanations are at least partially on target, because such explanations appeal essentially to the kinds of network properties that Fodor and Pylyshyn (1988) claim cannot explain systematicity.

1 Transformational Systematicity and Formal Thought Disorders

The linguistic kind of systematicity that Fodor and Pylyshyn (1988) are primarily concerned with can be broadly defined as follows:

Systematicity of cognitive representation: the capacity for having certain thoughts is "intrinsically connected" to the capacity to have certain other thoughts. (Aizawa 2003)

Fodor and McLaughlin (1990) would later define systematicity as follows:

As a matter of psychological law, an organism is able to be in one of the states belonging to the family only if it is able to be in many of the others. ... You don't find organisms that can think the thought that the girl loves John but can't think the thought that John loves the girl. (Fodor and McLaughlin 1990, 184)

The overriding assumption here is that linguistic systematicity requires compositionality. Furthermore, for thought to be systematic, its representations must be composed of syntactic components that have semantics, and those components must be recombinable to form other representations with the semantics of the components intact.

Transformational systematicity is much more than just linguistic systematicity, however. As Aydede (1995) notes, it also involves "inferential coherence" more than just compositionality:

Systematicity of thought is not restricted to the systematic ability to entertain certain thoughts. If the system of mental representations does have a combinatorial syntax, then there is a set of rules, syntactic formation rules, that govern the construction of well-formed expressions in the system. It is this fact that guarantees formative systematicity. But inferential thought processes are systematic too: the ability to make certain inferences is intrinsically connected to the ability to make many others. For instance, according to the classicist, you do not find minds that can infer "A" from "A&B" but cannot infer "C" from "A&B&C." Again, it is a nomological psychological fact that inferential capacities come in clusters that are homogeneous in certain aspects. How is this fact, which I will call inferential (or, sometimes, transforma-tional) systematicity, to be explained? (Aydede 1995, 4)

Following Aizawa (2003), we can then define systematicity of inference as follows: inferences of similar logical type generally ought to elicit corre-spondingly similar cognitive capacities, a capacity to perform one instance of a given type of inference given the capacity to perform another instance. However, transformational systematicity is more than this. It includes inferential coherence more broadly as well as temporal, causal, and narra-tive coherence.

The best way to appreciate this is to look at formal thought disorders in schizophrenia. Formal thought disorder (an abnormality in the form or structure of thought as opposed to the content) is considered diagnostic of schizophrenia in the absence of obvious brain trauma. Again, formal thought disorders are often diagnosed and discussed via linguistic and written output; therefore, both language and thought will be considered here. One well-known example is Dysexecutive syndrome: a breakdown in the unity of consciousness wherein subjects are not able to consider two obviously related things at the same time. As Raymont and Brook put it, "If the person has any representation of the second item at all, it is not unified with consciousness of the first one" (2009, 572). This particular syndrome is perhaps closest to the linguistic system. However, this is only one manifestation of a formal thought disorder.

Disturbances in the form of thought are disturbances in the logical process of thought—more simply, disturbances in the logical connections between ideas. Illogicality is present where, for example, we find erroneous conclusions, internal contradictions, or incoherence in one's thinking because of the loss of logical connections (McKenna 1997). Other specific and related examples of disorders of the form of thought include:

Derailment or loose association—Schizophrenics' ideas wander off track to another issue that is obliquely related or unrelated, deviating from the topic at hand. This is representative of a broader tendency in schizophrenic thought and language to include peculiar associations, the spreading and loosening of associations, expansions of meaning, tangents, neologisms, conceptual contamination, desymbolization of words, and so on (Tandon, Nasrallah, and Keshavan 2009). Sometimes the spreading of associations is exacerbated by the overwrought attention to the sound and many senses (connotations) of a word. In addition to phonetic issues, there are also problems with access to lexicon. Such language in extreme cases involves disorganized sentences, with no obvious logical or causal relationship between them. Such language is suggestive of conceptual disorganization and perhaps cognitive disorientation more generally.

Loss of goal—Schizophrenics often fail to take a train of thought to its natural conclusion. Sometimes this involves stereotyped thinking (empty repetition or inability to move beyond very rigid themes, which are returned to over and over). Such speech is often logically and semantically disordered, and filled with irrelevant information. In addition, relevant information is often ignored, conclusions are jumped to, and inductive inferences are poor (Tandon, Nasrallah, and Keshavan 2009). This failure to maintain a plan of discourse, or loss of voluntary control of speech, is perhaps indicative of larger cognitive deficits pertaining to action, planning, ordering, and sequencing (Mujica-Parodi, Malaspina, and Sackheim 2000).

Inflexible reasoning—Schizophrenic reasoning is often either overly concrete or overly abstract. Understanding the meaning of nonliteral utterances is often problematic, as is overcomplicating the meaning of literal utterances (Mujica-Parodi, Malaspina, and Sackheim 2000).

Thus, we know that transformational systematicity is real because it is deformed or absent in the case of disturbances in the form of thought. This begs the question of what causes such a disorder.

Before we explore the possible answers to that question, several caveats must be mentioned. First, reasoning is not a homogeneous cognitive function even within the categories of inductive and deductive reasoning there are many different types. While our concern here is primarily deductive reasoning, it is difficult to determine if the deficit in question is a general one or more task/type specific. Indeed, DSM-IV aside, it is not even a settled matter if schizophrenics have a general problem with deductive reasoning; the data on this issue are conflicting and inconclusive (Mujica-Parodi, Malaspina, and Sackheim 2000). For example, there is some evidence to suggest that only some types of logic problems elicit the deficits in reasoning in question. Some studies allege to show that the issues arise with logic problems involving sentences of a natural language (such as syllogisms) but not with purely symbolic problems (ibid.).

Second, in addition to reasoning and problem solving, schizophrenics exhibit several neurocognitive deficits with regard to executive functions, including all phases of memory, attention, affect, language, and much more. Given that each of these executive functions requires many of the others to be operating normally as at least supporting conditions, it is very hard to determine with any degree of certainty exactly where the problem lies—it is difficult to isolate specific causal variables. Going forward, then, the question is how shall we measure deficits in deductive reasoning and logic apart from all the deficits mentioned above (Mirian, Heinrichs, and McDermid Vaz 2011).

Given the preceding caveats, several possibilities remain open. (A) Assuming that deficits in deductive reasoning are indeed robust and well confirmed, are they evidence of more generalized impairments in abstract thinking or are they truly specific to deduction? (B) Are such deficits in reasoning really about the form of thought, that is, deductively invalid or inductively weak arguments, or are they caused by problems in the initial stages of reasoning such as encoding, acquisition and evaluation of premises, and so on (Mirian, Heinrichs, and McDermid Vaz 2011)? (C) Are the deficits in deduction instead symptomatic of the various abnormalities in language associated with the disease? For example, perhaps it is anomalies in the pragmatics of linguistic functions such as the inability to appreciate context driven rules and various local and global constraints on meaning and use that are the real problem (Marini et al. 2008). It is well known that such problems arise for analogical reasoning when schizophrenics attempt to initially map an analogue onto its target (Simpson and Done 2004). (D) Rather than deduction as such, are the deficits really a function of problems with memory, planning, attention, or some other executive function? (E) As there are delusional and nondelusional schizophrenics, hallucinating and nonhallucinating, paranoids and non-paranoids, and so on, are the problems with reasoning perhaps not about the neurocognitive disease driven defects as such, but about those other aspects of the disease? Maybe the deficits in logic are driven by the formation and maintenance of delusional beliefs, for instance. Some people have even suggested that schizophrenics have their own non-truth-preserving logic that we have yet to codify (Mujica-Parodi, Malaspina, and Sackheim 2000). (F) Finally, are deficits in reasoning caused by affective abnormalities such as to do with emotionally laden discourse that raises concerns about belief congruency or saliency (Marini and et al. 2008)?

Clearly, much more work needs to be done to resolve these questions. Better testing is required to isolate deficits in deductive reasoning from all the other possibilities, assuming for the moment that logical reasoning is in fact separable from these other functions. We need testing that goes beyond the usual methods of object-sorting tasks, syllogistic reasoning tasks, analogical reasoning tasks, assessing the interpretation of nonliteral sentences such as proverbs, metaphors, and so on. Logic and reasoning tests that approximate real-world reasoning in the wild would be helpful.

In short, more needs to be done to establish that deficits in the form of reasoning are correlated with or symptomatic of disorders of thought. For the purposes of what follows, however, we will assume that this connection is well documented.

2 Explaining Formal Thought Disorders: Dynamical Systems Theory and Systems Neuroscience

The question of what causes disorders in the form of thought is quite tricky, as we have just seen. Furthermore, schizophrenia has heterogeneous symptoms (both "positive" and "negative") in addition to formal thought disorders, affecting both central and peripheral systems. Positive symptoms are usually defined in terms of involving reality distortion, such as hallucinations and delusions, whereas the negative symptoms involve neurocognitive deficits. Interestingly, thought disorders could go either way in this classification system (Ventura, Thames, Wood, et al. 2010). The causes and determining factors of schizophrenia generally are widely regarded to be multicausal and interlevel, with a myriad of genetic, epigenetic, neurological, and environmental factors (leading to long period of abnormal neural development, i.e., neurodevelopmental pathogenesis [Spence 2009]). Some theorists hold that the brain abnormalities involved in schizophrenia derive from too much and/or too little synaptic pruning in development. For example, some people argue that while the positive symptoms in schizophrenia may be caused by the hyperconnectivity of certain brain regions ("pathological blooming"), the negative symptoms are caused by too much pruning—"aberrant pruning" (Rao et al. 2010). However, even if this is the case, the question remains as to the causes of the pruning abnormalities. There is definite genetic susceptibility, but no particular genes have been identified (Sun, Jia, Fanous, van den Oord, Chen, et al. 2010). While there are certainly biomedical models, epigenetic models, and biopsychosocial models of schizophrenia, at the present moment the biopsychosocial models seem the best supported (Spence 2009; Murphy 2006; Radden 2004). In short, schizophrenia seems to demand explanatory pluralism. Again, gene expression interacting with environmental causes (physical, biological, and social) at particular stages of development seems to be the best explanation. Nor do all the environmental causes or catalysts fall only during the time of brain development. For example, one well-known factor or correlate is social stress, such as being an immigrant to a new country (Eagleman 2011, 211). However, even restricting explanation to the brain, it is believed that the "mechanisms of schizophrenia," if you will, are multiply realized (Murphy 2006). While there are often structural abnormalities in the brains of schizophrenics such as thalamus volume deficits, there is no consensus about their universality, uniqueness, or explanatory force: there are no correlational laws as of yet and no localization involving such abnormalities that are widely regarded to explain schizophrenia (Spence 2009).

Models of schizophrenia range from the piecemeal to the global. Piecemeal models tackle things symptom by symptom, whereas global models seek some very general property of cognition with common effects on different capacities. Piecemeal models tend to seek local or "monotopical'" approaches that try and map particular symptoms or "subsyndromes" to structural abnormalities in certain brain regions in a one-to-one fashion. For example, thought disorders get mapped to hippocampal abnormalities (Volz et al. 2000). Cognitive dysmetria is a good example of a global model wherein it is

hypothesized that connectivity is disturbed among nodes located in the prefrontal regions, the thalamic nuclei and the cerebellum. A disruption in this circuitry might produce "cognitive dysmetria," resulting in difficulties to prioritize, to process and to coordinate, as well as to respond to information. This hypothesis is in accordance with the suggestion that schizophrenia is not caused by a single structural brain defect but by alterations of critical neuronal networks. (ibid., 46).

Theorists have been attempting to apply dynamical systems theory (focusing on nonlinear systems) and chaos theory in particular to schizophrenia at least since the 1990s; specifically through the study of recurrent neural networks. Central questions being posed in this literature include: (1) Are chaotic attractors diagnostic markers for schizophrenia? Some claim, for example, that when dynamical methods are applied to physiological brain signals one can show that complex dynamics are related to healthy mental states whereas simple, that is, chaotic dynamics are related to pathology. And (2) more generally, is schizophrenia a dynamical disease? In other words, can aspects of schizophrenia such as formal thought disorders be usefully and partially explained on the basis of the concepts from nonlinear dynamical systems in general or chaos theory in particular (Schmid 1991; Pezard and Nandrino 2001; Heiden 2006). The working assumption here is that changes in brain dynamics can be explanatorily correlated with changes in mental processes. It has been hypothesized, for instance, that abnormalities in cortical pruning during development could cause a decrease in neural network storage ability and lead to the creation of "spurious" attractors. Other attempts use the machinery of phase transitions and order-parameters to try and model various abnormalities in schizophrenia. For example, in one very simple model the order-parameter is the level of dopamine transmission wherein "the effects of dopamine are demonstrated by a mathematical model which can be interpreted in two ways: on the one hand as a model of typical excitatory-inhibitory circuits in the cortex, on the other hand of a negative feedback loop between thalamus, prefrontal cortex and striatum. The model exhibits different types of firing patterns and their bifurcation, from various kinds of periodicities up to erratic or chaotic behavior, corresponding to different levels of dopamine concentration" (Heiden 2006, 36). The problem with all these models has been that they are much too idealized and simplistic, and quite hard to compare with actual physiological data.

Fortunately, systems neuroscience has evolved further in the last few years (Sporns 2011; Silberstein and Chemero forthcoming[1]). It is an outgrowth of earlier attempts to apply dynamical systems theory to the brain. Systems neuroscience is attempting to determine how the brain engages in the coordination and integration of distributed processes at the various length and timescales necessary for cognition and action. The assumption is that most of this coordination represents patterns of spontaneous, self-organizing, macroscopic spatiotemporal patterns, which resemble the on-the-fly functional networks recruited during activity. This coordination often occurs at extremely fast timescales with short durations and rapid changes. There is a wide repertoire of models used to account for these self-organizing macroscopic patterns, such as oscillations, synchronization, metastability, and nonlinear dynamical coupling. Many explanatory models such as synergetics and neural dynamics combine several of these features, for example, phase-locking among oscillations of different frequencies (Sporns 2011).

Despite the differences among these models, we can make some important generalizations. First, dynamic coordination is often highly distributed and nonlocal. Second, population coding, cooperative, or collective effects prevail. Third, time and timing is essential in a number of ways. Fourth, these processes exhibit both robustness and plasticity. Fifth, these processes are highly context and task sensitive. Regarding the third point, there is a growing consensus that such integrated processes are best viewed not as vectors of activity or neural signals, but as dynamically evolving graphs. The evidence suggests that standard neural codes such as rate codes and firing frequencies are insufficient to explain the rapid and rapidly transitioning coordination. Rather, the explanation must involve "temporal codes" or "temporal binding," such as spike timing-dependent plasticity, wherein neural populations are bound by the simultaneity of firing and precise timing is essential. In these cases, neurons are bound into a group or functional network as a function of synchronization in time. The key explanatory features of such models then involve various time-varying properties such as the exact timing of a spike, the ordering or sequencing of processing events, the rich moment-to-moment context of real-world activity and immediate stimulus environment, an individual's history such as that related to network activation and learning, and so on. All of the above can be modeled as attractor states that constrain and bias the recruitment of brain networks during active tasks and behavior (Sporns 2011; Von der Malsburg, Philips, and Singer 2010).

There is now a branch of systems neuroscience devoted to the application of network theory to the brain. The formal tools of network theory are graph theory and dynamical systems theory, the latter to represent network dynamics—temporally evolving dynamical processes unfolding in various kinds of networks. While these techniques can be applied at any scale of brain activity, here we will be concerned with large-scale brain networks. These relatively new to neuroscience explanatory tools (i.e., simulations) are enabled by large data sets and increased computational power. The brain is modeled as a complex system: networks of both linear and nonlinear interacting components such as neurons, neural assemblies, and brain regions. In these models, rather than viewing the neurons, cell groups, or brain regions as the basic unit of explanation, brain multiscale networks and their large-scale, distributed, and nonlocal connections or interactions are the basic unit of explanation (Sporns 2011). The study of this integrative brain function and connectivity is primarily based in topological features (network architecture) of the network that are insensitive to, and have a one-to-many relationship with respect to, lower-level neurochemical and wiring details. At this level of analysis, all that matters is the topology: the pattern of connections. Different geometries (arrangement of nodes and edges) can instantiate the same topology. More specifically, a graph is a mathematical representation of some actual (in this case) biological many-bodied system. The nodes in these models represent neurons, cell populations, brain regions, and so on, and the edges represent connections between the nodes. The edges can represent structural features such as synaptic pathways and other wiring diagram type features, or they can represent more functional topological features such as graphical distance (as opposed to spatial distance).

Here we focus on the topological features, where the interest is in mapping the interactions (edges) between the local neighborhood networks, that is, global topological features—the architecture of the brain as a whole. While there are local networks within networks, the global connections between these are of the greatest concern in systems neuroscience. Graph theory is replete with a zoo of different kinds of network topologies, but one of perhaps greatest interest to systems neuroscience are small-world networks, as various regions of the brain and the brain as a whole are known to instantiate such a network. The key topological properties of small-world networks are:

Sparseness: Relatively few edges given a large number of vertices.

Clustering: Edges of the graph tend to form knots, for example, if X and Y know Z, there is a better than normal chance they know each other.

Small diameter: The length of the most direct route between the most distant vertices, for example, a complete graph, with $\frac{n^2}{2}$ edges, has a diameter of 1, since you can get from any vertex to any other in a single step. That is, most nodes are not neighbors of one another but can be reached through a short sequence of steps. In other words, the key topological properties of such networks are (A) a much higher clustering coefficient relative to random networks with equal numbers of nodes and edges and (B) short (topological) path length. That is, small-world networks exhibit a high degree of topological modularity (not to be confused with anatomical or cognitive modularity) and nonlocal or long-range connectivity. Small-world networks strike a balance between high levels of local clustering and short path lengths linking all nodes, even though most nodes are not neighbors of one another. Keep in mind that there are many different types of small-world networks with unique properties, some with more or less topological modularity, higher and lower degrees (as measured by the adjacency or connection matrix), and so on (Sporns 2011; Von der Malsburg et al. 2010).

The explanatory point is that such graphical simulations allow us to derive, predict, and discover a number of important things, such as mappings between structural and functional features of the brain, cognitive capacities, organizational features such as degeneracy, robustness and plasticity, structural or wiring diagram features, various pathologies such as schizophrenia, autism and other "connectivity disorders" when small-world networks are disrupted, and other essential kinds of brain coordination such as neural synchronization, and so on. In each case, the evidence is that the mapping between structural and topological features is at least many-one. Very different neurochemical mechanisms and wiring diagrams can instantiate the same networks and thus perform the same cognitive functions. Indeed, it is primarily the topological features of various types of small-world networks that explain essential organizational features of brains, as opposed to lower-level, local purely structural features. Structural and topological processes occur at radically different and hard (if not impossible) to relate timescales. The behavior and distribution of various nodes such as local networks are determined by their nonlocal or global connections. As Sporns puts it, "Heterogeneous, multiscale patterns of structural connectivity [small-world networks] shape the functional interactions of neural units, the spreading of activation and the appearance of synchrony and coherence" (2011, 259).

3   Schizophrenia in Graph Theory

There is growing evidence from systems neuroscience that the cognitive and behavioral deficits in schizophrenia are caused in part by a "disturbance of connectivity" involving "disorganized structural and functional connectivity" (Sporns 2011, 210). As we learned in the last section, graph theory predicts that different network architectures will present various advantages and vulnerabilities. Network disturbances cause disruption in the dynamics of circulating information or coupling, that is, a disruption to the large-scale structural and functional connectivity of the brain (topology). In particular, we are talking about disruptions of small-world network topologies whereby disturbances of connectivity are caused by the loss of specific nodes and edges (cells and their axonal connections), and nonlocal or global effects—changes in global network parameters (Sporns 2011, chap. 10). Both anomalous decreased and increased coupling are the result, leading to abnormal functional connectivity between regions of the frontal and temporal lobes, abnormal patterns of cortical synchronization within and across cortical areas during rest, sensory processing, and cognitive tasks (ibid.). Key network properties affected include the complementary properties of degeneracy/plasticity and graphical modularity.

Several neuroimaging studies have yielded evidence of dysfunctional connectivity between regions of the brain in schizophrenia. Recent studies seem to confirm that said dysfunctional connectivity causes disruption of the key topological properties of functional brain networks, that is, efficient small-world network properties (Yu et al. 2011; Liu et al. 2008). Utilizing the fMRI data from both healthy patients and schizophrenics, graphical analysis shows that the brain networks of healthy subjects had their small-world network properties intact, whereas those of patients with the disease did not (ibid.). The networks of schizophrenics exhibited disruptions in prefrontal, parietal, and temporal lobes, and disruptions of interregional connectivity in general, presumably as a result of dysfunctional integration (ibid.). Regarding the relationship between topological measures and clinical variables, studies also suggest that the degree and severity of dysfunctional integration is correlated with illness duration (Liu et al. 2008). Further studies provide even more connection between topological measures and clinical variables:

Interestingly, we found characteristic path length and global efficiency of the whole brain network were correlated with PANSS negative scale values in SZ [schizophrenia]. Higher negative scale scores were associated with longer character path length and lower global efficiency. These might indicate the more severe these symptoms, the lower information interactions among brain components. In addition, clustering coefficient, local efficiency of IC14 (occipital region) and shortest path length of IC32 (parietal region) were correlated with negative PANSS scores in SZ. Higher negative scale scores were associated with a higher clustering coefficient, higher local efficiency of IC14 (occipital region) and longer shortest path length of IC32 (parietal region). These findings are in line with studies which found psychopathology is associated with aberrant intrinsic organization of functional brain networks in schizophrenia and provide further evidence for this illness as a disconnection syndrome. (Yu et al. 2011, 10)

Much more work needs to be done focusing specifically on correlations between thought disorder symptoms and disruptions of small-world network topology.

There are several things to note here. First, systematicity may be conceived as a much more global and multifaceted feature of cognition than mere linguistic systematicity, which would be a subset at best—transformational systematicity is perhaps fundamental, and linguistic systematicity requires no separate explanation. In discussing abduction, Fodor concedes that the "globality" or holism of mental processes (problem solving, belief formation, etc.) is sufficient to kill the computational theory of mind (Fodor 2000, 47). However, he remains convinced that connectionism cannot account for "the causal consequences of logical form" (ibid.). Note however that graph/network theory has many more resources than straight connectionism. Furthermore, if one takes the network explanation seriously, there is no reason to believe that logical form has any special causal consequences. Nor is there any reason to believe that belief formation is about rules of deductive inference; it is rather perhaps more like model checking.

Second, the explanation for the failure of transformational systematicity and other symptoms does not refer to representations with a combinatorial structure, compositionality, LOT, and so on. Indeed, the dynamical and graphical explanation on offer here appears to be exactly the sort Fodor and Pylyshyn are skeptical of. What is doing the explaining in this case are topological and dynamical properties of neural networks. Of course, fans of the computational theory of mind could argue that networks merely implement or realize computational processes. There is not enough space to fully address this retort here, but again note that the explanatory properties on offer are essentially graphical and dynamical. In addition, recall the context- and task-sensitivity of network processes, the manner in which different and multiple components/networks are very quickly reconfigured to perform different tasks, and the many dimensions of the importance of timing in these network processes. None of this bespeaks of computationalism.

Third, there is no reason whatsoever to think that compositionality would be a feature of such graphical networks even if you believe that such networks support or realize (in some sense) representations (of some sort) in the brain. That is, there is no reason to believe that the brain is primarily or essentially a "logic-engine." The graphical explanation on offer is not obviously about the mechanism of computation in any standard sense at all. And although one might call such explanations mechanistic, they are certainly not mechanistic in the sense of localization and decomposition (Silberstein and Chemero forthcoming). Given what we know about networks and what we know about the widespread cognitive deficits of schizophrenics, perhaps what we are learning is that logical functions cannot really be separated in some modular fashion from other executive functions.

Fourth, given the plurality and spatiotemporally extended nature of the causes of schizophrenia and given the network perspective, there is no reason to think that the explanation of transformational systematicity, or its failure, resides solely in the brain. As Sporns says:

The operation of brain networks depends on a combination of endogenous patterns of neural activity and exogenous perturbations such as those generated by stimuli or tasks. For organisms situated and embodied in their natural environments, the nature and timing of these perturbations are strongly determined by bodily actions and behavior. By generating sequences of perturbations, embodiment generates sequences of network states and, thus, information in the brain. Hence, the activity of an organism in its environment contributes to neural processing and information flow within the nervous system. … The embodied nature of cognition is captured by dynamical system theory, which describes neural, behavioral, and environmental processes within a common formalism. (2011, 319)

Perhaps it isn't just disturbances of brain topology that are at issue, but disturbances to the topology and dynamics of brain, body, and environmental networks, per the dynamical-extended view of cognition (Silberstein and Chemero 2012). Thus, the topology of the brain isn't just a function of endogenous neural dynamics but is a function of the brain's coupling with body and environment, a much larger social topological network in which individuals themselves are hubs. As Sporns puts it, "Social interactions also give rise to networks among people. The dynamic coupling of the brains and bodies of interacting individuals blends into the complex networks that pervade social structures and organizations. Hence, social networks and brain networks are fundamentally interwoven, adjacent levels of a multi-scale architecture, possibly with some of the same network principles at play" (2011, 319).

If this idea strikes you as crazy, keep in mind that not only do schizophrenics have disorders of movement, thought, agency, and experience, but those disorders are often primarily about the temporal-causal coordination and the connectivity of these functions to one another. Several people have noted that symptoms of schizophrenia are about disturbances in temporal (i.e., dynamical) coordination (Kitcher and David 2003). One reason for this belief is their notoriously poor performance on various timing tests and related cognitive tests, which expose their abnormal temporal processing and cognition in all areas of life (ibid.). Phenomenologically, schizophrenics often report feeling out of spatial, emotional, and temporal sync with the environment, others, and their own bodies. They also report being unsure as to where they end and the environment begins. In addition, they also manifest several other disorders related to self-affection and self-caring (ibid.). So perhaps transformational systematicity of thought is really inextricably bound with or only one aspect of, a greater kind of systematicity involving brain-body-environment coordination. Just as Kant suggested, perhaps our coherent conceptions of self and world, object and action, space and time, are thoroughly interdependent. As many have noted, for example, successive experiences are not the same as experiences of succession; to have an apprehension of temporal order, one must apprehend the relevant period of time as a single act, and schizophrenics are not able to do so. However, rather than Kant's categories "in" the mind or brain, here we are talking about order-parameters operating over spatiotemporally extended graphical networks. If so, the interesting question now is what are the order parameters in question and how might we manipulate them to better help schizophrenics.

Note

1. The remainder of this section is taken from Silberstein and Chemero forthcoming, sect. 2.

References

Abrahamsen, A., and W. Bechtel. 2006. Phenomena and mechanisms: Putting the symbolic, connectionist, and dynamical systems debate in broader perspective. In *Contemporary Debates in Cognitive Science*, ed. R. J. Stainton, 159–185. Malden, MA: Blackwell.

Aizawa, K. 2003. *The Systematicity Arguments*. New York: Springer.

Aydede, M. 1995. Connectionism and the language of thought. http://csli -publications.stanford.edu/papers/CSLI-95-195.html.

Eagleman, D. 2011. *Incognito: The Secret Lives of the Brain*. New York: Vintage.

Fodor, J. 2000. *The Mind Doesn't Work That Way: The Scope and Limits of Computational Psychology*. Cambridge, MA: MIT Press.

Fodor, J., and B. McLaughlin. 1990. Connectionism and the problem of systematicity: Why Smolensky's solution doesn't work. *Cognition* 35:183–205.

Fodor, J., and Z. Pylyshyn. 1988. Connectionism and the cognitive architecture. *Cognition* 28:3–71.

Graham, G. 2010. *The Disordered Mind*. New York: Routledge.

Heiden, A. U. 2006. Schizophrenia as a dynamical disease. *Pharmacopsychiatry* 39 (Suppl. 1):S36–42.

Kitcher, T., and A. David. 2003. *The Self in Neuroscience and Psychiatry*. Cambridge: Cambridge University Press.

Lieberman, J. A., S. T. Stroup, and D. Perkins. 2006. *The American Psychiatric Publishing Textbook of Schizophrenia*. American Psychiatric Publishing.

Liu, Y., M. Liang, Y. Zhou, Y. He, Y. Hao, M. Song, C. Yu, H. Liu, Z. Liu, and T. Jiang. 2008. Disrupted small-world networks in schizophrenia. *Brain* 131:945–961.

Marini, A., I. Spoletini, I. A. Rubino, M. Ciuffa, P. Bria, G. Martinotti, G. Banfi, R. Boccascino, P. Strom, A. Siracusano, C. Caltagirone, and G. Spalletta. 2008. The language of schizophrenia: An analysis of micro and macrolinguistic abilities and their neuropsychological correlates. *Schizophrenia Research* 105:144–155.

Mirian, D. R., W. Heinrichs, and S. McDermid Vaz. 2011. Exploring logical reasoning abilities in schizophrenia patients. *Schizophrenia Research* 127:178–180.

McKenna, P. J. 1997. *Schizophrenia and Related Syndromes*, 14–15. Hove: Psychology Press.

Mujica-Parodi, L. R., D. Malaspina, and H. Sackheim. 2000. Logical processing, affect, and delusional thought in schizophrenia. *Harvard Review of Psychiatry* 8: 73–83.

Murphy, D. 2006. *Psychiatry in the Scientific Image*. Cambridge, MA: MIT Press.

Pezard, L., and J. L. Nandrino. 2001. Dynamic paradigm in psychopathology: "Chaos theory," from physics to psychiatry. *Encephale* 27 (3):260–268.

Radden, J. 2004. *The Philosophy of Psychiatry*. Oxford: Oxford University Press.

Rao, N. P., S. Kalmady, R. Arasappa, and G. Venkatasubramania. 2010. Clinical correlates of thalamus volume deficits in anti-psychotic-naïve schizophrenia patients: A 3-Tesla MRI study. *Indian Journal of Psychiatry* 52 (3): 229–235.

Raymont, P., and A. Brook. 2009. Unity of consciousness. In *The Oxford Handbook of Philosophy of Mind*, ed. B. McLaughlin, A. Beckermann, and S. Walter. Oxford: Oxford University Press.

Schmid, G. B. 1991. Chaos theory and schizophrenia: Elementary aspects. *Psychopathology* 24 (4):185–198.

Silberstein, M., and A. Chemero. 2012. Complexity and extended phenomenological-cognitive systems. *Topics in Cognitive Science* 4 (1):35–50.

Silberstein, M., and A. Chemero. Forthcoming. Constraints on localization and decomposition as explanatory strategies in the biological sciences. *Philosophy of Science*.

Simpson, J., and J. Done. 2004. Analogical reasoning in schizophrenic delusions. *European Psychiatry* 19:344–348.

Spence, S. 2009. *The Actor's Brain: Exploring the Cognitive Neuroscience of Free Will*. New York: Oxford University Press.

Sporns, O. 2011. *Networks of the Brain*. Cambridge, MA: MIT Press.

Sun, J., P. Jia, A. H. Fanous, E. van den Oord, X. Chen, B. P. Riley, R. L. Amdur, K. S. Kendler, and Z. Zhao. 2010. Schizophrenia gene networks and pathways and their applications for novel candidate gene selection. *PLoS ONE* 5 (6):e11351. doi:10.1371/journal.pone.0011351.

Tandon, R., H. Nasrallah, and M. Keshavan. 2009. Schizophrenia, "just the facts." 4. Clinical features and conceptualization. *Schizophrenia Research* 110:1–23.

Ventura, J., A. Thames, R. Wood, L. H. Guzik, and G. S. Hellemann. 2010. Disorganization and reality distortion in schizophrenia: A meta-analysis of the relationship between positive symptoms and neurocognitive deficits. *Schizophrenia Research* 121:1–14.

Volz, H.-P., C. Gaser, and H. Sauer. 2000. Supporting evidence for the model of cognitive dysmetria in schizophrenia—a structural Magnetic Resonance Imaging study using deformation-based morphometry. *Schizophrenia Research* 46:45–56.

Von der Malsburg, C., W. Philips, and W. Singer. 2010. *Dynamic Coordination and the Brain: From Neurons to Mind.* Cambridge, MA: MIT Press.

Yu, Q., J. Sui, S. Rachakonda, H. He, W. Gruner, Godfrey Pearlson, Kent A. Kiehl, and Vince D. Calhoun. 2011. Altered topological properties of functional network connectivity in schizophrenia during resting state: A small-world brain network study. *PLoS ONE* 6 (9):e25423. doi:10.1371/journal.pone.0025423.

Contributors

Ken Aizawa

Department of Philosophy
Rutgers University, Newark
USA

William Bechtel

Department of Philosophy
University of California, San Diego
USA

Gideon Borensztajn

Institute for Logic, Language and
Computation
University of Amsterdam
The Netherlands

Paco Calvo

Departamento de Filosofía
Universidad de Murcia
Spain

Anthony Chemero

Departments of Philosophy and
Psychology
University of Cincinnati
USA

Jonathan D. Cohen

Department of Psychology and
Princeton Neuroscience Institute,
Princeton University, Princeton,
New Jersey
USA

Alicia Coram

School of Historical and
Philosophical Studies
University of Melbourne
Australia

Jeffrey L. Elman

Department of Cognitive Science
University of California, San Diego
USA

Stefan L. Frank

Department of Cognitive,
Perceptual and Brain Sciences
University College London
UK

Antoni Gomila

Departament de Psicologia
Universitat de les Illes Balears
Spain

Seth A. Herd

Department of Psychology and
Neuroscience
University of Colorado Boulder
USA

Trent Kriete

Department of Psychology and
Neuroscience
University of Colorado Boulder
USA

Christian J. Lebiere

Psychology Department
Carnegie Mellon University
USA

Lorena Lobo

Facultad de Psicología
Universidad Autónoma de Madrid
Spain

Edouard Machery

Department of History and
Philosophy of Science
University of Pittsburgh
USA

Gary Marcus

Department of Psychology
New York University
USA

Emma Martín

Departamento de Filosofía
Universidad de Murcia
Spain

Fernando Martínez-Manrique

Departamento de Filosofía I
Universidad de Granada
Spain

Brian P. McLaughlin

Philosophy Department and
Cognitive Science Center
Rutgers University
USA

Randall C. O'Reilly

Department of Psychology and
Neuroscience
University of Colorado Boulder
USA

Alex A. Petrov

Department of Psychology
Ohio State University
USA

Steven Phillips

Human Technology Research
Institute
National Institute of Advanced
Industrial Science and Technology
Japan

William Ramsey

Department of Philosophy
University of Nevada, Las Vegas
USA

Michael Silberstein

Department of Philosophy
Elizabethtown College
USA

John Symons

Department of Philosophy
University of Kansas
USA

David Travieso

Facultad de Psicología
Universidad Autónoma de Madrid
Spain

William H. Wilson

School of Computer Science and
Engineering,
The University of New South
Wales,
Australia

Willem Zuidema

Institute for Logic, Language and
Computation
University of Amsterdam
The Netherlands

Index